POLYMER SCIENCE AND TECHNOLOGY

POLYMER-MATRIX COMPOSITES

TYPES, APPLICATIONS AND PERFORMANCE

Polymer Science and Technology

Additional books in this series can be found on Nova's website
under the Series tab.

Additional e-books in this series can be found on Nova's website
under the e-book tab.

POLYMER SCIENCE AND TECHNOLOGY

POLYMER-MATRIX COMPOSITES

TYPES, APPLICATIONS AND PERFORMANCE

RAKESH KUMAR
EDITOR

nova publishers
New York

For permission to use material from this book please contact us:
Telephone 631-231-7269; Fax 631-231-8175
Web Site: http://www.novapublishers.com

Additional color graphics may be available in the e-book version of this book.

Library of Congress Cataloging-in-Publication Data

Polymer-matrix composites (Nova Science Publishers)
Polymer-matrix composites : types, applications, and performance / editors, Rakesh Kumar (BIT Mesra, Patna Campus, Patna, India).
pages cm. -- (Polymer science and technology)
Includes bibliographical references and index.
ISBN: 978-1-61122-005-6 (hardcover)
1. Polymeric composites. I. Kumar, Rakesh, 1972- II. Title.
TA418.9.C6P657 2014
620.1'92--dc23

2014030324

Published by Nova Science Publishers, Inc. † New York

CONTENTS

PREFACE

The agricultural sector is the main source of raw materials for biopolymers such as soy protein isolates (SPI), natural fibers, polylactic acid (PLA), polyhydroxybutyrates (PHBs) and many more. Over the years, biopolymer sector has witnessed significant developments due to substantial investments that have been made in green technologies/processes. Apart from this, focus has also been given to improving the performance of the existing thermoplastic as well as thermoset polymers and synthetic fiber based composites. With the onset of green technology/process, agriculture has witnessed rapid transformation, which, in turn has opened new avenues for the development of commonly available synthetic polymers, such as polyethylene from sugarcane.

The properties of synthetic polymer based composites improve greatly with the incorporation of nanoparticles or nanommaterials. Carbon nanotubes, cellulosic nanowhiskers and naonosilica are widely used as a reinforcement material for polymer matrix. It is worth mentioning that nanosized devices may be a reality soon. Just one example of nanosized devices is the use of "E-DOG" which can detect bomb at a crowded place.

In fact, facilitating the commercialization of innovative processes/technologies developed in the laboratories is the first step. Second step is to take the products fabricated in the first step to the next level, i.e. market. Lastly, demonstrating these products in real-life conditions is necessary for terming it as marketable products. The truth is that the development of biopolymer or nanomaterial based polymer-matrix composites are progressing slowly. This book is written so as to address some of the types of polymer-matrix composites as well as performance of these composites through 11 original and very interesting Chapters. The use of biopolymer based composites has been greatly emphasized in Chapters 1 to 5 so as to reduce the carbon footprint. Chapters 6 to 9 are devoted to the performance of widely used thermoplastics, thermosets and conducting polymers. Lastly, the applications of natural fiber reinforced composites in automotive sectors as well as erosive wear performance have been discussed in Chapters 9 to 11.

The book starts with a discussion relating to the fabrication of carbon nanotubes (CNTs) incorporated PLA based composites. Functionalized CNTs are also incorporated in PLA. Factors affecting the structure and physicochemical properties, such as crystallinity, permeability and mechanical properties have been discussed in Chapter 1. Then the book continues with Chapter 2, where the incorporation of different types of nanoclay in regenerated cellulose based materials to produce nanocomposites has been discussed. It has been stated in Chapter 3 that the amorphous domains of starch, cellulose, chitin can be removed under acid hydrolysis to produce crystallites in nanoscale, which are called nanocrystals and these nanocrystals can be used as substitutes for inorganic nanoparticles in preparing polymer nanocomposites. In chapter 4, a discussion on soy protein isolate (SPI) and arylated SPI based nanocomposites is presented. SPI based nanocomposites have good mechanical properties but low water resistance which limits their applications. In the year 2008, it has been reported that arylated SPI films exhibited improved thermal stability and high water resistance. Thus arylated SPI is emerging as a new area of research and it is expected that the arylation of protein materials could serve as a general motif to fabricate water resistant protein based composites/nanocomposites. Chapter 5 addresses the relevant background of Alfa fiber. The preparation and performances of composites/nanocomposites based on Alfa fiber are discussed in this chapter.

Chapter 6 discusses the performance of polymer nanocomposites that is intimately related to the degree of dispersion of the nanofiller. In this chapter, it is reported that an alternative to conventional injection moulding (CIM) to achieve a higher dispersion of the nanofillers is the non-conventional shear control orientation in injection moulding (SCORIM) process. Glass fibre reinforced polymer (GFRP) has non corrosive, non conductive and non magnetic properties. In Chapter 7, GFRP composite as an alternative to steel have been reported. In Chapter 8, it has been stated that with the development of nanotechnology, various conducting polymers with metal oxide like ruthenium oxide, iridium oxide, manganese oxide, cobalt oxide, nickel oxide, tin oxide, iron oxide, pervoskites, ferrites *etc.* have been explored for energy storage up to ~1000 F/g specific capacity. The mentioned value of capacitance supports metal oxide and will play a crucial role in future supercapacitor technology development. Sulfur-polymer matrix composites as thermoplastic materials made of mineral aggregate and filler, with sulfur as a binder instead of cement and water is discussed in Chapter 9. The proportion of aggregate, filler and binder for the preparation of sulfur-polymer matrix composite mixture may vary depending on the application. In this chapter, various fillers were used for the production of sulfur- polymer matrix composite samples: talc, alumina, microsilica and fly ash.

Chapter 10 explores the use of natural fiber reinforced polymer products in automotive sectors. These products are inexpensive and more sustainable due to environmental concerns. Owing to their ability to create a positive environmental impact, natural fibres in composite materials are characterized by their biodegradability and decomposability at the end of their life cycle. Chapter 11 evaluates erosive wear performance of polymer matrix composites as a function of various operating and materials variables such as impingement angle, impact velocity, stand-off distance, erodent size, erodent flow rate, reinforcement content as well as their mechanical and thermal response. Various types of natural fibers such as bagasse, bamboo, chicken feather, coir, jute, lantana camera based polymer-matrix composites are studied for their erosive wear properties.

Finally, let me conclude this preface by thanking all the authors who have contributed to the realization of this book, without whom this project would have never been started. I thank them for their participation and patience during the preparation of this book and I am grateful that they have entrusted me to edit their contributions. I hope the readers will find this book useful. I am looking forward to receive comments and constructive feedbacks about contents.

In: Polymer-Matrix Composites
Editor: Rakesh Kumar

ISBN: 978-1-61122-005-6

Chapter 1

NANOCOMPOSITES BASED ON POLYLACTIC ACID (PLA) REINFORCED BY FUNCTIONALIZED CARBON NANOTUBES (CNT)

Lucía Famá *

LPMC, Dept of Physics, Faculty of Ciencias Exactas y Naturales and IFIBA-CONICET, University of Buenos Aires, Ciudad Universitaria, Buenos Aires, Argentina

ABSTRACT

Within biodegradable natural polymers, thermoplastic materials based on polylactic acid (PLA) have been studied over the last decades for their possible use in different kinds of industries such as packaging, textile, biomedicine, engineering, etc. However, PLA has certain disadvantages compared with synthetic materials, in particular low elongation at break and fairly fragility. To solve this problem some investigations have attempted to incorporate different micro/nano size particles in order to reinforce matrices developed by this polymer. Carbon nanotubes (CNT) are one of the fillers with high potential for industrial use; however, their use as reinforcing material has an important difficulty associated with their poor dispersion and bad interfacial interaction with polymer matrices when they are in their neat state. In this way, different types of treatment of carbon nanotubes have been studied in order to improve their dispersion in a PLA matrix. A review of some of those investigations will be presented in this chapter. In particular, the discussion will be focused on the description of different functionalization methods that were used to obtain those objectives. A comparative analysis of some covalent and non-covalent techniques and the variables involved in these processes will be performed. Factors affecting physical properties such as structure, crystallinity and mechanical behavior of PLA- functionalized CNT nanocomposites will be discussed. The most important results on mechanical, dynamic and quasi static behavior of these materials will be analyzed in separate sections. Finally, a conclusion about the importance of the functionalization of carbon nanotubes to improve the physical properties of PLA based composites will be presented.

---

* Dr. L. Famá, Dept of Physics, Faculty of Ciencias Exactas y Naturales and IFIBA-CONICET, University of Buenos Aires, Ciudad Universitaria, Pab. 1, (C1428EHA), Buenos Aires, Argentina. Phone: 54-11-45763300, ext. 255. Fax: 54-11-45763357. Email: lfama@df.uba.ar.

Keywords: Polylactic acid, Carbon nanotubes, Nanocomposites, Functionalization techniques, Physical properties

1. INTRODUCTION

Over the last decades, significant efforts have been achieved by important research groups in the study of environmentally friendly materials with the aim to replace the traditional synthetics such as nylon, polystyrene, polyvinyl chloride (PVC), polyethylene, polypropylene, etc., which come from oil sources, provoking environmental pollution and accumulation of many residues due to their extensive degradation time (their production involves high levels of CO_2 and they take hundreds of years to degrade). Other advantage of the use of biodegradable polymers is the low cost of their production.

Many polymers such as starch, polylactic acid, cellulose, chitosan, alginate, proteins, etc. [1-6], have been focused of several investigations with the purpose of making biodegradable and biocompatible materials, with the expectation of reducing the impact on the environment caused by the production and use of petrochemical products [7].

Among potential biodegradable polymers, polylactic acid (PLA) has been viewed as a good alternative in biodegradable formulations [8-11] because it is produced from renewable resources, making it largely available (large-scale production) in a wide range of grades. On the other hand, PLA has remarkable properties to fulfill different applications, such as packaging, textile, biomedicine, engineering, etc. [12-14].

Although PLA has many advantages over synthetic thermoplastic materials, this polymer has poor mechanical properties such as low elongation at break and fairly fragility, and it also has high water vapor permeation compared to other non-biodegradable materials [15]. The addition of micro- and nano- sized fillers into a polylactic acid matrix has been the topic of some studies in order to overcome those problems [4, 16]. It is well known that the size of reinforcements has an important influence on the properties of the resulting composites. For example, when nanofillers are used, the relationship between the area of the interface and the mass of the reinforcement is higher than in the case of micro or conventional reinforcements. This fact leads to the great interest of the use of nanoparticles in the development of composites.

The use of carbon nanotubes (CNT) to reinforce PLA matrices is currently investigated due to their high aspect ratio and great mechanical properties such as Young's modulus, tensile strength, as well as electrical conductivity [8, 9, 17, 18]. Generally, smaller size fillers have the capacity to improve several mechanical properties, such as storage modulus or tensile strength, when they are incorporated into a matrix with inferior properties; however, they can produce a detrimental effect on ductility [19, 20]. The interfacial adhesion between nanotubes and a polymer matrix plays an important role in improving the properties of nanocomposites based on polymers [21-24]. Furthermore, the addition of CNT affects material surface roughness and consequently, surface tension and wear properties.

Taking into account the great properties of both polylactic and carbon nanotubes, the development of PLA-CNT composites has been an attractive topic of study in order to use them in different applications including packaging, medicine, textile, military and energy, pharmaceuticals, cosmetics, etc.

Despite the lack of knowledge about the toxicity of carbon nanotubes, still controversial in the literature, CNT composites are being designed to interact with food, for example, emitting antimicrobials and antioxidants [25]. In particular, many packaging industries are trying to implement this new technology in bags, plates, cups, bowls, and coatings [26, 27].

In biomedicine, it has been shown that the addition of carbon nanotubes in a polylactic acid matrix can simulate bones formation with ability to generate a structural reinforcement [8].

This chapter will be focused on a review of some investigations about the influence of the incorporation of CNT functionalized by different techniques, on the structure, crystallinity and mechanical, dynamic and quasi static, properties of PLA based nanocomposites.

2. Poly (Lactic Acid) (PLA)

Polylactic acid (PLA) is linear aliphatic thermoplastic polyester mainly produced by ring-opening polymerization of lactic acid caused by the fermentation of annually renewable resources [8]. Lactic acid is one of the smaller optically active molecules and can be found in different forms: L(-)-lactic acid (most common) and D(+)-lactic acid, both produced by microorganisms. PLA is an organic acid that can be obtained from several natural sources such as corn starch (in United States), tapioca roots: chips or starch (mostly in Asia), and sugarcane (in the rest of the world) [28], allowing adapting to specific regional climates.

PLA is one of the few polymers in which its structure can easily be modified by the polymerization of a mixture of L and D isomers to obtain better molecular weight and amorphous or semi-crystalline polymers. Its properties can be modified through the variation of the isomers ratio (L/D) or of the relative contents of L and D. Depending on its distinct forms, poly (L-lactic acid), *PLLA*, or poly (D-lactic acid), *PDLA*, it has different properties. For example, PLLA has crystallinity of around 37 %, glass transition temperature between 60-65 °C, melting temperature between 173-178 °C and tensile modulus between 2.7-16 GPa [29, 30].

The heat resistant of the polylactic acid can be in temperatures around 110 °C [31], and it is soluble in chlorinated solvents, hot benzene, tetrahydrofuran and dioxane [32]. Its tensile strength and Young's modulus are comparable to those of some non-natural polymers such as polyethylene, but it is more hydrophilic and has lower density. PLA is also stable to ultra violet light (UV), resulting excellent to conform fabrics that resist fading. It can be formulated into rigid or flexible materials and can be copolymerized with other materials [33].

PLA has others important characteristics: is considered both as biodegradable (e.g., adapted for short-term packaging) and biocompatible in contact with living tissues (e.g., for biomedical applications). It complies with the rising worldwide concept of sustainable development and is classified as an environmentally friendly material [34]. Due to its status as biodegradable and biocompatible, once the life cycle of PLA products ends, they can be thrown into the field and, as they come from renewable resources, the raw material can be always available.

In recent years, PLA based materials were primarily different markets reference such as packaging (with short-term applications mainly on food), textile, biomedicine and engineering. For instance, some types of manufactured products are blow-molded bottles, injection-molded cups, spoons and forks, thermoformed cups and trays, paper coatings, fibers for the textile industry, sutures, bones simulators, films, and various molded articles [10].

The development of applications of PLA practically is related to environmental concerns and the adverse effects of using non-degradable or petrochemical polymers. Polylactic acid is a very useful polymer that has found applications in a wide range of industries because of its biodegradable and environmentally friendly characteristics.

The applications of this polymer in biomedical and pharmaceutical fields can be traced back to several decades, because of its bioresorbability and biocompatible properties in human body [10]. For example, due to its capacity to degrade into innocuous lactic acid, PLA is used as medical implants in the form of screws, pins, rods, and as a mesh [35]. Depending on the type of the polylactic, it can break down within the body between 6 months and 2 years. This gradual degradation is desirable for a support structure, because it gradually transfers the load to the body (e.g. to bones). PLA resorbs or degrades upon implantation into the body, but most of its mechanical properties are lost within a few weeks [36].

The use of PLA also has grown well in domestic market for general consumer in order to obtain good and biodegradable packaging. It is forecast that the development of this polymer will grow tremendously in the future, making its price comparable with that of the commodity plastics, but with benefit of being kind to the environment.

Although materials formed by PLA are still relatively new for their use in some plastic industries, they promise to become widely implemented throughout a broad range of applications. PLA can also serve as a compostable packaging material, either by casting, injection molding, or spun [35]. In particular, cups and bags have been made from this material. Market studies showed that polylactic acid and its derivatives are economically feasible products for the composition of packaging materials. Taking into account its current consumption, it is, at the present, the most important market in volume for biodegradable packaging [8, 10, 11]. During the last decade, the use of PLA in the industry of packaging has increased all across Europe, Japan, and United States; mainly in the area of fresh products, where polylactic is being used as food packaging for short shelf-life products, such as fruit and vegetables. Some applications in this area include drinking and sundae cups, salad trays, wrappings for sweets, lamination films, blister packages, and water bottles. However, to cater for a larger market, some drawbacks of PLA must be overcome, such as its limited mechanical properties, barrier behavior and heat resistance. With respect to meet market expectations, the world production of PLA must be increased. Some industries of food packaging such as Lindar, InnoWare Plastics and Carrefour Belgium actually use polylactic based materials due to their great qualities. This polymer is suitable to be used for light weight and transparent food packaging containers. It is highly glossy and can be easily printed as the existing materials, such as polystyrene, polyethylene, and polyethylene terephthalate. Other typical applications of PLA include yogurt pots, sandwich containers, and fresh food trays for fruits, pastas or cheeses [37-39]. Actually, some companies such as Frito-Lay (SunChip), Wal-Mart (salad packaging) and Naturally Iowa (EarthFirst-shrink sleeve label), use polylactic in the form of films, for bakery goods, confectionery, salads, wraps, envelope windows, laminated coatings, multilayer packaging, etc. [39, 40]. For example, PLA can be used to made biaxial oriented plastic films for packaging bags. These plastic bags take a few months to fully degrade. As already stated, PLA is one of the most studied bioplastics regarding recyclability, because it may biodegrade under certain conditions such as the presence of oxygen and moisture. It also can be degraded by abiotic degradation (simple hydrolysis) [34]. Every year, millions of synthetic plastic bags are disposed of, causing white pollution of ground and water, and many problems in the accumulation of residues. The substitution of petroleum based bags by PLA bags can make significant environmental savings [39].

As fibers and non-woven textiles, PLA also has many potential uses such as upholstery, disposable garments, awnings, feminine hygiene products, and diapers. Since 2009 the last trends for PLA applications have been in durable and biobased markets, such as automotive (e.g., seats, insulation) and building applications (e.g., acoustic and thermal insulation). The main advantages of PLA are its high bio-based content, good conditions to perform fibrous materials, and its corresponding physical and mechanical properties [34].

Currently, PLA is also used in compostable yard bags to promote national or regional composting programs. In addition, new applications such as cardboard or paper coatings are being pursued [11].

Despite the great characteristics of PLA, materials based on this polysaccharide have certain disadvantages compared to the synthetics. It is well known that due to the inferior physical properties and low molecular weight, PLA cannot replace petroleum based materials by itself. Therefore, different kinds of micro and nano reinforcements have been used in the development of composites based on this polymer.

3. Carbon Nanotubes (CNT)

Carbon nanotubes (CNT) have been focus of some investigations in the field of nanotechnology in recent years due to their unique physical properties [41-44]. CNT particles, discovered by Iijima in 1991 [45], are made up of thin sheets of benzene ring carbons rolled up into the shape of a seamless tubular structure. Their diameter is on the order of a few nanometers and can be up to several micrometers in length. Nanotubes belong to the family of fullerenes and are composed of sp^2 bonds, similar to those observed in graphite, and they naturally align themselves into ropes held together by van der Waals forces [46].

Based on their structure, CNT can be classified into two general categories: single-walled carbon nanotubes (SWCNT), which consists of one layer of cylindrical graphene, and multi-walled carbon nanotubes (MWCNT), that involves an array of such cylinders formed concentrically separated by 0.35 nm, similar to the basal plane separation in graphite [47]. Both types of carbon nanotubes have unique physical and chemical properties such as high aspect ratio, ultra-light weight, high mechanical properties, and great electrical and thermal conductivity [48, 49], which make them potentially useful in a wide variety of applications [50-52].

SWCNT usually have diameter around 1 nm and length that may be many thousands of times larger (up to the order of the centimeters). Theoretical and experimental investigations revealed that single-walled carbon nanotubes have extremely high tensile modulus [53] and tensile strength [54], demonstrating that they are stronger than steel, lighter than aluminum, and have more conductive than copper.

MWCNT have diameter from 2 to 100 nm and length of tens of microns [55]. Due to the combination of their high strength, flexibility and great values of aspect ratio (300-1000) [56-60], several investigations have focused on the properties of these structures [61]. The length and diameter of MWCNT differ greatly from those of SWCNT, thus, their properties are also very different. In a recent work, Nuriel et al. [62] studied the characteristics of MWCNT with only 15-30 mm of diameter, obtaining a surface tension of ~ 40-45 mJ/m^2.

Some of the properties of the different types of CNT are shown in Table 1.

Table 1. Properties of different structures of CNT [44]

Type of CNT	Property	Quantity
SWCNT	Density	~ 0.6 g/cm^3
	Tensile modulus	~ 1 TPa
	Gas-adsorption capacity	5-10 wt%
	Type of semiconducting material	as-grown of p-type
MWCNT	Density	~ 1-2 g/cm^3
	Tensile modulus	~ 1.8 TPa
	Heat conductivity	~ 1200 W/mk
	Long-term current carrying capacity	10^9-10^{10} A cm^{-2}

The main advantages of carbon nanotubes over the micrometric reinforcements lie in their high aspect ratio (k = L/d) which makes them promising to transfer more effectively their excellent properties to a polymer matrix [63-65].

It is well known that the final properties of composites containing CNT strongly depend on the aspect ratio of the filler. It has been shown that the modulus increment of composite materials is maximized with the use of multi-walled nanotubes. This material tends to have a large surface area per unit volume which results in very good load transfer to the nanotube network [66]. The small size of carbon nanotubes is comparable to the size of polymer molecular chains, resulting in a considerably higher interaction surface area between the filler and matrix than that of the conventional micro-sized particles.

Intense research has been developed on CNT composite materials with enhancements in mechanical [67], electrical [68], and thermal properties [69].

Up to date, much research has focused on increasing mechanical parameters such as Young's modulus, tensile strength and toughness of composite materials, because of the demand of these properties in many technological applications; and, in this way, CNT have been considered ideal candidates for this purpose [21, 70, 71].

The very important properties of carbon nanotubes make them potentially useful in a wide variety of applications such as nanotechnology, electronics, optics and other fields of materials science [72]. In particular, there is a great interest in the application of carbon nanotubes in the development of nano reinforced polymer films. The application of CNT in polymers such as starch, chitosan, PLA, etc., has also been studied in the last years [4, 73-75]. Depending on the polymer used and its processing conditions, the final properties of the film seem to vary. The low density of the nanotubes is clearly an advantage for based polymer composites. Carbon nanotubes offer multifunctionality and better performance during compression due to their flexibility and low propensity to fracture.

Although composites developed with polymers and nanotubes are materials that should be applied on different areas, there have not been many successful experiments that show this fact. The main problem is creating a good interface between nanotubes and polymer matrix to attain good load transfer from the filler to the matrix, during loading. Nanotubes are almost always organized into aggregates. To maximize the advantage of nanotubes as reinforcing structures in composites, the aggregates need to be broken up and dispersed or cross-linked to prevent slippages. In addition, the surfaces of nanotubes have to be chemically modified to achieve strong interfaces between them and the surrounding polymer chains [72].

It is very difficult to observe a good dispersion of carbon nanotubes in a polymer matrix when they are directly mixed with the polysaccharide, resulting in a deterioration of mechanical and permeability properties of the composite. Recently, some research suggested that CNT had to be functionalized to acquire great dispersion in a biodegradable polymer matrix, and thus, the final composites could have better properties for some applications as replacement of synthetic materials [4, 73, 74].

Some of different types of functionalization methods of carbon nanotubes are set out below.

3.1. Carbon Nanotubes Functionalization

The successful use of composite materials depends largely on their final properties, such as toughness, resistance, Young's modulus, tensile strength, gas and water vapor permeability, that are strongly linked to the dispersion of the filler throughout the polymer matrix and to the interfacial adhesion between the phases. In general, weak interfaces provide good energy absorption mechanisms and create composites with low stiffness and strength but with high toughness; whereas strong interfaces lead to hard composites with high tensile strength but brittle [75-77].

In the case of carbon nanotubes, it is seen that there are many difficulties in transforming them into soluble materials that can be easily manipulated. Their effectiveness strongly depends of their homogeneous dispersion throughout the polymer matrix without destroying their integrity and the adequate interfacial adhesion between the phases [78-80]. A good dispersion not only makes available more surface area of the filler, but also prevents their aggregation [81]. If CNT should evenly be distributed in the polymer matrix, some challenges have to be faced, like the length of the tubes, their entanglement, volume fraction and the high viscosity of the matrix.

The dispersion of the nanotubes can be initially done through various methods like sonication, stirring, calendaring, etc.; however, these methods are only suitable for very low viscous matrix materials and small volumes [82]. In the particular case of composites based on PLA, it was found that direct blending of CNT and this polymer provokes a tendency to aggregate fillers, and their non-uniform dispersion in the matrix often results in deleterious effects [46, 83, 84]. Obtaining a homogeneous dispersion of carbon nanotubes in PLA matrix is not easily accomplished, primarily by their very low solubility in most solvents [85].

Many studies have proposed different methods of functionalization of carbon nanotubes in order to solve the problem of their poor dispersion in a PLA matrix. For this purpose, generally, it has used two major types of functionalization: covalent, with which it has generated a multitude of structures based on structural modification of the filler using small molecules and/or polymers [86-90], and non-covalent functionalization, where π-π interaction, π-cation interaction and ionic interaction between CNT and polymers enable the absorption of polymers onto CNT surfaces [91-94].

In the covalent functionalization, CNT are solubilized or derivatized with inorganic or organic functional groups to facilitate a strong bonding and selectivity through either hydrogen or covalent bonds. For example, the generation of hydrophilic substituent surfaces such as carboxylic, hydroxyl or sulfonic acid groups by suitable chemical method [57, 58, 95, 96] is frequently used, in particular, in medical and biological applications [97]. However, these methods significantly perturb the conjugated π system of the nanotube due to the dramatic changes in its structure and, consequently, varying its electronic properties [98, 99]. Accordingly, these modifications are not useful in applications where it is mostly needed the important electrical or mechanical properties of carbon nanotubes.

Moreover, non-covalent functionalization method, which involves the physical adsorption of molecules on the walls of the carbon nanotubes through π interactions or van der Waals type, is actually effective to enhance the properties of CNT. In many cases, this strategy preserves the structural and electronic integrity of the fillers, allowing the use of both electrical and mechanical properties for different applications. To date, numerous systems based on non-covalent interaction are successful in provocation of some changes in solubility and/or electronic properties of nanotubes [94].

The following are some methods of functionalization of carbon nanotubes reported from different research groups that have achieved an effective dispersion of CNT in PLA matrix [4, 101, 104, 105, 107-111, 114. 117].

To improve the solubility of CNT within PLA matrix, and therefore their dispersion, generally functional groups such as hydroxyl (OH) and carboxyl (COOH) are introduced at the surface of CNT via oxidation processes in the presence of oxygen, air, concentrated sulfuric acid, nitric acid, aqueous hydrogen peroxide, or different acid mixtures [100]. For example, Wu et al., [101] investigated the effect of several methods of functionalization of MWCNT on the rheology and thermal stability of PLA based nanocomposites prepared by melt-compounding. These authors used carboxylic and hydroxyl groups to functionalize the filler and compared modified MWCNT-PLA materials with composites reinforced by purified MWCNT (as model). They published that the best dispersion extent of MWCNT within PLA matrix had been achieved with carboxylic-functionalized MWCNT [102].

In order to introduce carbon nanotubes into a PLA matrix, Seligra et al., [4] developed a technique to functionalize the filler involving some different reactions. The first one was a Fenton reaction to generate hydroxyl and carboxyl groups on the walls of MWCNT. Then, carboxyl groups were lengthened by reacting with thionyl chloride to generate acid chloride groups. Finally a reaction with triethylene glycol was performed to leave an OH terminal distanced from the filler wall, thereby increasing their availability. Fenton reaction was performed using H_2O_2 and $FeSO_4$ solutions. Then, HCl was added to redissolve the iron that had not reacted. For reaction with thionyl chloride and ethylene glycol, functionalized MWCNT by Fenton and thionyl chloride ($SOCl_2$) were used to produce acid chloride groups and, then, triethylene glycol was added.

An alternative way to increase the hydroxyl groups on the nanotube wall is through diazotization reactions, using isoamyl nitrite and p-aminobenzoic acid [103]. This methodology is softer, so it would yield derivatized nanotubes of greater length.

Pillai et al., [104] described a comparison between two different chemical strategies for the functionalization of multi-walled carbon nanotubes before incorporating them into a PLA matrix. In the first procedure, MWCNT were functionalized in a direct amination process that avoided the use of strong acids or acid chloride unlike conventional functionalization methods; whereas the second method consisted in a two steps procedure involving mild acid treatment followed by amidation. Both procedures not only allowed control the amine content on CNT surfaces but also the obtaining of a remarkable degree of fictionalization. Moreover, in the work published by Novais et al., [105], carbon nanotubes were functionalized by 1,3-dipolar cycloaddition reaction, generating pyrrolidine groups at the nanotube surface [106]. These authors further studied the reaction of functionalized CNT with PLA in solution and in polymer melt. Their results revealed evidence of polymer bonded to the nanotubes surface, only when reaction was carried out in polymer melt. Other authors such as Georgakilas et al., [107] and Kordatos et al., [108] also reported successful surface functionalization of carbon nanotubes via 1,3 dipolar cycloaddition.

In order to enhance the interfacial interactions between PLA and SWCNT, Chiu et al., [109] did functionalization treatments to purified SWCNT using nitric acid/sulfuric acid mixture, and grafted with (3-isocyanatopropyl)triethoxysilane. When functionalized SWCNT were used to reinforce PLA matrix, the fractured surface of composites did not present pullout phenomenon.

Martinez and Salavagione, [110] proposed two CNT functionalization strategies: a direct esterification of filler, and its functionalization through acyl chloride derivative to obtain soluble nanocomposites in water and dimethyl sulfoxide, allowing the conformation of PLA-CNT films easily.

Chen et al., [111] reported attachment of long-chain alkylamines onto SWCNT treated with acid. Many others authors also adopted similar strategies to the surface-functionalization of MWCNT [112, 113]. To prevent serious damages on the surface of the graphene layers, Viswanathan et al., [114] developed carbanion initiated polymerization process to surface graft polymer chains onto SWCNT. Several researches also proposed the surface-functionalization of CNT by short chain alkyl radicals [115-117]. For example, Holzinger et al., [117] functionalized SWCNT using nitrenes or carbons at reaction temperatures of 160 °C and 60 °C, respectively, for 3 h.

Other authors reported that a stable and uniform suspension of nanotubes in PLA polymer matrix could be achieved by using various surfactants; however, this might be inconvenient from the standpoint of removal and separation of the surfactants that may lead to undesirable effects [118].

4. PLA - Functionalized CNT Nanocomposites

Since 1994, when the first polymer nanocomposites using carbon nanotubes as filler were reported by Ajayan et al., [119], there has been progress in the study of bionanocomposites formed by different biodegradable polymers as matrix and reinforced with functionalized CNT, to be used in various industries such as packaging, hygiene, cleaning, textile, biomedical, engineering, etc. [120-122]. An important factor for the implementation of these bionanocomposites in mentioned industries is the conscience of environmental impacts. The inherent properties of CNT enhance composite materials with the addition of the advantages of its eco-friendly nature that have been effectively used to create new class of materials. They also provide additional advantages such as ease to process, transparent, recyclable and their low density [123].

Nanocomposites have been an important proposal, for example, for delivery systems used in medicine, and particularly, the use of PLA as matrix has been interesting due to its hydrolytic degradability and low toxicity. The most important properties of PLA-CNT are drug release rate and matrix degradation rate, which are affected by the particle design and material properties [124].

Next, the description of different techniques used by some scientist to develop PLA-CNT composites, and the influence of the functionalization of this filler in the physicochemical properties of final nanocomposites, will be exposed.

4.1. PLA - Functionalized CNT Performance

As it is well known, the properties of composite materials depend on the structure formed during the processing steps [125]. There are several techniques reported for the manufacture of PLA-CNT composites, such as solution mixing, melt blending, in-situ polymerization, etc. [46, 126-138].

A brief review of some preparation methods of PLA-CNT composites are given here.

The solution mixing is one of the most commonly techniques used for preparing composites based on a polymer-matrix and reinforced with CNT, particularly on a laboratory scale. This method generally involves three main steps. The first is the dispersion of the filler in a suitable solvent. Then, the solution is mixed with the polymer. And finally, the mixture is casted in a mold and the solvent is evaporated [138, 139]. The choice of the adequate solvent is determined by the solubility of the polymer. Difficulties in dispersing pristine CNT in a solvent by simple stirring often require the use of high-power ultrasonication. An important issue in solution casting systems is the speed at which the solvent is removed [140]. Many PLA-MWCNT composites have been successfully prepared by the solution mixing method [4, 129, 141, 142]. However, for a successfully solution mixing some treatments such as heat-treated [143], acid-treated [132] or functionalization [144, 145] of carbon nanotubes are often necessary to improve the dispersion of the filler.

There have been extensive works that published the melt blending process of CNT with polymers [142, 146-149]. This technique generally involves heat processing of the polymer and CNT in mixing type of equipment such as screw extruder or batch mixer. The mixers impart shear and elongation stress to the process, helping to break apart CNT agglomerates and dispersing them uniformly in the polymer matrix. The extruder is much more versatile because by simply changing the screw configuration (in a twin-screw system), a better control of shear and mixing is obtained [138]. The dispersion of the filler depends on process conditions. Some variables to consider are: proper balance of dispersing mixing and length of the different zones, design of screw, melt temperature, residence time of melt, shear, etc. The melt blending is convenient, especially, by its cost effectiveness, fast production and environmental benefits [150].

Another advantage of the melt methods is that it does not require the use of organic solvents during processing [138]. However, compared with the solution mixing, the melt blending is generally less effective at dispersing nanotubes into polymers, and is limited to lower concentrations of filler due to the high viscosity of the composites with higher contents.

In-situ polymerization involves dispersion of carbon nanotubes in a monomer followed by polymerization [138, 151]. This method enables covalent bonding between functionalized nanotubes and polymer matrix using various chemical reactions. In-situ polymerization process is very attractive for polymers that are thermally unstable or insoluble in solvents. With this technique it is possible to have a high nanotube loading [127].

4.2. Physical Properties of PLA - Functionalized CNT

Some important functional properties of nanocomposites are provided by the nanostructures of the reinforcements added to the matrices due to their larger surface to volume ratio that improves the surface reactivity at nano regime and their molecular level distribution in the polymeric matrices [123, 152]. Many of these properties are precisely controlled by the size and the morphology of the nanoreinforcements, which allows the development of materials in accordance with the demands of consumers and the needs for some applications including biomedical and sensors or transducers, with relatively low concentrations [153]. In the particular case of nanocomposites formed by a PLA matrix and reinforced by carbon nanotubes, some factors can influence in the final properties of these materials; in particular, the class of functionalization of the filler. As suggested along this chapter, the incorporation of functionalized CNT in a PLA polymer matrix leads to changes in the physicochemical properties of the final nanocomposites. The most relevant results reported in the literature about the influence of the functionalization of this filler on the structure, crystallization behavior, permeability and mechanical (dynamic and quasi static) properties, of PLA based nanocomposites are given below.

4.2.1. Structure

In order to achieve improvements in the properties of PLA-CNT composites for their successful implementation on several applications, there are some key issues to be resolved. One is the good dispersion of the filler in the matrix, as well as a major adhesion between both, to ensure low percolation thresholds combined with high mechanical performance. A good interfacial bonding between CNT and matrix, and a perceptible reinforcement effect when the nanotubes are added into the matrix, can bring enhancements in most of the properties of composites. For these reasons it is essential the study the filler dispersion in the matrix. Most published research has investigated this phenomenon by performing extensive studies of the structure of the developed composites.

Several cases where MWCNT were functionalized by the incorporation of carboxyl and/or hydroxyl on their walls resulted with a great dispersion of the filler in a PLA matrix. In this regard, Wu et al., [101] investigated the effect of various types of functionalized MWCNT on the dispersion of the filler in a PLA matrix, and on the final structure of composites prepared by melt-compounding. These authors used as models MWCNT functionalized by carboxylic and hydroxyl groups to compare with purified filler. They revealed that the best dispersion extent of MWCNT within the PLA matrix had been achieved with carboxylic-functionalized MWCNT compared with that of hydroxyl and purified filler (rheological measurements and transmission electron microscopy, TEM, analyses). The authors attributed these results to the nice affinity between PLA chain and the carboxylic groups on the surface of MWCNT, showing that most of MWCNT had been fully dispersed and MWCNT had been randomly oriented, locally forming interconnecting structure.

Other example is the investigation of Xu et al., [154], who published the influence of functionalized multi-walled carbon nanotubes (f-MWCNT) prepared by covalent grafting polylactide. Optical microscope (OM), field-emission scanning electron microscope (FE-SEM) and TEM observations, demonstrated that nanocomposites with f-MWCNT had improved the filler dispersion into the matrix and the interfacial adhesion between the fillers and the polymer, in comparison with the case of PLA reinforced with pristine MWCNT.

Li et al., [155] introduced carboxyl groups to MWCNT treated by mixed acids, and developed PLA/c-MWCNT composites via in-situ polymerization. These authors reported an improvement in the compatibility between the filler and PLA, when MWCNT had been functionalized. There was an interaction force between PLA and c-MWCNT, and even certain PLA segments were grafted on the filler. They showed SEM images of pure MWCNT, carboxylated MWCNT, and PLA/c-MWCNT composites, reveling that pure fillers had assumed hollow cylinder with long caliber and random state, and they had been severely agglomerated. After modification by mixed acids, carboxylated MWCNT presented as short segments with open port, and dispersed well, suggesting oxidization and carboxylation of MWCNT. SEM analysis tested by these authors, finally demonstrated that carboxylated MWCNT had presented an even dispersion into the matrix.

In a recent publication, Seligra et al., [4] studied the structural properties of films based on PLA and reinforced with functionalized carbon nanotubes (fMWCNT) via Fenton reaction (to generate hydroxyl and carboxyl groups on the walls of the filler) and subsequently by a reaction with thionyl chloride and with triethylene glycol. They used three concentrations of functionalized carbon nanotubes to prepare PLA-fMWCNT nanocomposites: C_1 (0.026 wt%), C_2 (0.10 wt%), and C_3 (0.18 wt%). When the authors performed SEM tests, excellent dispersion of the filler into the PLA matrix for composites with filler concentrations till 0.10 wt%, was observed. The cryogenic fracture surface of the composite C_2 showed functionalized carbon nanotubes isolated and well adhered to PLA matrix without fissures and apparent aggregations, preserving the neat filler length (~ 1 μm). They demonstrated that the functionalization of carbon nanotubes did not produced breakage of fillers, probably due to the increased polarity by the functional groups formed on the surfaces of fMWCNT, as well as to the good interactions of –COOH groups with C=O esters from PLA matrix. What is more, they observed a great adhesion between the functionalized filler and PLA.

Chen et al., [156] attached poly (L-lactic acid) to MWCNT functionalized with hydroxyl groups by surface-initiated ring-opening polymerization. They tested the nanostructures of PLLA-MWCNT by Fourier transform infrared spectroscopy (FTIR), TEM, and SEM analysis. The results showed that PLLA layers on filler surface were not uniform but covered the surface like stains. The amount of PLLA grafted increased when these authors used two kinds of solvents (N,N-dimethylformamide, *DMF*, and toluene). They showed that grafted PLLA layer on MWCNT was more uniform with DMF than with toluene, and some parts of PLLA-grafted MWCNT surface prepared in toluene were bare. Finally they reported that MWCNT-g-PLLA prepared for a longer reaction time was more readily dispersed in organic solvents such as chloroform and DMF. The authors attributed this fact not only to the higher PLLA content but also to the smaller area of the bare MWCNT surface.

Others studies showed that functionalization of multi-walled carbon nanotubes when PLA was covalently grafted onto the filler resulted in the conformation of nanocomposites with optimal filler dispersion. For example, Song et al., [17] developed PLA-grafted-MWCNT using biocompatible poly (L-lactic acid) covalently grafted onto the convex surfaces and tips of carbon nanotubes by in-situ polycondensation of the commercially available L-lactic acid monomers. The studies of high-resolution transmission electron microscopy (HRTEM) images of PLA-grafted-MWCNT and of the purified CNT revealed that the outmost layer of purified CNT was not very smooth, which has been attributed to strong acid oxidization and its morphology, unlike to that of the polymer-functionalized CNT materials. They also observed a core/shell structure with CNT at the center for PLA-g-CNT, which has indicated that CNT core was evenly covered with a layer of PLA chains. Finally, the authors concluded that grafting reaction had taken place not only on the convex sidewalls but also at the tips of the oxidized CNT, and moreover the soft polymer shells on the walls and on the tips of CNT had been totally uniform.

In the particular case of the work of Yoon et al., [133], the dispersion of MWCNT-g-PLA and MWCNT-COOH in a PLA matrix has been investigated by SEM test. For PLA/MWCNT-COOH nanocomposites, MWCNT-COOH formed aggregates in PLA and there has been no good adhesion between the filler and the matrix (each MWCNT-COOH was pulled out from PLA). They have attributed this behavior to the van der Waals interaction among MWCNT as well as to the poor interfacial adhesion between MWCNT-COOH and matrix. In the case of PLA/MWCNT-g-PLA, the authors showed that the material containing MWCNT-g-PLA with longer PLA chains displayed better dispersion of the filler in the matrix and relatively smoother surfaces with less pull-out of MWCNT from PLA, demonstrating that longer polylactic chains of MWCNT-g-PLA had better compatibility with the matrix leading to improve the dispersion of this filler in the nanocomposites.

When PLLA/f-MWCNT was prepared using commercial fillers, a fine dispersion of f-MWCNT was achieved throughout the PLLA matrix without any apparent aggregation, for concentrations lower than 2 wt% of filler. However, for PLLA/f-MWCNT with that concentration of filler, slight aggregation of f-MWCNT was observed, indicating that the functionalized carbon nanotubes could be uniformly distributed within the PLLA matrix in low concentrations, but slight aggregations would occur with further increasing the filler content [157].

Papageorgiou et al., [158] worked with poly (L-lactic acid) and 2.5 wt% of oxidized multi-walled carbon nanotubes (o-MWCNT) nanocomposites, prepared by solvent evaporation method. For the uniform dispersion of these nanoparticles into the matrix the authors used a mixture of dichloromethane/tetrahydrofuran 50/50 (w/w), finding that nanoparticles were well dispersed avoiding precipitation. They have attributed the fine dispersion to the interactions between the carboxyl groups of oxidized MWCNT and the hydroxyl end groups of PLA.

Han et al., [159] investigated biodegradable composites based on poly (L-lactic acid) and recycled carbon fiber (RCF), prepared through melting extrusion. In their studies the authors assured that the modification of the surface of RCF was essential to achieve a good reinforcing effect, due to the weak interfacial adhesion of the filler studied. When they observed the surface characteristics of RCF before and after the treatment with a silane coupling agent of 3-glycidoxy-propyltrimethoxysilane, some residues on RCF surfaces were found. The authors attributed this fact to the carbonaceous deposition that resulted from the thermo oxidative decomposition during the recycling process of RCF, which would cause a reduction in the capability of interfacial bonding between RCF and PLA matrix. In the image of RCF after their modification, a textured surface was observed. Si and O elements appearing on the RCF surface were indicatives of the presence of silane coupling agent. The surface-treatment of RCF with silane coupling agent enhanced the interfacial adhesion between RCF and PLLA, and thus PLLA-RCF composites achieved significant improvements in some properties. The morphology of the fracture surface of these nanomaterials indicated that RCF obtained a homogeneous dispersion into PLLA matrix due to the good interfacial interaction.

Taking into account previous reviews, the functionalization process of carbon nanotubes is critical to achieve their good dispersion in a PLA matrix. However, some research of PLA-CNT nanocomposites with non-functionalized filler has been reported. Kobashi et al., [160] investigated the dispersion of non-functionalized MWCNT immersed in PLA with a solvent using melt-process, in order to use them as liquid sensor based on the change of the MWCNT network structure during the solvent contact. Through TEM images, the authors suggested the existence of a three-dimensional MWCNT network like structure within the PLA matrix. They showed an image with single carbon nanotubes and small MWCNT clusters with typical diameters of about 500 nm. Individual nanotubes were interconnected to form these clusters, leading to a conductive network.

To avoid the functionalization of carbon nanotubes, Villmow et al., [80] investigated the influence of twin screw extrusion process conditions on the dispersion of MWCNT in a PLA matrix. Screw and temperature profile, as well as rotation speed were systematically varied during the production of PLA-CNT nanocomposites to determine the impact on the dispersion and distribution of the filler within the matrix. In that work the authors prepared master-batches rather than direct mixing. In all the cases, no large unfilled areas were observed. TEM investigations revealed good distribution of MWCNT with a very dense arrangement of the filler in all master-batches, optimizing the systems.

4.2.2. Crystallinity

It is well known that crystallinity plays an important role in the physicochemical properties and in the biodegradability of nanocomposites. What is more, it affects the structure and morphology of the polymer matrix and the final composites. One important point for the use of composites based on PLA in order to replace synthetic materials is the knowledge of the crystallinity of this polymer. Studying its crystallization behavior is particularly crucial to control the degradation rate, heat resistance, optical, mechanical and barrier properties of PLA composites [161-163]. The crystallization conditions of PLA depend from the different crystal structures presented in this material [164].

Quan et al., [165] investigated the influence of the addition of non-functionalized MWCNT into two different polilactic acid matrices on the crystallinity of the final nanomaterial. They incorporated the filler in PLLA and PDLA through a directly melt mixing process. Through X-ray diffraction, non-isothermal and isothermal crystallization, the authors showed that the formation of stereo complex greatly increased the crystallinity of composites and, meanwhile, MWCNT acted as heterogeneous nucleating agent, which significantly accelerated nucleation and spherulite growth. Therefore, PLLA-MWCNT and PDLA-MWCNT composites presented very low percolations threshold due to the volume exclusion effect.

Other critical factors that affect the behavior of the crystallinity of a composites material are the type of filler used and the treatment received. Regarding PLA-CNT materials, unmodified and modified CNT were investigated in a number of works before incorporating them into the PLA matrix and it was emphasized in the influence of the filler on the crystallinity of the resulting composites [135, 155, 157-159, 164-168].

One example is the investigation by Saeidlou et al., [164], who showed that when very low concentration of multi-wall carbon nanotubes were mixed with PLLA the resulting composites did not produce significant changes in their crystallinity [166, 169, 170], concluding that carbon nanotubes could not play a significant nucleating role for PLA melt processing. Similar conclusions have been found by Shieh et al., [166] in wide angle X-ray diffraction (WAXD) of PLLA based composites reinforced by CNT functionalized via grafting with PLLA (CNT-g-PLLA). The authors demonstrated that modified nanotubes could assist the disorder-to-order transition in PLLA crystals, but didn't lead to a more compact chain packing of the crystal lattice in PLLA composites than in pure PLLA. On the other hand, Wu et al., [136] studied melt and cold crystallization of PLA/CNT composites, obtaining that the addition of the filler played dual roles: as nucleating agent and physical barrier. Their results indicated that the incorporation of CNT accelerated the melt crystallization and the cold crystallization [152].

Zhao et al., [157] reported some results about the effect of the presence of commercial f-MWCNT on no isothermal melt crystallization, overall isothermal melt crystallization kinetics and spherulitic morphology of neat PLLA and its nanocomposites with low filler concentrations. They observed that the crystallization of PLLA in nanocomposites enhanced compared with that of neat PLLA. The authors attributed that behavior to the heterogeneous nucleation of the f-MWCNT used. Furthermore, the overall crystallization rate of PLLA increased with increasing the f-MWCNT contents in the PLLA matrix.

Kuan et al., [135] have prepared materials from poly (L-lactic acid) and multi-walled carbon nanotubes using water-crosslinking technique, in order to improve some physical characteristics of PLA-MWCNT nanocomposites. In particular, in X-ray diffraction investigations they found that a crystalline PLLA could be transformed to a semicrystalline polylactic acid after water-crosslinking reaction, leading to a change in the crystallinity of the final composite.

In other investigations of the crystallinity of PLA-fMWCNT nanocomposites, González Seligra et al., [167] obtained a slightly increase in the crystallinity of the materials when small concentrations of fMWCNT were added. They first showed that PLA pattern seemed to be similar to one of an amorphous material [166, 171. When they calculated the crystallinity of the matrix they obtained less than 1 %. With the addition of functionalized carbon nanotubes appreciable peaks at about $2\ \theta = 16.5\ ^{\circ}$, $19\ ^{\circ}$ and $29.5\ ^{\circ}$ appeared, leading to an increase in the degree of crystallinity of composites. The authors explained this fact in function of the effect of the functionalization of CNT. Nanoparticles act as nucleating agents, settling inside the material that allows to the crystallization of PLA, transforming their composites in semi-crystalline [154].

Nanocomposite films based on carbon nanostructures and PLLA have been also investigated by Armentano et al., [168]. In particular, they studied materials based on PLLA reinforced with pure SWCNT and carboxylated SWCNT, at 1 wt%. These authors demonstrated that different PLLA crystallites were formed and a fraction interface polymer was organized around the nanotube sidewalls. Their investigations also revealed that SWCNT and their COOH groups created heterogeneous nucleation on the carbon nanotube sidewalls.

At the carboxylated nanotubes polymer interface, chemical affinity was modulated and the crystal order was enhanced [130].

The crystallinity of biodegradable poly (D,L-lactide)/carboxyl-functionalized multi-walled carbon nanotubes (c-MWCNT) composites, achieved via in-situ polymerization, have been evaluated by Li et al., [155]. All the characteristic peaks observed by these authors demonstrated great accomplishment of carboxylated MWCNT, and good chemical bonding of PLA and the modified MWCNT. They showed non differences in X-ray diffractions patterns between pure and functionalized fillers, obtaining diffraction peaks at 26.06°, 42.97° and 53.63°. However, the authors detected higher intensity of diffraction peaks and more crystallization peaks in the carboxylated MWCNT composite than in that of pure MWCNT, and the full width at half maximum of the diffraction peak at 26.06° changed from 3.41° to 3.46°. In XRD pattern of PLA/c-MWCNT, not detected diffraction peaks of MWCNT, and the only peak in the observed curve demonstrated an amorphous sample, suggesting the embedding of c-MWCNT by PLA.

Papageorgiou et al., [158] investigated nanocomposites of poly (L-lactic acid) with oxidized multi-walled carbon nanotubes (o-MWCNT) by WAXD tests. The result of matrix showed that it was essentially amorphous, since only the amorphous halo of PLA could be seen in its pattern. The authors demonstrated that nanoparticles were effective nucleating agents. The cold-crystallization was affected by the presence of the filler. On cooling, the melt crystallization was accelerated by the presence of MWCNT, and the effective activation energy was decreased. However, that phenomenon began at lower temperatures and this resulted in the formation of an imperfect crystalline structure which reduced macromolecular chain mobility of the remaining amorphous polymer, limiting the ultimate crystallinity of nanocomposites and consequently, increasing the tendency for recrystallization. Just as previously, in this case the authors attributed the behavior to a nucleation due to the presence of the nanofiller.

When Han et al., [159] studied PLLA-RCF composites, they demonstrated that the addition of RCF with a large surface area greatly influenced on the crystallinity of PLLA. The polymer used was initially semi-crystalline. The neat PLLA exhibited well-developed spherulites with a perfect structure and a large size, indicating the fine crystallinity of the polymer. However, the spherulitic size of PLLA dramatically decreased with the addition of RCF and declined with increasing the filler. The authors deduced that the incorporation of RCF enhanced the capability of nucleation for PLLA due to the heterogeneous nucleating effect. They also showed that the density of the spherulites of PLLA-RCF composites was much higher than that of the neat PLLA. Owing to the nucleating effect of RCF, the nucleation density of PLLA spherulites increased with increasing RCF concentration. Finally, their conclusions were that the presence of RCF in the PLLA matrix significantly had affected not only the spherulitic size but also the overall crystallization process of PLLA.

4.2.3. Mechanical Properties

The excellent mechanical properties of carbon nanotubes suggest that the incorporation of small amounts of this filler into a PLA matrix can lead to nanoconposites materials with significant improvements in their mechanical properties. Some important advancements have been made for improve these properties in PLA-CNT composites in the recent years. As is well known, the dispersion of the filler into the matrix and the adhesion between both are fundamental factors to improve the mechanical properties of PLA-CNT materials, and as has been discussed above, the functionalization of CNT leads to the enhancement of their dispersion and adhesion to the PLA matrix. Here, some relevant results of the most important mechanical, dynamic and quasi static, properties of PLA- functionalized CNT composites after different types of filler functionalization reported in the last years are present.

4.2.3.1. Dynamic Behavior

The dynamic mechanical analysis is a useful and very sensitive technique for the investigation of the microstructure of macromolecular chain conformations and movements during the exposure to a variety of temperatures of the materials conformed by polymers.

Papageorgiou et al., [158] studied the storage modulus and the loss tangent of nanocomposites of polylactic acid with oxidized multi-walled carbon nanotubes (o-MWCNT) as a function of temperature. They observed the behavior of the neat PLA material and nanocomposites containing 2.5 wt% of filler. PLA matrix presented one transition (T_g) around 49 °C, while this parameter slightly shifted to higher values in the case of the nanocomposites. Usually T_g of the polymeric based composites tends to increase with the addition of nanoparticles, due to the interactions between the polymer chains and the fillers and by the reduction of the macromolecular chain mobility at the zone surrounding the nanoparticles [172, 173]. When these authors investigated the intensity of the most important peak, they observed a frequently reported behavior in the literature: this parameter was drastically decreased after the addition of MWCNT. The glass transition is a complex phenomenon which depends on some number of factors such as chain flexibility, molecular weight, branching, cross-linking, intermolecular interactions, and steric effects. The authors explained that the small increase of the temperature of the transition after the addition of MWCNT has been due to the existence of rigid fillers incorporated into the matrix [174]. However, the chain segment mobility of the PLA phase was not influenced drastically by the introduction of carbon nanotubes.

The viscoelastic behavior of PLA based composites is very sensitive not only to the presence of MWCNT but also to their dispersion into the matrix. In this way, Wu et al., [101] studied the dynamic storage modulus (G') and loss modulus (G'') of samples with PLA and carbon nanotubes. They investigated these parameters in function of different types of functionalized MWCNT (carboxylic, hydroxyl and purified fillers), all the cases with identical dimension parameters and special surface area. In the case of the carboxylic fillers, several concentrations were studied. The authors noted that G' increased monotonously with increase of MWCNT content due to the reinforcement effect of the MWCNT and, at relative lower concentration (less than 2 wt%), all the composites with carboxylic filler showed higher G'' than G' at low-frequency region, indicating that the relaxation behavior was still dominated by local PLA dynamics. When the authors explored the influence of the functionalization of the filler on the mechanical behavior of composites, they observed that PLA with hydroxyl CNT presented highest values of G'' compared with those of PLA-carboxylic CNT and PLA-purified filler samples. As all samples had identical MWCNT concentrations, same dimension parameters and special surface area, the differences in the modulus were indicative of the different dispersion states of MWCNT as a result of the functionalization methods. Finally they concluded that the carboxylic functionalization yielded composites with better dispersed fillers than those of the purified filler. The non-functionalized filler had poor dispersion due to the large difference in its affinity with the PLA matrix.

Other work that investigated the effects of the addition of carboxylated carbon nanotubes on dynamic mechanical properties of nanocomposites based on PLLA was the published by Armentano et al., [168]. In their study, it is reported that the addition of only 1 wt% of both SWCNT and carboxylated SWCNT into the matrix increased storage modulus, possibly due to a reduction in polymer chain molecular mobility at the PLLA/SWCNT interface [175]. However, when these authors compared the dynamical mechanical parameters between the composites with non-functionalized SWCNT and modified fillers, they observed the highest storage modulus in the case of PLLA/SWCNT-COOH nanocomposites and also a shift in their glass transition temperature, exhibiting a better interaction with the polymer matrix than that of the non-functionalized SWCNT material. The authors attributed this last behavior to the partial decrease in PLLA chain mobility due to the presence of SWCNT and COOH groups.

When González Seligra et al., [167] evaluated the dynamic mechanical curves of PLA-fMWCNT bionanofilms, they revealed that the loss tangent of all the materials studied presented one important thermal transition in the range of 50-55 °C, corresponding to the predominant material, PLA. The matrix had its transition at 50 °C, and when the filler was added, the temperature of the transition slightly shifted to higher temperatures. The authors attributed these results to the higher surface area promoted by the incorporation of the nanotubes and the very good adhesion between the matrix and the functionalized fillers [73, 176-178]. The relaxation observed in the curves of loss tangent corresponded to the sharp drop in storage modulus. Finally, the authors reported an increment in storage modulus at room temperature with the filler increase, becoming almost 50 % with just 0.55 wt% of fMWCNT.

Zhao et al., [157] also studied PLLA based composites with very small quantity of functionalized CNT. In their investigation the authors used commercial f-MWCNT. They observed that G' of PLLA/f-MWCNT nanocomposites improved significantly compared to that of the neat PLLA. They explained such results by some physical interactions between f-MWCNT and the PLLA matrix due to the high surface area of f-MWCNT and by the stiffening effect of the filler in the matrix. It should be noted that the increment of G' was significant with filler concentrations from 0 to 0.5 wt%; however, the difference was very slight when f-MWCNT increased from 0.5 to 2 wt%, indicating that such effect was more pronounced at lower filler content.

Han et al., [159] also demonstrated that the incorporation of their functionalized filler (RCF) into PLLA improved significantly the storage modulus of composites. The fact of this behavior was the rigidity effect and the orientation distribution of the filler in the matrix as well as the interfacial interaction between RCF and matrix due to the great functionalization. When these authors evaluated the loss tangent curves as function of the temperature, only one peak corresponding to the glass transition temperature of PLA was observed, which seemed to shift slightly to higher temperatures with the incorporation of the filler, due to the fact that RCF, as rigid load, confined the motion of the PLLA segments. They also noted that the values of tan δ of PLLA-RCF composites were larger than that of neat PLLA. This was attributed to the high rigidity of the PLLA chains as well as to the friction loss between the interfaces of PLLA and RCF. Finally, the authors mentioned that a surface treatment of RCF with a coupling agent improved the interfacial adhesion between the filler and the matrix, and thus a significant enhancement of mechanical properties happened.

On the other hand, Shieh et al., [166] examined the effect of the functionalization of CNT via grafting modification with PLLA (CNT-g-PLLA) on dynamic mechanical properties of PLLA/CNT-g-PLLA composites. The authors showed that both storage modulus and glass transition temperature of composites were greater for higher CNT-g-PLLA concentrations. The storage modulus at 30 ºC (below T_g) of pure PLLA increased about 2.3 times compared to that of PLLA/CNT-g-PLLA composite with 5 wt% of filler. The enhancement of this parameter in the composite suggested that filler-matrix interaction was strong, concluding that CNT via grafting modification with PLLA could be well dispersed in the PLLA matrix resulting in an effective reinforcing effect. This great dispersion presented a single T_g in all the materials, giving the highest values in the case of the composites. This indicated that the mobility of PLLA chains in matrix was reduced by the functionalization of CNT. In contrast, when composites with unmodified CNT were evaluated, shifts of T_g towards lower temperatures with respect to matrix were observed, indicating a poor dispersion of the filler in the PLLA matrix.

Xu et al., [154] used a covalent functionalization to modify MWCNT by a reaction of five-armed star polylactide with MWCNT–COCl to prepare PLA nanocomposites. They found that the values of G' and G'' for composites containing 0.5 and 1.0 wt% of f-MWCNT did not differ much from each other, but they increased from 2.0 wt% of filler. In those cases the slopes of the modulus curves changed significantly. The transition behaviors at low frequencies demonstrated that the long-range chains motion of polymer was restrained significantly by the f-MWCNT network. Meanwhile, the frequency dependence of G'' also showed similar tendency than G'. Frequency dependence of loss tangent was dependent on filler concentration. For composites containing 0.5 and 1.0 wt% of f-MWCNT, tan δ decreased with increasing frequency. When the concentration of filler reached 2.0 wt% or above it, a moderate increase of tangent appeared with increasing frequency, so the f-MWCNT concentration of percolation was between 1.0 and 2.0 wt%. In that study the functionalization of MWCNT enhanced the compatibility between f-MWCNT and PLA, which contributed to the low percolation concentration for the composites.

The mentioned examples in this part of the chapter have shown that dynamic mechanical behavior of PLA based composites reinforced by functionalized CNT when fillers are well dispersed, typically behaves as follows: with the increment of CNT, a shift of transition peak slightly towards higher temperatures happens, and storage modulus calculated at room temperature increases. This behavior is the result of the hindered chain and segmental mobility of polymer due to the addition of the filler. However, as also was shown, to improve these dynamic mechanical parameters is necessary the functionalization of the filler.

4.2.3.2. Quasi Static Behavior

It is well established in the literature that mechanical effectiveness of reinforced materials depends on the stress transfer from the external filler to the reinforcing phase [63]. This efficiency is determined by the available interfacial area between the filler and the matrix and also by the strength of this interface, which enhances when the area-to-volume ratio of fillers and the interfacial adhesion, increase. The interfacial interaction between the filler and the matrix is an important factor that affects not only physicochemical properties of composites, also their static mechanical behavior. In the case of poor adhesion, the interfacial layer can't transfer stress. If the adhesion between matrix and filler isn't good, the loading stresses can't be transferred to the filler and a reduction in the effective surface area can be obtained. Thus, ultra-high strength of PLA-CNT composites demands uniform dispersion of the nanofillers into the polymer matrix and, consequently, strong interaction between CNT and PLA [130, 133, 135, 156, 159, 166, 179-187].

In the literature there are several researches on the behavior of tensile properties of composites formed by PLA and reinforced by functionalized CNT. In most cases it can be noted that stress-strain curve presents a slight strain range where stress increases linearly with strain (at low strains), followed by a non-linear behavior, arriving finally to a plastic deformation region till a breaking strain [47, 179, 182].

An example of this behavior has been studied by Lin et al., [179]. These authors investigated the influence of the functionalization of MWCNT by carboxyl acid and hydroxyl groups on tensile curves. They observed that stress increased sharply during the initial stage of stretching and, subsequently, it reduced until finally fracture occurred. The tensile strength of PLGA/c-MWCNT films was enhanced nearly three-fold compared to that of the pure PLGA materials and approximately two-fold compared to that of PLGA/MWCNT composites. PLGA/c-MWCNT presented its elastic modulus around 82 % higher than that of the pure PLGA, while PLGA/MWCNT acquired a value of G' almost 73 % major compared to that of the matrix. Through tensile tests, the authors demonstrated that the presence of the functionalized MWCNT in PLGA nanocomposites could efficiently enhance the reinforcement by improving their dispersion in this polymer. This reinforcement effect depends on the interfacial adhesion and interface area between CNT and polymer. The acidic treatment process with carboxyl and hydroxyl groups formed hydrogen bonds with the polymer to enhance the interfacial adhesion, increasing the interface area of MWCNT and the matrix. The homogeneous distribution of c-MWCNT in PLGA helped in the effective large aspect ratio of filler and strong interaction between CNT and the polymer [130, 180, 181]. Hence, the authors could conclude that c-MWCNT with homogeneous dispersion in the PLGA matrix were the most effective reinforcement material.

Recently, Yoon et al., [133] also studied the influence of carboxylic acid-functionalized MWCNT (MWCNT-COOH) in PLLA based composites prepared by ring-opening polymerization. They observed that PLA/MWCNT-COOH nanocomposites had slightly higher storage modulus and tensile strength than the matrix when they were measured at room temperature, exhibiting values of about 1928 MPa and 49.3 MPa, respectively. On the other hand, these authors reported that the surface resistivity of the composites was much lower than that of the neat PLA.

When Wu and Liao, [13] measured the tensile properties of PLA/multihydroxyl-functionalized MWCNT composites, they showed that all the materials presented linear range at very low strain, and then a plastic deformation. PLA matrix yielded a breaking strain up to 4 %, which decreased with the increase of the filler content (happening at approximately 2.5 % for the composite with 3 wt% of MWCNT), followed by a stiffening of the material. They attributed their results to the orientation of the layer of the filler, as well as to the interfacial adhesion between polymer and the modified MWCNT.

Chen et al., [156] also evaluated mechanical properties of composites formed by MWCNT modified by the introduction of hydroxyl groups in order to obtain better disperse of the filler in a PLA matrix. Besides, they modified the L-lactide by the surface-initiated ring-opening polymerization. These authors showed that upon incorporation of only 1 wt% of filler, the tensile modulus and tensile strength of PLLA composites enhanced 91 % and 52 %, respectively, compared to the matrix. They explained their results in terms of a strong adhesion between PLLA and carbon nanotubes as a result of their high miscibility due to the great process of functionalization of the filler.

Seligra et al., [182] observed the stress-strain curves of PLA based composites reinforced with MWCNT functionalized by –COOH groups. They showed the following behavior: a linear range till ~ 1 % for the matrix and ~ 0.8 % for the composites, followed by a non-linear zone until failure. Important improvements in the tensile strength values have been found by these authors with the incorporation of very small amounts of functionalized carbon nanotubes into the PLA matrix, reaching almost 20 % in the case of the nanocomposites with only 0.10 wt% of fMWCNT, accompanied by a reduction in ε_b and in the toughness. Significant improvements in tensile strength could be due to the high modulus of the fillers as well as to their homogeneous dispersion in the PLA matrix and good interfacial adhesion achieved [4]. The authors also attributed the important and positive effect on tensile strength of the composites with the addition of very small amount of functionalized carbon nanotubes to various factors: the stiffness of the filler, which contributed to the presence of partially immobilized polymer phases [188], their high aspect ratio and surface area, as well as to the nano scale dispersion of the filler in the matrix by the great functionalization process.

Another study on the influence of the functionalization of CNT on the static mechanical properties of PLA-CNT composites has been reported by Mina et al., [183]. These authors compared the tensile strength and the storage modulus of pure PLA films and PLA based composites with different concentrations of non-functionalized CNT and functionalized filler by nitric acid treatment (0.5 wt%, 1.0 wt% and 1.5 wt% of filler). They observed low increments in mechanical properties of the composites with both untreated and modified nanotubes concentration, probably because of crystalline structures of PLA and the interaction of its molecules with the nanotubes. However, the best results have been found in the case of composites with functionalized MWCNT. It is known that acid treatment can increase surface roughness on the filler layers, facilitating the hydrogen bonding between the fillers and the PLA molecules [13], and lowering tensile values can be attributed to the poor dispersion of MWCNT due to the formation of small agglomerations.

Kuan et al., [135] investigated MWCNT/PLLA composites using two different matrices (with low- and high-crystalline poly (L-lactic) acid), and water-crosslinking technique, on physical properties. They observed a trend of increasing in both tensile and flexural strength, with MWCNT content. In the case of MWCNT/low-crystalline PLLA materials, the tensile strength increased 14.4 % with only 0.5 wt% of filler. The homogeneous distribution of MWCNT in the low-crystalline matrix (LC-PLA) increased the interface bonding and the shear stress between MWCNT and the polymer. Without interfacial adhesion between fillers and matrix, the tensile strength would decrease as the filler content increases [129, 189]. In the same work, these authors modified MWCNT by surface-grafted maleic anhydride (MA-g-MWCNT) and blended them with LC-PLA to improve the bonding between the filler and PLA, and, thereby, enhance mechanical properties of composites. Materials with coupling agents exhibited greater tensile strength than untreated ones due to the improvement of interfacial bonding. Finally, the reported results indicated that the modification of the surface of carbon nanotubes enhanced the compatibility between the polymer and the filler, increasing tensile properties, and, what is more, this parameter improved when covalent or hydrogen bonding between MA-g-MWCNT and LC-PLA were made.

PLA based nanocomposites with expanded graphite, conformed by melt-blending technique have been prepared by Fukushima et al., [184]. On mechanical properties of those materials a significant enhancement in stiffness with the addition of nanofillers was observed. However, the tensile strength remained practically constant and was similar to that of matrix. Significant increments in Young's modulus in all the composites also were reported, reaching more than 20 % in composites with 3 wt% of filler. The authors attributed this last result to the good dispersion and reinforcement effect of the fillers used.

Feng et al., [185] investigated PLLA composites reinforced by two kinds of multi-walled carbon nanotubes: functionalized (Fe_3O_4/MWCNT) and non-functionalized (commercial MWCNT), prepared by solution. When these authors compared the mechanical variables of the different composites with those of the PLLA matrix, they showed that the tensile strength values of both composites slightly increased as the filler concentration increased; however, this parameter was higher in the case of the addition of Fe_3O_4/MWCNT. On the other hand, Young's modulus and elongation at break also markedly improved in this last composite (from 211 MPa to 321 MPa and from 38 % to 73 %, respectively), due to the good dispersion of the functionalized filler into the polymer matrix.

Another example of the improvement in tensile properties of PLA based composites with increasing CNT when they are well dispersed has been reported by Han et al., [159], who showed results of the tensile and flexural measurements of PLLA/RCF composites with the variation of the RCF concentration. A significant improvement in both tensile and flexural strength was achieved with the addition of only 3 wt% of filler. The trend to the increment of the mechanical strength was maintained with increasing the filler content. The authors attributed the significant enhancement in the mechanical properties of the composites to the improved interfacial shear strength between the filler and the polymer matrix [190].

Among the many studies reviewed of nanocomposites based on PLA and reinforced by non-functionalized carbon nanotubes, Park et al., [186] described the stress-strain curves of materials formed through a melt process using a twin-screw extruder. The tensile strength and modulus of PLA/CNT nanocomposites were enhanced as the filler content increased because of the nano reinforcing effect of CNT by their high aspect ratio. They also reported a gradual decrement in elongation at break with increasing CNT concentration, indicating that PLA/CNT materials were more brittle by the incorporation of CNT. These authors explained the increases of the storage modulus of PLA/CNT to the great nanotube-nanotube and nanotube-matrix interactions. However, the first interaction became more dominant with increasing the CNT content and led to interconnected structures of CNT in PLA/CNT nanocomposites, resulting in agglomerations that disturbed the efficient transfer from the filler. Therefore, a less uniformly dispersion and a highly concentration of CNT could disturb efficient CNT transfer to the PLA matrix at higher filler content [187]. Finally, the authors concluded that just low contents of carbon nanotubes could uniformly be dispersed in the matrix because, at higher filler amounts, some agglomerations appeared, resulting in stress concentrations.

CONCLUSION

In this chapter it shows the discussion of the most important recent results on the physical properties of nanocomposites based on polylactic acid and reinforced by functionalized carbon nanotubes. Some different methods of functionalization used in order to obtain optimum dispersion of CNT in a PLA matrix were exposed. Important attention on covalent and non-covalent techniques and in the variables involved in such functionalization methods on the behavior of the final properties of composites has been paid. In particular, factors that affect the structure, crystallinity and mechanical (dynamic and quasi static) properties of PLA- functionalized CNT nanocomposites have been presented. It was found that most of the studies reviewed are agreed on the need to functionalize carbon nanotubes prior to incorporate them into a PLA matrix for their better dispersion and great adhesion between them and the matrix. This improvement has been attributed to the modification of the surface of carbon nanotubes that would enhance the compatibility between the polymer and the fillers, increasing the physical properties of the conformed composites.

In the case of the crystallinity, some researchers reported increments in the intensity of the peaks presented in the X-ray diffraction studies of PLA-CNT composites when functionalized carbon nanotubes were added. It summarized this fact as a function of the effect of the modification of the filler, and it concluded that these nanoparticles act as nucleating agents, settling inside of the material allowing to the crystallization of PLA, and transforming the composites in semi-crystalline.

When mechanical properties were reviewed, in practically all the investigations similar behavior was observed. The dynamic mechanical behavior of PLA-CNT when the filler was well dispersed, typically was as follows: one transition was detected in tangent curves corresponding to the PLA domination and, with the increment of the functionalized CNT, a shift of the transition peak to higher temperatures happened, as well as the storage modulus at room temperature increased. This behavior is the result of the interaction between the polymer chains and the filler as well as to the reduction of the macromolecular chains mobility at the zone surrounding the nanoparticles. In the tensile properties of PLA-CNT composites it was found that the stress-strain curves presented a little linear strain range, at the beginning, followed by a non-linear behavior, and finally arriving to a plastic deformation till a breaking strain. When functionalized carbon nanotubes were incorporated, increasing trends of tensile strength were observed. This enhancement was attributed to the efficiency of the stress transfer from the external fillers to the reinforcing phase, due to the available interfacial area between the filler and matrix and by the strength of the interface that increased as the area-to-volume ratio of CNT and the interfacial adhesion increased. When the adhesion was poor, the interfacial layer could not transfer stress.

From the results presented above, it was found that the reviewed research in this chapter provided coherent and plausible arguments on improving physical properties of composites based on PLA and reinforced by carbon nanotubes when the filler is functionalized.

ACKNOWLEDGMENTS

The author wishes to thank the National Research Council of Argentina (CONICET) (Project: PIP 11220090100699), University of Buenos Aires (UBACyT 2010-2011, 2011-2014 and 2012-2015, Projects: 20020090300055, 20020100100350 and 20020110200196, respectively), Lic. Paula González Seligra, and Drs. Melisa Lamanna and Silvia Goyanes, for their support.

REFERENCES

[1] Famá, L., Goyanes, S., Gerschenson, L., (2007). Influence of storage time at room temperature on the physicochemical properties of cassava starch films. *Carbohydr. Polym.* 70, 265.

[2] García, N.L., Famá, L., Dufresne, A., Aranguren, M., Goyanes, S., (2009). A comparison between the physico-chemical properties of tuber and cereal starches. *Food Res. Int.* 42, 976.

[3] Siqueira, G., Bras, J., Dufresne, A., (2009). A: Cellulose whiskers versus microfibrils: influence of the nature of the nanoparticle and its surface functionalization on the thermal and mechanical properties of nanocomposites. *Biomacromolecules* 10, 425.

[4] Seligra, P., Nuevo, F., Lamanna, M., Famá, L., (2013). Covalent grafting of carbon nanotubes to PLA in order to improve compatibility. *Compos. Part B- Eng.* 46, 61.

[5] Rudnik, E., (2013). *Compostable polymer properties and packaging applications.* In S. Ebnesajjad, (Eds.), *Plastic Films in Food Packaging*, Elsevier: Pennsylvania, US, pp 217.

[6] Zhang, Z., Ortiz, O., Goyal, R., Kohn, J., (2014). *Biodegradable polymers.* In R. Lanza, R. Langer, J.P. Vacanti, (Eds.), *Principles of Tissue Engineering*, Academic Press: US, pp 441.

[7] Yu, L., Dean, K., (2006). Polymer blends and composites from renewable resources. *Prog. Polym. Sci.* 31, 576.

[8] Auras, R., Harte, B., Selke, S., (2004). An overview of polylactides as packaging materials. *Macromol. Biosci.* 4, 835.

[9] Zhang, J.F., Sun, X., (2005). *Poly(lactic acid) based bioplastics.* In R. Smith, (Eds.), *Biodegradable Polymers for Industrial Applications*, Woodhead Publishing Limited: London, UK, pp 251.

[10] Doi, Y., Steinbuchel, A., (2002). Biopolymers, Volume 4, Polyesters III - Applications and Commercial Products, Wiley-VCH: Berlin, DE, pp 410.

[11] Domenek, S., Courgneau, C., Ducruet, V., (2011). *Characteristics and applications of poly (lactide).* In S. Kalia, L. Avérous, (Eds.), *Biopolymers: Biomedical and Environmental Applications*, Wiley-Scrivener: New York, US, pp 183.

[12] Harrison, B.S., Atala, A., (2007). Carbon nanotube applications for tissue engineering. *Biomaterials* 28, 344.

[13] Wu, C.-S., Liao, H.-T., (2007). Study on the preparation and characterization of biodegradable polylactide/multi-walled carbon nanotubes nanocomposites. *Polymer* 48, 4449.

[14] Sorrentino, A., Gorrasi, G., Vittoria, V., (2007). Potential perspectives of bio-nanocomposites for food packaging applications. *Trends Food Sci. Technol.* 18, 84.

[15] Hansen, N.M., Plackett, D., (2008). Sustainable films and coatings from hemicelluloses: A review. *Biomacromolecules* 9, 1494.

[16] Petinakis, E., Liu, X., Yu, L., Way, C., Sangwan, P., Dean, K., Bateman, S., Edward, G., (2010). Biodegradation and thermal decomposition of poly(lactic acid)-based materials reinforced by hydrophilic fillers. *Polym. Degrad. Stabil.* 95, 1704.

[17] Song, W., Zheng, Z., Tang, W., Wang, X., (2007). A facile approach to covalently functionalized carbon nanotubes with biocompatible polymer. *Polymer* 48, 3658.

[18] Sundqvist, P., Garcia-Vidal, F.J., Flores, F., Moreno, M.M., Navarro, C.G., Bunch, J.S., Herrero, J.G., (2007). Voltage and length-dependent phase diagram of the electronic transport in carbon nanotubes. *Nano Lett.* 7, 2568.

[19] Wisse, E., Govaert, L.E., Meijer, H.E., Meijer, E.W., (2006). Unusual tuning of mechanical properties of thermoplastic elastomers using supramolecular fillers. *Macromolecules* 39, 7425.

[20] So, H.H., Cho, J.W., Sahoo, N.G., (2007). Effect of carbon nanotubes on mechanical and electrical properties of polyimide/carbon nanotubes nanocomposites. *European Polym. J.* 43, 3750.

[21] Zilli, D., Goyanes, S., Escobar, M.M., Chiliotte, C., Bekeris, V., Cukierman, A.L., Rubiolo, G.H., (2007). Comparative analysis of electric, magnetic, and mechanical properties of epoxy matrix composites with different contents of multiple walled carbon nanotubes. *Polym. Composite* 28, 612.

[22] Choudhary, V., Gupta, A., (2011). *Polymer/carbon nanotube nanocomposites.* In S. Yellampalli, (Eds.), *Carbon Nanotubes - Polymer Nanocomposites*, Intech: Rijeka, HR, pp 65.

[23] Suhr, J., Zhang, W., Ajayan, P., Koratkar, N., (2006). Temperature activated interfacial friction damping in carbon nanotube polymer composites. *Nano Lett.* 6, 219.

[24] Kim, J.Y., Han, S., Hong, S., (2008). Effect of modified carbon nanotube on the properties of aromatic polyester nanocomposites. *Polymer* 49, 3335.

[25] Lyons, K., (2010). Nanotechnology: Transforming food and the environment. *Food First Backgrounder* 16, 1.

[26] Avérous, L., Boquillon, N., (2004). Biocomposites based on plasticized starch: thermal and mechanical behaviours. *Carbohydr. Polym.* 56, 111.

[27] Wang, Y., Rakotonirainy, M., Papua, W., (2003). Thermal behavior of Zeína-base biodegradable film. *Starch/Stärke* 55, 25.

[28] http://www.ceresana.com/en/market-studies/plastics/bioplastics/

[29] Södergård, A., Stolt, M., (2002). Properties of lactic acid based polymers and their correlation with composition. *Prog. Polym. Sci.* 27, 1123.

[30] Middleton, J.C., Tipton, A.J., (2000). Synthetic biodegradable polymers as orthopedic devices. *Biomaterial* 21, 2335.

[31] Fiore, G., Jing, F., Young, V.Jr., Cramer, C., Hillmyer, M., (2010). High Tg aliphatic polyestersby the polymerization of spirolactide derivatives. *Polym. Chem.* 1, 870.

[32] Garlotta, D., (2001). A literature review of poly(lactic acid). *J. Polym. Environ.* 9, 63.

[33] http://www.eis.uva.es/~macromol/curso08-09/pla/Pag%20web/acido%20polilactico.html

[34] Avérous, L., (2013). *Syntesis, properties, environmental and biomedical applications of polylactic acid.* In S. Ebnesajjad, (Eds.), *Handbook of Biopolymers and Biodegradable Plastics: Properties, Processing and Applications*, Wiley & Srivener Publishing: Oxford, UK, pp 171.
[35] Suzuki, S., Ikada, Y., (2010). *Medical applications.* In R. Auras, L.-T. Lim, S. Selke, H. Tsuji, (Eds.), *Poly(Lactic Acid): Synthesis, Structures, Properties, Processing, and Applications*, Wiley & Sons: New York, US, pp 443.
[36] Ramakrishna, S., Mayer, J., Wintermantel, E., Leong, K.W., (2001). Biomedical applications of polymer composite materials: A review. *Compos. Sci. Technol.* 61, 1189.
[37] http://www.natureworkllc.com
[38] http://www.natureworksllc.com/Product-and-Applications/Fresh-Food-Packaging.as px
[39] Sin, L.T., Rahmat, A.R., Rahman, W.A., (2013). *Applications of poly(lactic acid).* In S. Ebnesajjad, (Eds.), *Handbook of Biopolymers and Biodegradable Plastics: Properties, Processing and Applications*, Wiley & Srivener Publishing: Oxford, UK, pp 55.
[40] http://www.natureworksllc.com/Product-and-Applications/Films.aspx
[41] Lu, X., Chen, Z.X., (2005). Curved pi-conjugation, aromaticity, and the related chemistry of small fullerenes (< C-60) and single-walled carbon nanotubes. *Chem. Rev.* 105, 3643.
[42] Ajayan, P.M., (1999). Nanotubes from carbon. *Chem. Rev.* 99, 1787.
[43] Baughman, R.H., Zakhidov, A.A., de Heer, W.A.A., (2002). Carbon nanotubes - the route toward applications. *Sci.* 279, 787.
[44] Komarov, F.F., Mironov, A.M., (2004). Carbon nanotebes: Present and future, *Phys. Chem. Solid State* 5, 411.
[45] Iijima, S., (1991). Helical microtubules of graphitic carbon. *Nature* 354, 56.
[46] Moniruzzaman, M., Winey, K.I., (2006). Polymer nanocomposites containing carbon nanotubes. *Macromolecules* 39, 5194.
[47] Awasthi, K., Srivastava, A., Srivastava, O.N., (2005). Ballmilled carbon and hydrogen storage. *J. Nanosci. Nanotechnol.* 5, 1616.
[48] Sinha, N., Yeow, J.T.-W., (2005). Carbon nanotubes for biomedical applications. *IEEE Trans. Nanobiosci.* 4, 180.
[49] Goyanes, S., Bernal, C., Famá, L., (2013). *Application of Carbon Nanotubes (CNTs) in the Development of Nanocomposites Based on Biodegradable Polymers.* In A.K. Mishra, (Eds.), *Carbon Nanotubes: Synthesis and Properties*, Nova publisher: Hardcover, US, pp 157.
[50] Baughman, R.H., Cui, C.X., Zakhidov, A.A., Iqbal, Z., Barisci, J.N., Spinks, G.M., Wallace, G.G., Mazzoldi, A., De Rossi, D., Rinzler, A.G., Jaschinski, O., Roth, S., Kertesz, M., (1999). Carbon nanotube actuators. *Sci.* 284, 1340.
[51] Dresselhaus, M.S., Dresselhaus, G., Eklund, P.C., (1999). Science of Fullerenes and Carbon Nanotubes, Academic Press: San Diego, CA, pp 965.
[52] Harris, P.J., (1999). Carbon Nanotube and Related Structures. New Materials for the 21st Century, Cambridge University Press: Cambridge, EN, pp 287.
[53] Uchida, T., Kumar, S., (2005). Single wall carbon nanotube dispersion and exfoliation in polymers. *J. Appl. Polym. Sci.* 98, 985.

[54] Louri, O., Wagner, H.D., (1998). Transmission electron micros- copy observations of fracture of single-wall carbon nanotubes under axial tension. *Appl. Phys. Lett.* 73, 3527.

[55] Coleman, J., Khan, U., Blau, W., Gunko, Y., (2006). Small but strong: A review of the mechanical properties of carbon nanotube–polymer composites. *Carbon* 44, 1624.

[56] Matayabas, J.C., Turner, S.R., (2000). *Nanocomposite technology for enhancing the gas barrier of polyethylene terephthalate.* In T.J. Pinnavaia, G.W. Beall, (Eds.), *Polymer-Clay Nanocomposites*, Wiley & sons: New York, US, pp 207.

[57] Sung, Y.T., Kum, C.K., Lee, H.S., Byon, N.S., Yoon, H.G., Kim, W.N., (2005). Dynamic mechanical and morphological properties of polycarbonate/multi-walled carbon nanotube composites. *Polymer* 46, 5656.

[58] De Azeredo, H.M., (2009). Nanocomposites for food packaging applications. *Food Res. Int.* 42, 1240.

[59] Bose, S., Khare, R.S., Moldenaers, P., (2010). Assessing the strengths and weaknesses of various types of pre-treatments of carbon nanotubes on the properties of polymer/carbon nanotubes composites: A critical review. *Polymer* 51, 975.

[60] Zapata-Solvas, E., Gomez-García, D., Poyato, R., Lee, Z., Castillo-Rodroguez, M., Dominguez-Rodriguez, A., Radmilovic, V., Padrure, N.P., (2010). Mircrostructural effects on the creep deformation alimuna/single-wall carbón nanotubes composites. *Am. Ceram. Soc.* 59, 1464.

[61] Dresselhaus, M., Dresselhaus, G., Avorius, P., (2001). Carbon nanotubes: synthesis structure, properties and applications, Springer: US, pp 448.

[62] Nuriel, S., Liu, L., Barber, A.H., Wagner, H.D., (2005). Direct measurement of multiwall nanotube surface tension. *Chem. Phys. Lett.* 404, 263.

[63] Qian, D., Dickey, E.C., Andrews, R., Rantell, T., (2000). Load transfer and deformation mechan- isms in carbon nanotube-polystyrene composites. *Appl. Phys. Lett.* 76, 2868.

[64] Jin, L., Bower, C., Zhou, O., (1998). Alignment of carbon nanotubes in a polymer matrix by mechanical stretching. *Appl. Phys. Lett.* 73, 1197.

[65] Wang, Y., Wu, J., Wei, F., (2003). A treatment method to give separated multiwalled carbon nanotubes with high purity, high crystallization and a large aspect ratio. *Carbon* 41, 2939.

[66] Cadek, M., Coleman, J.N., Ryan, K.P., Nicolosi, V., Bister, G., Fonseca, A., Nagy, J.B., Szostak, K., Beguin, F., Blau, W.J., (2004). Reinforcement of polymers with carbon nanotubes: The role of nanotube surface area. *Nano Lett.* 4, 353.

[67] Sandler, J K., Kirk, J.E., Kinloch, I.A., Shaffer, M.S., Windle, A.H., (2003). Development of a dispersion process for carbon nanotubes in an epoxy matrix and the resulting electrical properties. *Polymer* 44, 5893.

[68] Wei, C., Srivastava, D., Cho, K., (2002). Thermal expansion and diffusion coefficients of carbon nanotube-polymer composites. *Nano Lett.* 2, 647.

[69] De Falco, A., Fascio, M.L., Lamanna, M.E., Corcuera, M.A., Mondragon, I., Rubiolo, G.H., D'Accorso, N.B., Goyanes, S., (2009). Thermal treatment of the carbon nanotubes and their functionalization with styrene. *Phys. B: Condens. Matter* 404, 2780.

[70] Ruan, S.L., Gao, P., Yang, X.G., Yu, T.X., (2003). Toughening high performance ultrahigh molecular weight polyethylene using multiwalled carbon nanotubes. *Polymer* 44, 5643.

[71] Ajayan, P.M., Zhou, O.Z., (2001). Applications of carbon nanotubes. Carbon nanotubes: Synthesis, structure, properties, and applications. *Topics Appl. Phys.* 80, 391.

[72] Famá, L.M., Pettarin, V., Goyanes, S.N., Bernal, C.R., (2011). Starch based nanocomposites with improved mechanical properties. *Carbohydr. Polym.* 83, 1226.

[73] Famá, L., Gañán Rojo, P.G., Bernal, C., Goyanes, S., (2012). Biodegradable starch based nanocomposites with low water vapor permeability and high storage modulus. *Carbohydr. Polym.* 87, 1989.

[74] Deng, P., Xu, Z., Li, J., (2013). Simultaneous determination of ascorbic acid and rutin in pharmaceutical preparations with electrochemical method based on multi-walled carbon nanotubes-chitosan composite film modified electrode. *J. Pharmaceut. Biomed.* 76, 234.

[75] Mamedov, A.A., Kotov, N.A., Prato, M., Guldi, D.M., Wicksted, J.P., (2002). Hirsch, A. Molecular design of strong single-wall carbon nanotube/polyelectrolyte multilayer composites. *Nat. Mater.* 1, 190.

[76] Chen, J., Rao, A.M., Lyuksyutov, S., Itkis, M.E., Hamon, M.A., Hu, H., Cohn, R.W., Eklund, P.C., Colbert, D.T., Smalley, R.E., Haddonet, R.C., (2001). Dissolution of full-length single-walled carbon nanotubes. *J. Phys. Chem. B* 105, 2525.

[77] Frankland, S.J., Caglar, A., Brenner, D.W., Griebel, M., (2002). Molecular simulation of the influence of chemical cross-links on the shear strength of carbon nanotube-polymer interfaces. *J. Phys. Chem. B* 106, 3046.

[78] Breuer, O., Sundararaj, U., (2004). Big returns from small fibers: a review of polymer/carbon nanotube composites. *Polym. Compos.* 25, 630.

[79] Xia, H., Song, M., Jin, J., Chen, L., (2006). Poly(propylene glycol)-grafted multi-walled carbon nanotube polyurethane. *Macromol. Chem. Phys.* 207, 1945.

[80] Villmow, T., Potschke, P., Pegel, S., Haussler, L., Kretzschmar, B., (2008). Influence of twin-screw extrusion conditions on the dispersion of multi-walled carbon nanotubes in a poly(lactic acid) matrix. *Polymer* 49, 3500.

[81] Liu, L., Wagner, H.D., (2005). Rubbery and glassy epoxy resins reinforced with carbon nanotubes. *Compos. Sci. Technol.* 65, 1861.

[82] Lu, K.L., Lago, M., Chen, Y.K., Green, M.L., Harris, P.J., Tsang, S.C., (1996). Mechanical damage of carbon nanotubes by ultrasound. *Carbon* 34, 814.

[83] Watts, P.C., Hsu, W.K., Chen, G.Z., Fray, D.J., Kroto, H.W., Walton, D.R., (2001). A low resistance boron-doped carbon nanotube–polystyrene composite. *J. Mater. Chem.* 11, 2482.

[84] Kis, A., Csanyi, G., Salvetat, J-P., Lee, T-N., Couteau, E., Kulik, A.J., Benoit, W., Brugger, J., Forro, L., (2004). Reinforcement of single-walled carbon nanotube bundles by inter- tube bridging. *Nat. Mater.* 3, 153.

[85] Ausman, K.D., Piner, R., Lourie, O., Rouff, R.S., Korobov, M., (2000). Organic solvent dispersions of single-walled carbon nanotubes: Toward solutions of pristine nanotubes. *J. Phys. Chem. B* 104, 8911.

[86] Liu, Y.Q., Yao, Z.L., Adronov, A., (2005). Functionalization of single-walled carbon nanotubes with well-defined polymers by radical coupling. *Macromolecules* 38, 1172.

[87] Liu, Y.Q., Adronov, A., (2004). Preparation and utilization of catalyst-functionalized single-walled carbon nanotubes for ring-opening metathesis polymerization. *Macromolecules* 37, 4755.

[88] Li, H.M., Cheng, F.O., Duft, A.M., Adronov, A., (2005). Functionalization of single-walled carbon nanotubes with well-defined polystyrene by "click" coupling. *J. Amer. Chem. Soc.* 127, 14518.

[89] Strano, M.S., Dyke, C.A., Usrey, M.L., Barone, P.W., Allen, M.J., Shan, H.W., Kittrell, C., Hauge, R.H., Tour, J.M., Smalley, R.E., (2003). Electronic structure control of single-walled carbon nanotube functionalization. *Sci.* 301, 1519.

[90] Zurek, E., Autschbach, J., (2004). Density functional calculations of the ^{13}C NMR chemical shifts in single-walled carbon nanotubes. *J. Amer. Chem. Soc.* 126, 13079.

[91] Gilje, S., Han, S., Wang, M., Wang, K.L., Kaner, R.B., (2007). A chemical route to graphene for device applications. *Nano Lett.* 7, 3394.

[92] Bunch, J.S., van der Zande, A.M., Verbridge, S.S., Frank, I.W., Tanenbaum, D.M., Parpia, J.M., Craighead, H.G., McEuen, P.L., (2007). Electromechanical resonators from graphene sheets. *Sci.* 315, 490.

[93] Li, D., Muller, M.B., Gilje, S., Kaner, R.B., Wallace, G.G., (2008). Highly conducting graphene sheets and Langmuir-Blodgett films. *Nat. Nanotechnol.* 3, 101.

[94] Seligra, P., Nuevo, F., Ribba, L., García, N.L., Lamanna, M., Famá, L., (2011). Biodegradable PLLA-MWCNTs nanocomposites, *The 3rd International Conference on Biodegradable and Biobased Polymers, BIOPOL*, Strasbourg, France.

[95] Veedu, A., Cao, X., Li, K.M., Soldano, C., Kar, S., Ajayan, P.M., Ghasemi-Nejhad, M.N., (2006). Multifunctional composites using reinforced laminae with carbon nanotube forests. *Nature Mater.* 5, 457.

[96] Liu, P., (2005). Modifications of carbon nanotubes with polymers. *Eur. Polym. J.* 41, 2693.

[97] Pantarotto, D., Partidos, C.D., Hoebeke, J., Brown, F., Kramer, E., Briand, J-P., Muller, S., Prato, M., Bianco, A., (2003). Synthesis, structural characterization, and immunological properties of carbon nanotubes functionalized with peptides. *Chem. Biol.* 10, 961.

[98] Chen, R.J., Zhang, Y., Wang, D., Dai, H., (2001). Noncovalent sidewall functionalization of single-walled carbon nanotubes for protein immobilization. *J. Amer. Chem. Soc.* 123, 3838.

[99] Nakashima, N., Tomonari, Y., Murakami, H., (2002). Water-soluble single-walled carbon nanotubes via noncovalent sidewall-functionalization with a pyrene-carrying ammonium ion. *Chem. Lett.* 6, 638.

[100] Sahoo, N.G., Rana, S., Cho, J.W., Li, L., Chan, S.H., (2010). Polymer nanocomposites basedon functionalized carbon nanotubes. *Prog. Polym. Sci.* 35, 837.

[101] Wu, D., Wu, L., Zhang, M., Zhao, Y., (2008). Viscosity and thermal stability of polylactide composites with functionalized carbon nanotubes. *Polym. Degrad. Stab.* 93, 1577.

[102] Raquez, J.-M., Habibi, Y., Murariu, M., Dubois, P., (2013). Polylactide (PLA)-based nanocomposites. *Prog. Polym. Sci.* 38, 1504.

[103] Huang, W.J., Lin, Y., Taylor, S., Gaillard, J., Rao, A.M., Sun, Y.P., (2002). Sonication-assisted functionalization and solubilization of carbon nanotubes. *Nano Lett.* 2, 231.

[104] Pillai, S.K., Ramontja, J., Ray, S.S., (2011). *Amine functionalization of carbon nanotubes for the preparation of CNT based polylactide composites – A comparative study*. In S. Mathur, S.S. Ray, S. Widjaja, D. Singh, (Eds.), *Nanostructured Materials and Nanotechnology V: Ceramic Engineering and Science Proceedings*, Hoboken: New Jersey, US, pp 43.
[105] Novais, R.M., Simon, F., Pötschke, P., Villmow, T., Covas, J.A., Paiva, M.C., (2013). Poly(lactic acid) composites with poly(lactic acid)-modified carbon nanotubes. *J. Polym. Sci., Part A: Polym. Chem.* 51, 3740.
[106] Paiva, M.C., Simon, F., Novais, R., Ferreira, T., Proença, M., Xu, W., Besenbacher, F., (2010). Controlled functionalization of carbon nanotubes by a solvent-free multicomponent approach. *ASC Nano* 4, 7379.
[107] Georgakilas, V., Kordatos, K., Prato, M., Guldi, D.M., Holzinger, M., Hirsch, A., (2002). Organic functionalization of carbon nanotubes. *J. Am. Chem. Soc.* 124, 760.
[108] Kordatos, K., Da Ros, T., Bosi, S., Vazquez, E., Bergamin, M., Cusan, C., Pellarini, F., Tomberli, V., Baiti, B., Pantarotto, D., (2001). Novel versatile fullerene synthons. *J. Org. Chem.* 66, 4915.
[109] Chiu, W.-M., Kuo, H.-Y., Tsai, P.-A., Wu, J.-H., (2013). Preparation and properties of poly (lactic acid) nanocomposites filled with functionalized single-walled carbon nanotubes. *J. Polym. Environ.* 21, 350.
[110] Martinez, G., Salavagione, H., (2011). Nanocompuestos poliméricos de grafeno: Preparación y propiedades. *Rev. Iberoam. Polím.* 12, 53.
[111] Chen, J., Hamon, M.A., Hu, H., Chen, Y., Rao, A.M., Eklund, P.C., Haddon, R.C., (1998). Solution properties of single-walled carbon nanotubes. *Sci.* 282, 95.
[112] Czerw, R., Guo, Z., Ajayan, P.M., Sun, Y.P., Carrol, D.L., (2001). Functionalization of carbon nanotubes by free radicals. *Nano. Lett* 1, 423.
[113] Fu, K., Huang, W., Lin, Y., Riddle, L.A., Carroll, D.L., Sun, Y.P., (2001). Defunctionalization of functionalized carbon nanotubes. *Nano Lett.* 1, 439.
[114] Viswanathan, G., Chakrapan, N., Yang, H., Wei, B., Chung, H., Cho, K., Ryu, C.Y., Ajayan, P.M., (2003). Single-step in situ synthesis of polymer-grafted single-wall nanotube composites. *J. Am. Chem. Soc.* 125, 9258.
[115] Peng, H.Q., Alemany, L.B., Margrave, J.L., Khabashesku, V.N., (2003). Sidewall carboxylic acid functionalization of single-walled carbon nanotubes. *J. Am. Chem. Soc.* 125, 15174.
[116] Ying, Y.M., Saini, R.K., Liang, F., Sadana, A.K., Billups, W.E., (2003). Functionalization of carbon nanotubes by free radicals. *Org. Lett.* 5, 1471.
[117] Holzinger, M., Vostrovsky, O., Hirsch, A., Hennrich, F., Kappes, M., Weiss, M., Jellen, R., Angew, F., (2001). Sidewall functionalization of carbon nanotubes. *Chem. Int. Ed. Engl.* 40, 4002.
[118] Jeong, J.S., Moon, J.S., Jeon, S.Y., Park, J.H., Alegaonkar, P.S., Yoo, J.B., (2007). Mechanical properties of electrospun PVA/MWNTs composite nanofibers. *Thin Solid Films* 515, 5136.
[119] Ajayan, P.M., Stephan, O., Colliex, C., Trauth, D., (1994). Aligned carbon nanotubes. Arrays formed by cutting a polymer resin-nanotube composite. *Sci.* 265, 1212.
[120] Lagaron, J.M., Lopez-Rubio, A., (2011). Nanotechnology for bioplastics: Opportunities, challenges and strategies. *Trends Food Sci. Tech.* 22, 611.

[121] Petersson, L., Oksman, K., (2006). Biopolymer based nanocomposites: Comparing layered silicates and microcrystalline cellulose as nanoreinforcement. *Compos. Sci. Technol.* 66, 2187.

[122] Liu, Y., Chakrabartty, S., Gkinosatis, D.S., Mohanty, A.K., Lajnef, N., (2007). Multi-walled Carbon Nanotubes/Poly(L-lactide) Nanocomposite Strain Sensor for Biomechanical Implants, *Biomedical Circuits and Systems Conference, BIOCAS IEEE*, Montreal, Canadá.

[123] Samir, M.A., Alloin, F., Dufresne, A., (2005). Crosslinked nanocomposite polymer electrolytes reinforced with cellulose whiskers. *Biomacromolecules* 6, 612.

[124] De Jong, W.H., Borm, P.J.A., (2008). Drug delivery and nanoparticles: Applications and hazards. *J. Nanomedicine* 3, 133.

[125] Bokobza, L., (2007). Multiwall carbon nanotube elastomeric composites: A review. *Polymer* 48, 4907.

[126] Harris, P.J., (2004). Carbon nanotube composites. *Int. Mater. Rev.* 49, 31.

[127] McClory, C., Chin, S.J., McNally, T., (2009). Polymer/carbon nanotube composites. *Aust. J. Chem.* 62, 762.

[128] Martinez-Hernandez, A.L., Velasco-Santos, C., Castaño, V.M., (2010). Carbon nanotubes composites: processing, grafting and mechanical and thermal properties. *Curr. Nanosci.* 6, 12.

[129] Moon, S.I., Jin, F., Lee, C.J., Tsutsumi, S., Hyon, S.H., (2005). Novel carbon nanotube/poly(L-lactic acid) composites: Their modulus. Thermal stability and electrical conductivity. *Macromol. Symp.* 224, 287.

[130] Zhang, D., Kandadai, M.A., Cech, J., Roth, S., Curran, S.A., (2006). Poly(L-lactide) (PLLA)/multi-walled carbon nanotube (MWCNT) composite: Characterization and biocompatibility evaluation. *J. Phys. Chem. B* 110, 12910.

[131] Shieh, Y.T., Liu, G.L., (2007). Effects of carbon nanotubes on crystallization and melting behavior of poly(L-lactide) via DSC and TMDSC studies. *J. Polym. Sci. Pol. Phys.* 45, 1870.

[132] Kwon, J.Y., Kim, H.D., (2005). Preparation and properties of acid-treated multiwalled carbon nanotube/waterborne polyurethane nanocomposites. *J. Appl. Polym. Sci.* 96, 595.

[133] Yoon, J.T., Lee, S.C., Jeong, Y.G., (2010). Effects of grafted chain length on mechanical and electrical properties of nanocomposites containing polylactide-grafted carbon nanotubes. *Compos. Sci. Technol.* 70, 776.

[134] Barrau, S., Vanmansart, C., Moreau, M., Addad, A., Stoclet, G., Lefebvre, J.M., (2011). Crystallization behavior of carbon nanotube-polylactide nanocomposites. *Macromolecules* 44, 6496.

[135] Kuan, C.F., Chen, C.H., Kuan, H.C., Lin, K.C., Chiang, C.L., Peng, H.C., (2008). Multi-walled carbon nanotube reinforced poly (l-lactic acid) nanocomposites enhanced by water-crosslinking reaction. *J. Phys. Chem. Solids* 69, 1399.

[136] Wu, D., Wu, L., Zhou, W., Zhang, M., Yang, T., (2010). Crystallization and biodegradation of polylactide/carbon nanotube composites. *Polym. Eng. Sci.* 50, 1721.

[137] Supronowicz, P.R., Ajayan, P.M., Ullmann, K.R., Arulanandam, B.P., Metzger, D.W., Bizios, R., (2002). Novel-current conducting composite sub- strates for exposing osetoblasts to alternating current stimulation. *J. Biomed. Mater. Res.* 59, 499.

[138] Bhattacharya, M., Seong, W.-J., (2013). *Carbon nanotubes – based materials – preparation, biocompatibility, and applications in dentistry*. In K. Subramani, W. Ahmed, J.K. Hartsfield, (Eds.), *Nanobiomaterials in Clinical Dentistry*, Elsevier Inc.: US, pp 37.

[139] Safadi, B., Andrews, R., Grulke, E.A., (2002). Multiwalled carbon nanotube polymer composites: Synthesis and characterization of thin films. *J. Appl. Polym. Sci.* 84, 2660.

[140] Bergin, S.D., Nicolosi, V., Streich, P.V., Giordani, S., Sun, Z., Windle, A.H., Ryan, P., Niraj, N.P., Wang, Z.-T., Carpenter, L., Blau, W.J., Boland, J.J., Hamilton1, J.P., Coleman, J.N., (2008). Towards solution of single-walled carbon nanotubes in common solvents. *Adv. Mater.* 20, 1876.

[141] Kumar, B., Castro, M., Feller, J.F., (2012). Poly(lactic acid)–multi-wall carbon nanotube conductive biopolymer nanocomposite vapour sensors. *Sensor. Actuat. B-Chem.* 161, 621.

[142] Kaminsky, W., (2011). *Polyolefin carbon nanotube composites by in-situ polymerization*. In T. McNally, P. Pötschke, (Eds.), *Polymer Carbon Nanotube Composites: Preparation Properties and Applications*, Woodhead: Cambridge, UK, pp 3.

[143] Chang, T.E., Kisliuk, A., Rhodes, S.M., Brittain, W.J., Sokolov, A.P., (2006). Conductivity and mechanical properties of well-dispersed single-wall carbon nanotube/polystyrene composite. *Polymer* 47, 7740.

[144] Zhu, B.K., Xie, S.H., Xu, Z.K., Xu, Y.Y., (2006). Preparation and properties of polyimide/aluminum nitride composites. *Compos. Sci. Technol.* 66, 548.

[145] Liu, Y.L., Chang, Y.H., Liang, M., (2008). Poly(2,6-dimethyl-1,4-phenylene oxide) (PPO) multi-bonded carbon nanotube (CNT): Preparation and formation of PPO/CNT nanocomposites. *Polymer* 49, 5405.

[146] Lee, S.S., Park, C.Y., Lee, D.S., (2008). Properties of nanocomposites based on sulfonated poly(styrene-b-ethylenebutylene-b-styrene) and multiwalled carbon nanotubes. *Colloid. Surf. A: Physicochem. Eng. Aspects* 239, 313.

[147] Xie, X.L., Aloys, K., Zhou, X.P., Zeng, F.D., (2003). Ultrahigh molecular mass polyethylene/carbon nanotube composites: Crystallization and melting properties. *J. Therm. Anal Calorim.* 74, 317.

[148] Jin, Z., Pramoda, K.P., Goh, S.H., Xu, G., (2002). Poly(vinylidene fluoride)-assisted melt-blending of multi-walled carbon nano-tube/poly(methyl methacrylate) composites. *Mater. Res. Bull.* 37, 271.

[149] Shih, Y.-F., Wang, Y.-P., Hsieh, C.-F., (2011). Preparation and properties of PLA/long alkyl chain modified multi-walled carbon nanotubes nanocomposites. *J. Polym. Eng.* 31, 13.

[150] Han, Z., Fin, A., (2011). Thermal conductivity of carbon nanotubes and their polymer nanocomposites: A review. *Prog. Polym. Sci.* 36, 914.

[151] Yang, Z., Chen, X., Pu, Y., Zhou, L., Chen, C., Li, W., Xu, L., Yi, B., Wang, Y., (2007). Facile approach to obtain individual-nanotube dispersion at high loading in carbon nanotubes/polyimide composites. *Polym. Adv. Technol.* 18, 458.

[152] Reddy, M.M., Vivekanandhan, S., Misra, M., Bhatia, S.K., Mohanty, A.K., (2013). Biobased plastics and bionanocomposites: Current status and future opportunities. *Prog. Polym. Sci.* 38, 1653.

[153] Darder, M., Aranda, P., Ruiz-Hitzky, E., (2007). Bionanocomposites: A new concept of ecological, bioinspired, and functional hybrid materials. *Adv. Mater.* 19, 1309.

[154] Xu, Z., Niu, Y., Yang, L., Xie, W., Li, H., Gan, Z., Wang, Z., (2010). Morphology, rheology and crystallization behavior of polylactide composites prepared through addition of five-armed star polylactide grafted multiwalled carbon nanotubes. *Polymer* 51, 730.

[155] Li, Q.-H., Zhou, Q.-H., Deng, D., Yu, Q.-Z., Gu, L., Gong, K.-D., Xu, K.-H., (2013). Enhanced thermal and electrical properties of poly (D,L-lactide)/multi-walled carbon nanotubes composites by in-situ polymerization. *Trans. Nonferrous Met. Soc. China* 23, 1421.

[156] Chen, G.-X., Kim, H.-S., Park, B.H., Yoon, J.-S., (2007). Synthesis of poly(L-lactide)-functionalized multiwalled carbon nanotubes by ring-opening polymerization. *Macromol. Chem. Phys.* 208, 389.

[157] Zhao, Y., Qiu, Z., Yang, W., (2009). Effect of multi-walled carbon nanotubes on the crystallization and hydrolytic degradation of biodegradable poly(L-lactide). *Compos. Sci. Technol.* 69, 627.

[158] Papageorgiou, G.Z., Achilias, D.S., Nanaki, S., Beslikas, T., Bikiaris, D., (2010). PLA nanocomposites: Effect of filler type on non-isothermal crystallization. *Thermochim. Acta* 511, 129.

[159] Han, H., Wang, X., Wu, D., (2012). Preparation, crystallization behaviors, and mechanical properties of biodegradable composites based on poly(L-lactic acid) and recycled carbon fiber. *Compos. Part A Appl. Sci. Manuf.* 43, 1947.

[160] Kobashi, K., Villmow, T., Andres, T., Pötschke, P., (2008). Liquid sensing of melt-processed poly(lactic acid)/multi-walled carbon nanotube composite films. *Sens. Actuators B: Chem.* 134, 787.

[161] De Santis, P., Kovacs, A.J., (1968). Molecular conformation of poly (S-lactic acid). *Biopolymers* 6, 299.

[162] Eling, B., Gogolewski, S., Pennings, A.J., (1982). Melt-spun and solution spun fibers. *Polymer* 23, 1587.

[163] Cartier, L., Okihara, T., Ikada, Y., Tsuji, H., Puiggali, J., Lotz, B., (2000). Epitaxial crystallization and crystalline polymorphism of polylactides. *Polymer* 41, 8909.

[164] Saeidlou, S., Huneault, M.A., Li, H., Park, C.B., (2012). Poly(lactic acid) Crystallization. *Prog. Polym. Sci.* 37, 1657.

[165] Quan, H., Zhang, S.J., Qiao, J.L., Zhang, L.Y., (2012). The electrical properties and crystallization of stereocomplex poly(lactic acid) filled with carbon nanotubes. *Polymer* 53, 4547.

[166] Shieh, Y.T., Liu, G.L., Twu, Y.K., Wang, T.L., Yang, C.H., (2010). Effects of carbon nanotubes on dynamic mechanical property, thermal property, and crystal structure of poly(l-lactic acid). *Polymer* 48, 145.

[167] González Seligra, P., Lamanna, M., Famá, L., (2014). PLA-fMWCNT bionanofilms with high modulus and great properties to apply in packaging and biomedicine *Procedia Mater. Sci. In press.*

[168] Armentano, I., Dottori, M., Fortunati, E., Mattioli, S., Kenny, J.M., (2010). Biodegradable polymer matrix nanocomposites for tissue engineering: A review. *Polym. Degrad. Stabil.* 95, 2126.

[169] Xu, H.S., Dai, X.J., Lamb, P.R., Li, Z.M., (2009). Poly(L-lactide) crystallization induced by multiwall carbon nanotubes at very low loading. *J. Polym. Sci. Part B Polym. Phys.* 47, 2341.

[170] Li, Y., Wu, H., Wang, Y., Liu, L., Han, L., Wu, J., Xiang, F., (2010). Synergistic effects of PEG and MWCNTs on crystallization behavior of PLLA. *J. Polym. Sci. Part B Polym. Phys.* 48, 520.

[171] Yasuniwa, N., Sakamo, K., Ono, Y., Kawahara, W., (2008). Melting behavior of poly(l-lactic acid): X-ray and DSC analyses of the melting process. *Polymer* 49, 1943.

[172] Chrissafis, K., Paraskevopoulos, K.M., Papageorgiou, G.Z., Bikiaris, D.N., (2008). Thermal and dynamic mechanical behaviour of bionanocomposites: Fumed silica nanoparticles dispersed in poly(vinyl pyrrolidone), chitosan and poly(vinyl alcohol). *Appl. Polym. Sci.* 110, 1739.

[173] Chen, K., Wilkie, C.A., Vyazovkin, S., (2007). Nanoconfinement revealed in degradation and relaxation studies of two structurally different polystyrene-clay systems. *J. Phys. Chem. B* 111, 12685.

[174] Yoon, J.T., Jeong, Y.G., Lee, S.C., Min, B.G., (2009). Influences of poly(lactic acid)-grafted carbon nanotube on thermal, mechanical, and electrical properties of poly(lactic acid). *Polym. Adv. Technol.* 20, 631.

[175] Kwiatkowska, M., Broza, G., Sculte, K., Roslaniec, Z., (2006). The in-situ synthesis of poly(butylene terephthalate)/ carbon nanotubes composites. *Rev. Adv. Mater. Sci.* 12, 154.

[176] Ma, X., Jian, R., Chang, P., Yu, J., (2008). Fabrication and characterization of citric acid-modified starch nanoparticles/plasticized-starch composites. *Biomacromolecules* 9, 3314.

[177] Ma, X., Yu, J., Wang, N., (2008). Glycerol plasticized-starch/multiwall carbon nanotube composites for electroactive polymers. *Compos. Sci. Technol.* 68, 268.

[178] Rao, Y., Pochan, J., (2007). Mechanics of polymer-clay nanocomposites. *Macromolecules* 40, 290.

[179] Lin, C., Wang, Y., Lai, Y., Yang, W., Jiao, F., Zhang, H., Ye, S., Zhang, Q., (2011). Incorporation of carboxylation multiwalled carbon nanotubes into biodegradable poly(lactic-co-glycolic acid) for bone tissue engineering. *Colloid. Surface. B* 83, 367.

[180] Shi, X.F., Sitharaman, B., Pham, Q.P., Liang, F., Wu, K., Billups, W.E., Wilson, L.J., Mikos, A.G., (2007). Fabrication of porous ultra-short single-walled carbon nanotube nanocomposite scaffolds for bone tissue engineering. *Biomaterials* 28, 4078.

[181] Wang, S.F., Shen, L., Zhang W.D., Tongs, Y.J., (2005). Preparation and properties of chitosan/ carbon nanotubes composites. *Biomacromolecules* 6, 3067.

[182] González Seligra, P., Lamanna, M., Famá, L., (2013). Nanofilms PLA-MWCNT con mejoras en las propiedades de tracción y en la permeabilidad al vapor de agua, *13º Congreso Internacional en Ciencia y Tecnología de Metalurgia y Materiales, SAM-CONAMET*, Iguazú, Argentina.

[183] Mina, M.F., Beg, M.D.H., Islam, M.R., Alam, A., Nizam, A., Younus, R.M., (2013). Characterization of biodegradable nanocomposites with poly (lactic acid) and multi-walled carbon nanotubes. *World Acad. Sci. Eng. Tech.* 73, 1019.

[184] Fukushima, K., Murariu, M., Camino, G., Dubois, P., (2010). Effect of expanded graphite/layered-silicate clay on thermal, mechanical and fire retardant properties of poly(lactic acid). *Polym. Degrad. Stabil.* 95, 1063.

[185] Feng, J., Sui, J., Cai, W., Wan, J., Chakoli, A.N., Gao, Z., (2008). Preparation and characterization of magnetic multi-walled carbon nanotubes-poly(l-lactide) composite. *Mater. Sci. Eng. B* 150, 208.

[186] Park, S.H., Lee, S.G., Kim, S.H., (2013). Isothermal crystallization behavior and mechanical properties of polylactide/carbon nanotube nanocomposites. *Compos. Part A* 46, 11.

[187] Yao, Z., Zhu, C.C., Cheng, M., Liu, J., (2001). Mechanical Properties of Carbon Nanotube by Molecular Dynamics Simulation. *Comput. Mater. Sci.* 22, 180.

[188] Szymczyk, A., Roslaniec, Z., Zenker, M., García-Gutiérrez, M.C., Hernández, J.J., Rueda, D.R., Nogales, A., Ezquerra, T.A., (2011). Preparation and characterization of nanocomposites based on COOH functionalized multi-walled carbon nanotubes and on poly(trimethylene terephthalate). *Express Polym. Lett.* 5, 977.

[189] Huda, M.S., Drzal, L.T., Misra, M., (2005). Effect of fiber surface treatment on the properties of laminated biocomposites from poly(lactic acid) and kenaf fibers. *Ind. Eng. Chem. Res.* 44, 5593.

[190] Keener, T.J., Stuart, R.K., Brown, T.K., (2004). Maleated coupling agents for natural fibre composites. *Compos. Part A: Appl. Sci. Manuf.* 35, 357.

In: Polymer-Matrix Composites
Editor: Rakesh Kumar

ISBN: 978-1-61122-005-6

Chapter 2

RECENT DEVELOPMENTS IN MONTMORILLONITE AND SEPIOLITE FILLED REGENERATED CELLULOSE NANOCOMPOSITES: CHARACTERIZATIONS AND PROPERTIES

Mohammad Soheilmoghaddam[1], Mat Uzir Wahit[1,2*] and Harintharavimal Balakrishnan[2]

[1]Enhanced Polymer Research Group (EnPRO),
Department of Polymer Engineering, Faculty of Chemical Engineering,
Universiti Teknologi Malaysia (UTM), Johor, Malaysia
[2]Centre for Composites (CfC),
Universiti Teknologi Malaysia (UTM), Johor,Malaysia

ABSTRACT

In recent years, the development of environmentally friendly materials obtained from renewable resources has attracted enormous attention due to the new sustainable development policies. Cellulose is a readily available, naturally occurring biodegradable and biocompatible linear polysaccharide. Recently, room temperature ionic liquids have been used as solvents to produce regenerated cellulose (RC) due to their attractive properties such as good chemical and thermal stability, low flammability, low melting point and ease of recycling. Polymer/nanofiller nanocomposites are believed to have strong potential to widen polymer applications due to its enhanced performance. It is also widely accepted that the incorporation of small amount of nanofiller (less than 5 wt.%) into bio-based matrices results in nano-biocomposite materials with enhanced mechanical, permeability and thermal properties. Montmorillonite (MMT) has a nano-sized layered structure with large surface area thus providing sufficient interfacial regions in polymer nanocomposite. Besides, the needle like fibers based, natural hydrous magnesium silicate sepiolite has also been investigated due to its high surface area, unique geometry and its ability to form the hydrogen bonding with polymers as well as to

* Corresponding Author address: Centre for Composites (CfC), Universiti Teknologi Malaysia (UTM), 81300 Skudai, Johor Bahru, Johor. Tel: +607-5535909, Email: mat.uzir@cheme.utm.my.

disperse well in the matrix. This chapter aims to highlight the effect of the addition of two different types of nanofillers such as organically modified MMT and sepiolite to produce RC nanocomposites, on selected properties.

Keywords: Regenerated cellulose, Montmorillonite, Sepiolite, Nanocomposites

1. INTRODUCTION

1.1. Regenerated Cellulose

Biopolymers from renewable resources have gained significant attention over the past decade. Due to their biodegradability, biopolymers are ideally positioned to reduce the over-exploitation of our limited fossil resources [1]. Cellulose is the most abundant naturally occurring linear polymer in nature, which consists of β 1–4 glucopyranose units (Figure 1) exhibiting many unique properties and an applicability to a variety of uses such as paper, textiles, membranes, artificial fibers and the food industry [2].

Cellulose wall has the linear structure which is made up of layers upon layers of microfibrils with approximately 10 to 20 nm in diameter and may consists of up to 40 cellulose chains. Cellulose fiber has high tensile strength and is associated with other natural polymers within the plant cell wall. Hemicellulose wraps the microfibril's cellulose core, a branched polymer composed of a mixture of primarily sugars (xylose, arabinose), and some hexoses (mannose, galactose, glucose) (Figure 1) [3, 4].

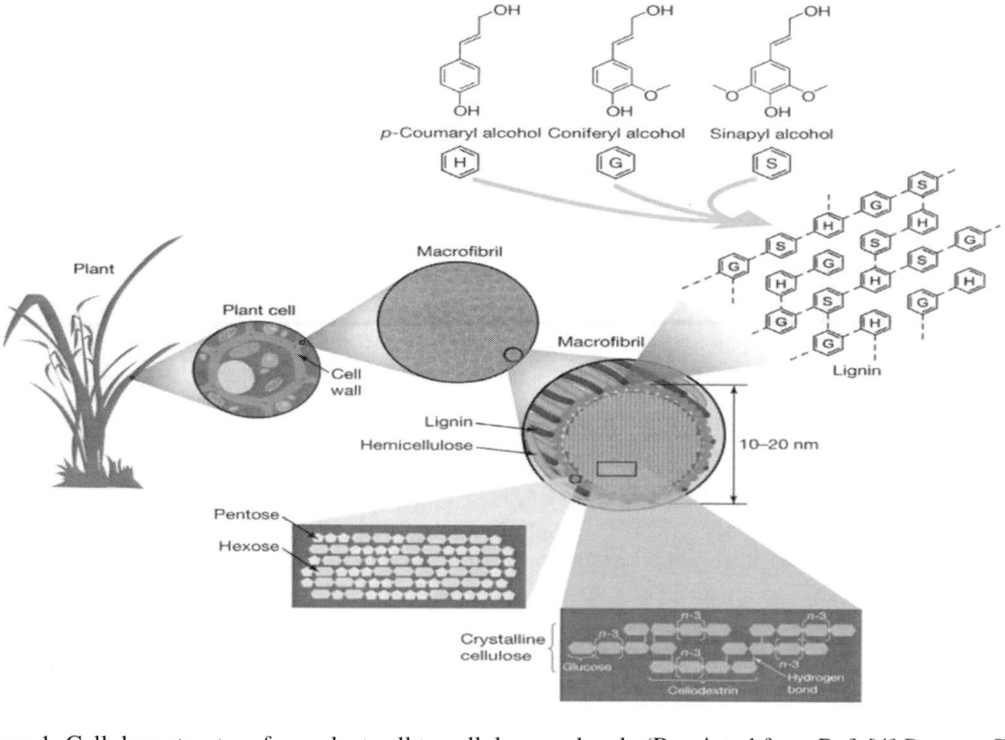

Figure 1. Cellulose structure from plant cell to cellulose molecule (Reprinted from Ref. [4] Potters, G., Goethem, V.D., Schutte, F., (2010). Nature Education 3,14).

The crystalline nature of cellulose was revealed almost a century ago when Nishikawa and Ono recorded the first X-ray diffraction patterns from fiber bundles from various plants. The combination of X-ray diffraction with model building and conformational analyses in the 1970s provided "modern" crystal and molecular structures for most of the cellulose polymorphs [5]. Cellulose is known to exist in at least four polymorphic crystalline forms, of which the structure and properties of cellulose I (native cellulose) and cellulose II (regenerated cellulose and mercerized cellulose) have been most extensively studied. As a first approximation, the crystal structure of cellulose I determined by X-ray diffraction can be described by monoclinic unit cell which contains two cellulose chains in a parallel orientation with a twofold screw axis [2].

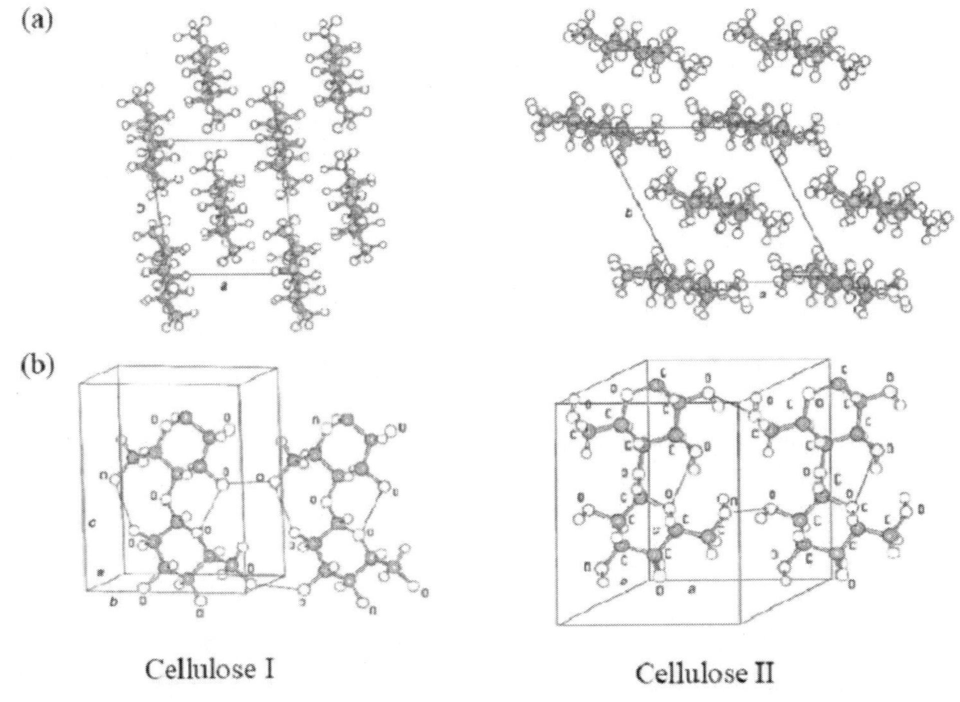

Figure 2. Crystal structure of cellulose I and cellulose II: (a) projection of the unit cell (UC) along the a-b plane; (b) projection of the UC parallel to the (100) lattic plane (cellulose I) and the (010) latic plane (cellulose II) (Reprinted from Ref. [2] Klemm, D., Heublein, B., Fink, H-P., Bohn, A., (2005). Angew. Chem. Int. Ed. 44, 3358).

Cellulose II can be formed from cellulose I by treatment with aqueous sodium hydroxide (mercerizing) or by dissolution of the cellulose and subsequent precipitation/regeneration, as it is done in the formation of regenerated cellulose fibers and films. This monoclinic crystal structure with two antiparallel chains in the unit cell is characterized by specific unit cell geometry with a modified H-bonding system. The alkalization of cellulose is of considerable importance to commercial scale cellulose production as a method for increasing the reactivity (activation) of subsequent reaction as well as for mercerizing of cotton. It is not yet understood how the parallel chain arrangement of cellulose I undergoes transition into the antiparallel orientation of cellulose II without an intermediate dispersion of cellulose molecules. There are currently few reports on the structure of noncrystalline random cellulose chain segments. More knowledge in this area is required, as these structure elements have a significant influence on the accessibility and reactivity of cellulose, as well as on the properties of regenerated cellulose fiber and films [6, 2]. Figure 2 illustrates the crystalline structure of cellulose I and cellulose II.

The cellulose special crystal and molecular structure imparts significant properties to cellulose such as: hydrophilicity, chirality, degradability and broad chemical variability initiated by the high donor reactivity of the -OH groups. The solvent resistance of cellulose has promoted several researches for its dissolution and modification [7-9].

Over the years, several solvent systems such as, NaOH/urea aqueous solution [10, 11], liquid ammonia/ammonium thiocyanate [12, 13], lithium chloride (LiCl)/ N,N-Dimethylacetamide (DMAc) [14], LiCl/1,3-dimethyl-2-imidazolidinone (DMI) [15], N-methylmorpholine-N-oxide (NMMO)/water [16-18] for the preparation of regenerated cellulose materials have been developed. Due to environmental concerns, (NMMO)/water has emerged as the most commercially desirable process. However, the NMMO/water process still has some environmental issues to contend with including several adverse side reactions, the formation of undesired byproducts and expensive solvent recovery. These issues need to be overcome prior to efficient and environmentally-friendly production of regenerated cellulose [19,20].

Recently, a green solvent system known as ionic liquid (IL) has been gaining interest. IL is a salt which traditionally known as the molten salts or fused salts with melting points below 100 °C and often lower temperature [21]. The ILs usage is first reported in 1914 with the preparation of ethylammonium nitrite as reported by Paul Walden [22]. ILs attract notable importance as a salt electrolytes for thermal batteries in the early 1960s at U.S. Air Force Academy [23] and the given compounds had alkylpyridinium cations. This salt had problem because it had tendency to be reduced easily so the new salt had been developed which was 1-alkyl-3-methylimidazolium. Some of these materials are liquid at room temperature and most of them melt below 100 °C. They are considered as the green solvents due to their low pressure and recycling possibility. These water-free systems consist fully of ions so ILs can be the solvent for variety of synthesis.

Ionic liquids are characterized by excellent dissolubility, low toxicity, thermal stability, near zero volatility and recyclability [24]. The desirable properties of ILs can also be tailored since they offer great flexibility in the design of cationic and anionic structures and as such they are termed the "designers solvent" [25]. Literature has shown a variety of ionic liquids (ILs) to be good cellulose and biomass non-derivating solvents [26-29]. The most promising anions for cellulose dissolution are acetate and chloride, whereas different cations can be employed [28]. Swatloski et al., [30] reported for the first time that ionic liquid 1-butyl-3-methylimidazolium chloride (BMIMCl) has a good dissolving power for cellulose which opened up a new way of developing a class of cellulose solvent systems and promoted several other groups to test a variety of other ILs on their ability to dissolve cellulose (Figure 3) [31-33].

Cellulose with variety of DP was dissolved in BMIMCL directly without other derivative reaction occurring under certain conditions. After cellulose dissolution in IL it was regenerated by treatment with water, cellulose structure changed from cellulose I to cellulose II and while the degree of crystallinity and thermal stability decreased more than the original, the application value of the products did not affect by this changes in addition, ionic liquid is recoverable [34].

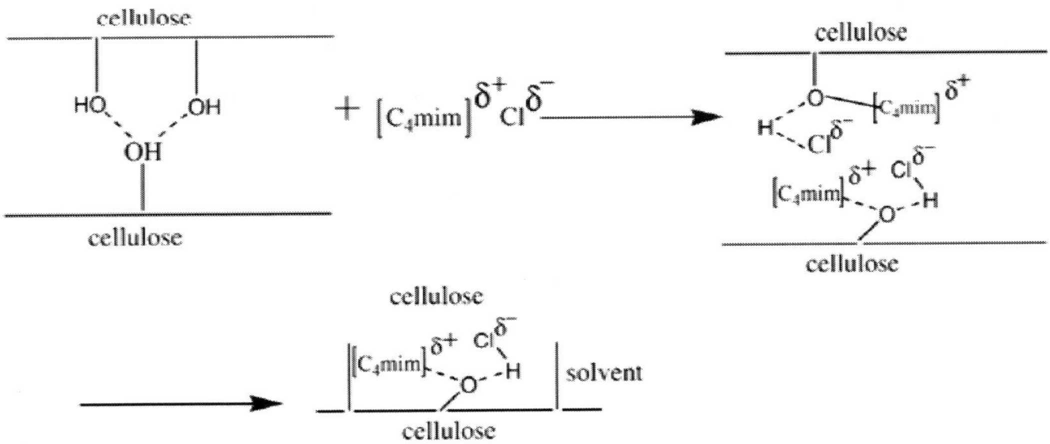

Figure 3. Mechanism of dissolution of cellulose with BMIMCl and transformation of cellulose I to cellulose II (Reprinted from Ref. [32] Feng, L., Chen, Z., (2008). Journal of Molecular Liquids 142, 1).

As a result of the cellulose solution in BMIMCL ionic liquid a flexible gel material was prepared by Kadokawa et al., [35]. Cellulose was dissolved at 100 ºC for 24h then cooled to the room temperature and kept it 7 days between two glass plates to obtain the gel material and when the material was heated to 120 ºC it became soft at 150 ºC converted to fluid. The thermogravimetric analysis (TGA) and X-ray diffraction (XRD) results showed the crystalline structure of cellulose was significantly disrupted in the material and by keeping the material in room temperature for two days a new gel with more transparency compared with the original one was regenerated [35].

1.2. Nanofillers

A new category of hybrid composite materials based on the ultra incorporation of nano-sized fillers (nanofillers) into a polymer matrix has been investigated. Nanotechnology is the ability to work on a scale of about 1-100 nm in order to understand, create, characterize and use material structure, devices and system with new properties derived from their based on the nanostructures. Nanocomposites could exhibit unique modifications in their properties, in comparison with conventional composites in terms of physical properties, including barrier, flammability resistance, thermal and environmental stability, solvent uptake, and rate of biodegradability of biodegradable polymers [3].

1.2.1. Montmorillonite (MMT)

In recent years, organically modified layered silicate such as montmorillonite (MMT) has been widely used as reinforcement to prepare polymer nanocomposites due to its easy availability, low cost and more importantly environmental friendliness. Polymer/MMT offer tremendous improvement in wide range of physical and engineering properties at low filler (MMT) loadings. MMT with nanosized layered structure has a large surface area providing sufficient interfacial regions in polymer nanocomposites for enhancement in thermal and tensile properties with low loading into the polymer matrix [36, 37].

MMT consists of layered structure which are made up of two tetrahedrally coordinated silicon atoms combined with an edge-shared octahedral sheet of either magnesium or aluminum hydroxide (Figure 4). Each layered sheet is about 1 nm thick and its length varies from tens of nanometers to more than one micron, depending on the layered silicate. MMT nanoclays are unique since they have a platy structure with a unit thickness of one nanometer or less. With an aspect ratio in 1000:1 range, MMT nanoclays are easily available. The general formula of MMT is $M_x(Al_{4-x}\ Mg_x)Si_8O_{20}(OH)_4$, where M is either Na or Ca [38, 39].

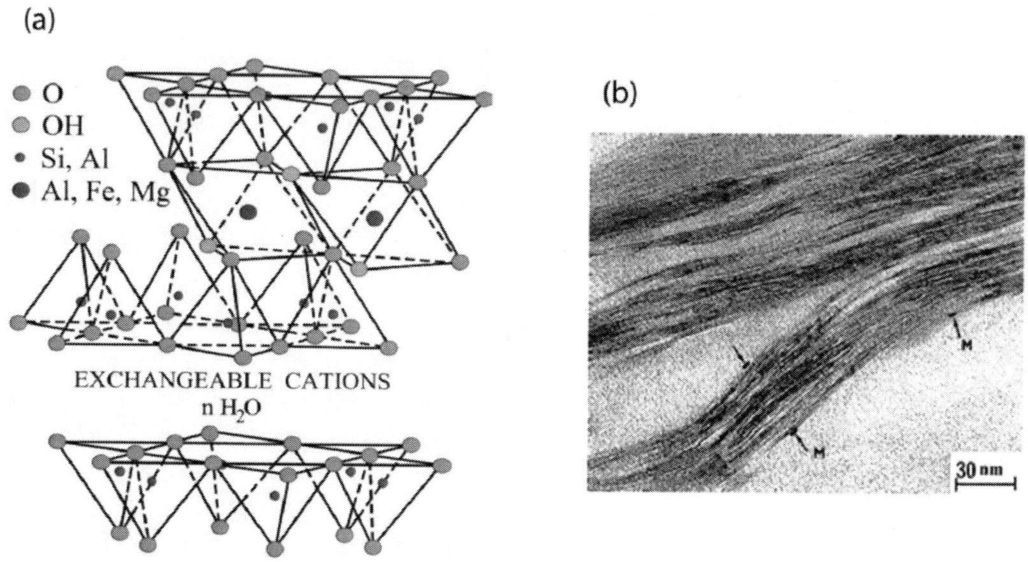

Figure 4. Layered silicate structure of MMT (a) schematic and (b) transmission electron microscope (TEM) (Reprinted from Ref. [39] Bordes, P., Pollet, E., Avérous, L., (2009). Progress in Polymer Science 34, 125. Reprinted from Ref. [40] Merlin, F., Lombois, H., Joly, S., Lequeux, N., Halary, J-L., Van Damme, H., (2002). J. Mater. Chem. 12, 3308).

Isomorphous substitution of Si^{4+} and Al^{3+} in tetrahedral and Al^{3+} and Mg^{2+} on octahedral sheet cause an excess of negative charges within the MMT layers. These negative charges are counterbalanced by cation such as Ca and Na situated between layers. Due to high hydrophilicity of layered silicate, water molecules are also present between the layers. Stacking of the layers leads to regular Van der Waals gap called interlayers or galleries [38].

1.2.2. Sepiolite

The majority of researchers have focused on two-dimensional (2D) clay materials. However, these nanofillers are not the only type of nanoclay that can be used for nanocomposite preparation. One-dimensional (1D) nano-scale fillers such as sepiolite are believed to disperse well due to their special needle like geometry [41-43]. Sepiolite is a magnesium hydrosilicate composed of two tetrahedral silica sheets sandwiching a central sheet of magnesium oxide hydroxide with $Si_{12}O_{30}Mg_8(OH)_4(H_2O)_4{\bullet}8H_2O$ as the unit cell formula [44-46]. Sepiolite has a surface area of about 300 m^2/g with a very high density of surface silanol groups (Figure 5) that are easily available for coupling reactions, resulting in a strong interaction between the inorganic particles and organic polymer matrix [47, 46].

Consequently, sepiolite filled polymer nanocomposites exhibit extraordinary performance on mechanical, thermal and barrier properties [48-50].

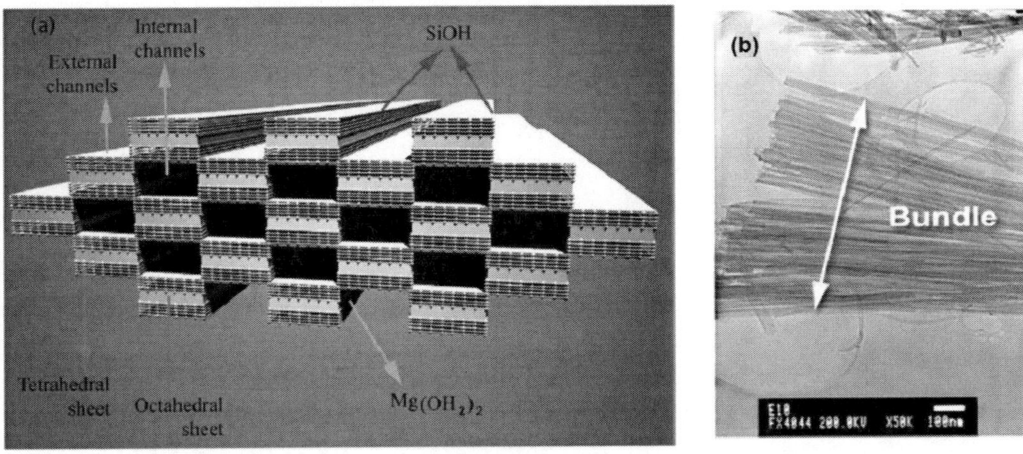

Figure 5. (a) Schematic and (b) TEM structure of a sepiolite (Reprinted from Ref. [51] Soheilmoghaddam, M., Wahit, M.U., Yussuf, A.A., Al-Saleh, M.A., Whye, W.T., (2014). Polymer Testing 33, 121 Reprinted from Ref. [52] García-Romero, E., Suárez, M., (2013). Appl. Clay Sci. 86, 129).

2. Regenerated Cellulose Nanocomposites

2.1. Regenerated Cellulose/MMT

Incorporation of nanofillers into polymer matrix has been proved to be a powerful tool in order to increase the polymer properties. Of particular interest are polymer nanocomposites reinforced with organically modified layered silicate such as MMT. MMT with nano-sized layered structure has a large surface area providing sufficient interfacial regions in polymer nanocomposites. Polymer/MMT nanocomposite materials are proven to have strong potential to widen polymer nanocomposites applications [36, 37, 53].

Recently, White et al., [54] reported preparation of regenerated cellulose/MMT nanocomposites. Cotton linter was used as cellulose source and two different solvents include, lithium chloride/*N*,*N*-dimethylacetamide (LiCl/DMAc) and N-Methylmorpholine N-oxide (NMMO)/water system were used as the solvent for preparation of cellulose/MMT nanocomposites. Unmodified MMT and organo-modified MMT were used. They reported that LiCl/DMAc method was unsuccessful. It was also reported that unmodified MMT was not exfoliated and the method that resulted in superior cellulose/MMT nanocomposites used an organo-modified MMT with NMMO as cellulose solvent. The disappearance of MMT peak in Figure 6 as shown in XRD result revealed that MMT was intercalated or exfoliated in cellulose/MMT nanocomposites.

A significant improvement in thermal properties of cellulose/MMT was also reported by White et al., [54] when compared with cellulose control source. As shown in Figure 7, the degradation temperature of the nanocomposites increased by 45 °C and the char yield increased between 5 % and 31 % depending on the amount of MMT added to regenerated cellulose.

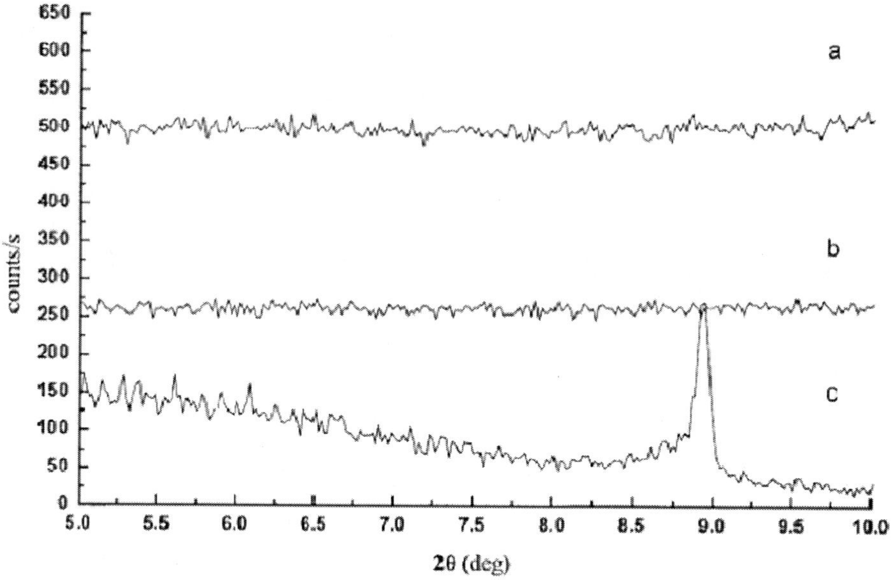

Figure 6. XRD patterns of (a) the control cotton sample, (b) the nanocomposite with 15 wt.% MMT, and (c) organo- modified MMT (Reprinted from Ref. [54] White, L.A., (2004). J. Appl. Polym. Sci. 92, 2125).

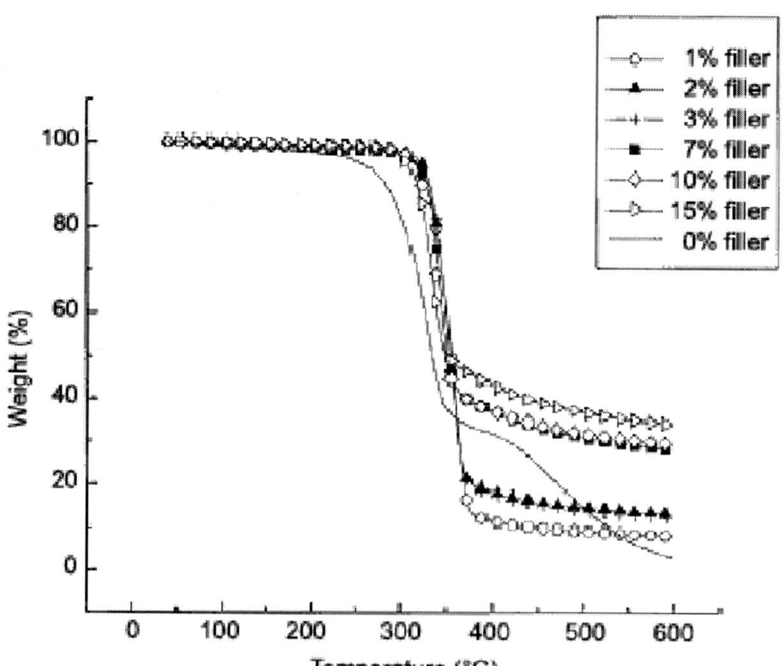

Figure 7. TGA thermograms comparing regenerated cellulose nanocomposites with various MMT contents (Reprinted from Ref. [54] White, L.A., (2004). J. Appl. Polym. Sci. 92, 2125).

Cerruti et al., [55] prepared cellulose montmorillonite nanocomposite using N-Methylmorpholine N-oxide (NMMO) as the solvent. Using XRD analysis and TEM observation, they found that MMT was intercalated and exfoliated in the pure cellulose matrix. XRD pattern for nanomer (1.28nm) and the nanocomposites are shown in Figure 8. They reported that in the case of the nanocomposite i.e., MMT-A poor intercalation of cellulose within montmorillonite stacks was observed, yielding a material with both micro-composite and nanocomposite morphological features. For nanocomposite i.e., MMT-B, which was obtained by reacting the polymer/MMT mixture for the longer time after the complete dissolution of cellulose, different morphology was observed. The XRD patterns showed that the peak at 2θ = 3.4 was shifted to smaller angle and clay plates exfoliation operated by the polymer.

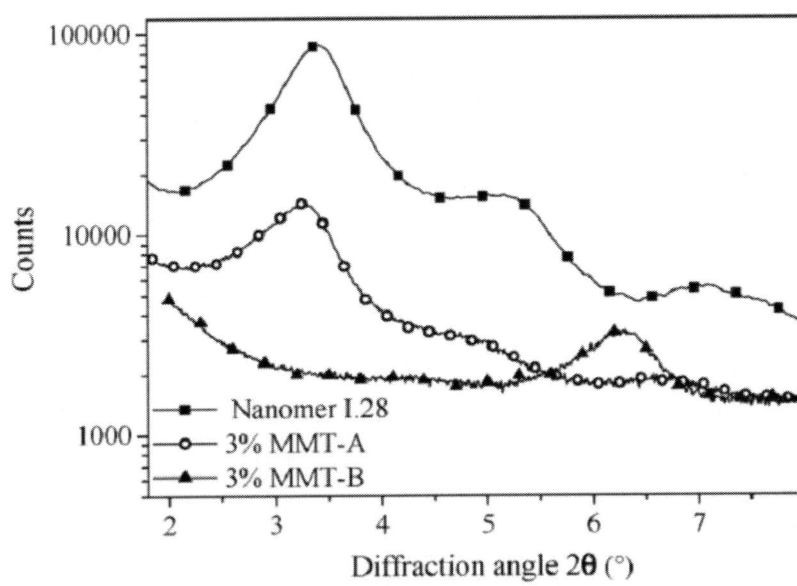

Figure 8. XRD spectra organo-modified montmorillonite nanomer (1.28) and the cellulose based nanocomposites. (Reprinted from Ref. [55] Cerruti, P., Ambrogi, V., Postiglione, A., Rychlý, J., Matisová-Rychlá, L., Carfagna, C., (2008). Biomacromolecules 9, 3004).

The regenerated cellulose/MMT nanocomposites via NMMO as solvent were prepared by Lee et al., [56]. This research studied the effect of MMT hydrophobicity on the nanocomposite properties. It was reported that the surface hydrophobicity of the MMT played a dominant role in MMT interaction and exfoliation within the cellulose matrix. The researchers observed that NaMMT was intercalated and modified MMT i.e. 10A, which was more hydrophilic than other kinds of tested modified MMT, was totally exfoliated in cellulose matrix. For regenerated cellulose/NaMMT and regenerated cellulose/MMT10A nanocomposite films, a 43 % and 18 % increase in the tensile strength were observed, respectively. However, the tensile strength in other nanocomposites decreased. This decrement was probably attributed to the poor compatibility between cellulose and the aggregation formation of hydrophobic MMT particles in the composite matrix.

The work of Mahmoudian et al., [20] investigated the effect of different MMT contents on regenerated cellulose/montmorillonite (RC/MMT) nanocomposite films prepared in ionic liquid, 1-butyl-3-methylimidazolium chloride (BMIMCl) using solution casting method. It was revealed that the Young's modulus of RC/MMT films increased linearly with increasing MMT loadings (Figure 9). The RC/MMT-6 (6 wt.% MMT content) nanocomposite film exhibited about 40% increase in modulus compared to the pure RC films. The improvement in modulus of RC is due to the high rigidity exerted by the MMT as it does not deform or relax and therefore the cellulose's chain movement was suppressed by MMT particles. This improvement in modulus may also be ascribed to the dispersion and alignment of MMT platelets within cellulose. As when MMT particles disperse in intercalated or exfoliated form, it will provide higher interfacial region which will make stress transfer to MMT layers more effective. For RC/MMT-8 (8 wt.% MMT content), the rise in modulus is less significant. Reductions in modulus enhancement at higher level of MMT loading was ascribed to reduced level of dispersion.

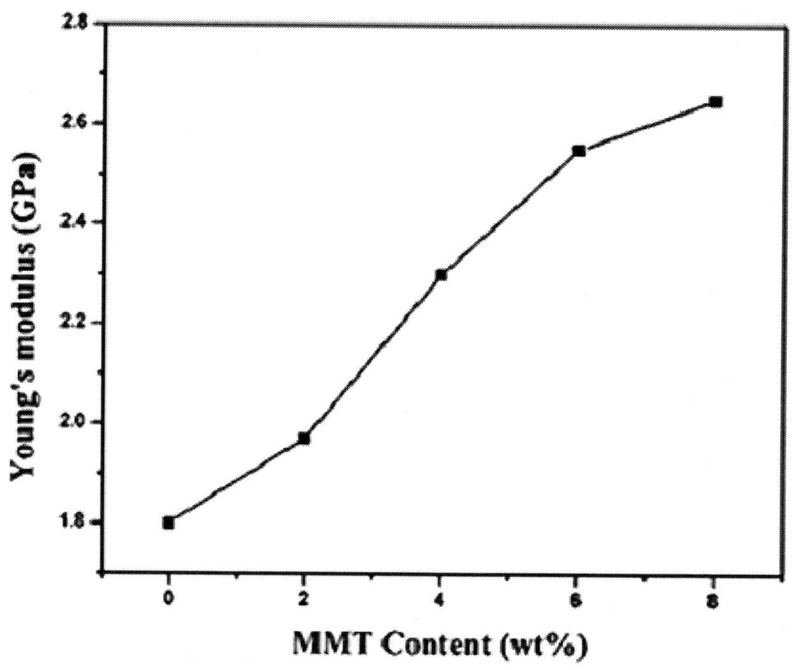

Figure 9. Effect of MMT contents on Young's modulus of RC/MMT nanocomposite films. (Reprinted from Ref. [20] Mahmoudian, S., Wahit, M.U., Ismail, A.F., Yussuf, A.A., (2012). Carbohydr. Polym. 88, 1251).

Similarly, addition of MMT increased tensile strength of the films. The RC nanocomposite with 6 wt.% MMT content exhibited about 12% increase in the tensile strength (Figure 10). As the MMT loading further increased, the tensile strength decreased, which is attributed to the aggregation of MMT particles. However, the elongation at break values of RC/MMT nanocomposites decreased with the presence of MMT. The elongation at break values of the films decreased from 7.5% to 4.5% as the MMT content increased from 0 to 8 wt.%. This could be attributed to the presence of MMT which restrains the slippage movement of RC chains during deformation and therefore resulting in decreased elongation at break values (Figure 10).

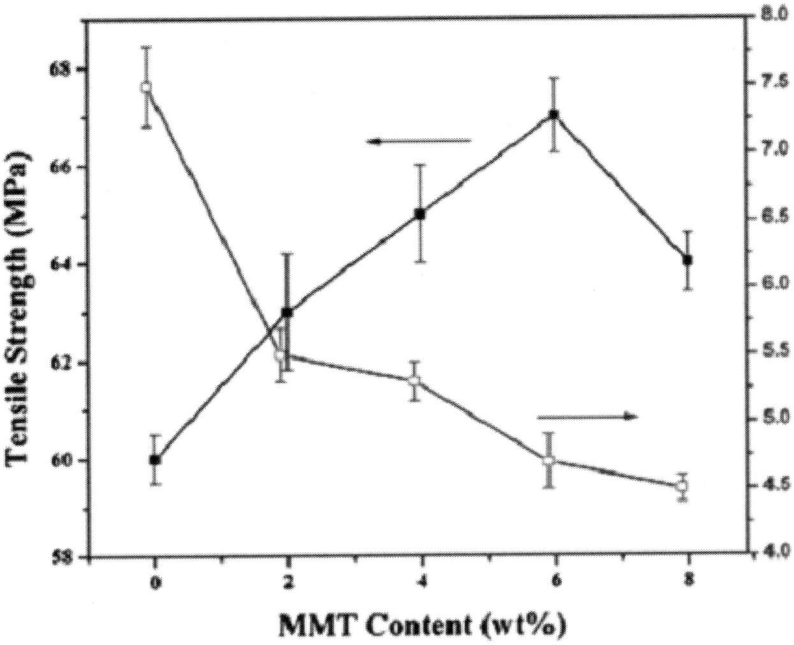

Figure 10. Effect of MMT contents on tensile strength and elongation at break of RC/MMT nanocomposite films (Reprinted from Ref. [20] Mahmoudian, S., Wahit, M.U., Ismail, A.F., Yussuf, A.A., (2012). Carbohydr. Polym. 88, 1251).

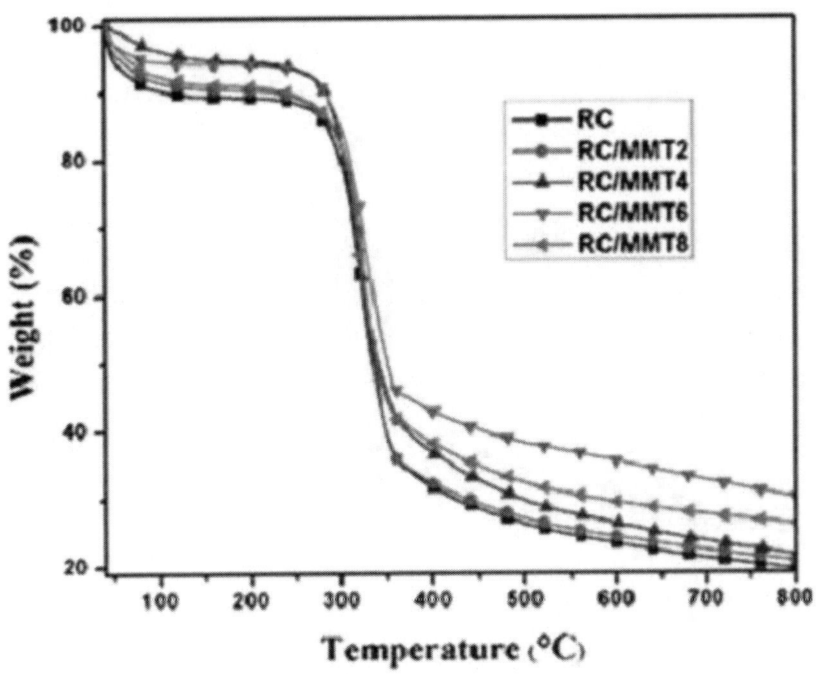

Figure 11. TGA curves of RC/MMT nanocomposites with various MMT loadings (Reprinted from Ref. [20] Mahmoudian, S., Wahit, M.U., Ismail, A.F., Yussuf, A.A., (2012). Carbohydr. Polym. 88, 1251).

Thermal-degradation of RC and RC/MMT nanocomposites was studied by Mahmoudian et al., [20] using thermogravimetric analysis under nitrogen to determine the effect of MMT on thermal stability of RC. The traces recorded at 10 °C/min for RC and RC/MMT nanocomposites are reported in Figure 11. The thermal stability is characterized by the temperatures at 10%, 20% and 80% weight losses occurred, referred to as T_{10}, T_{20} and T_{80} (data are shown in Table 1). It can be observed that the incorporation of MMT improved the decomposition temperature in RC/MMT nanocomposites. The thermal stability of nanocomposites increased with increasing MMT, up to a loading of 6 wt.%. It can be seen that, the T_{20} of RC films improved around 10 °C from 299.9 °C to 309.4 °C with 6 wt.% MMT content. It is believed that MMT platelets (impermeable) increased the thermal stability of cellulose, by providing hindrance to the diffusion of heat mass transfer (outmigration of degraded volatiles) to the surface, thus retarding the decomposition rate.

Table 1. TGA results of RC and RC/MMT nanocomposites films (Reprinted from Ref. [20] Mahmoudian, S., Wahit, M.U., Ismail, A.F., Yussuf, A.A., (2012). Carbohydr. Polym. 88, 1251)

Samples	T_{10} (°C)	T_{20} (°C)	Residue (%) at 650 (°C)
RC	111.9	299.9	22.1
RC/MMT-2	229.3	305.3	23.2
RC/MMT-4	283.7	304.0	24.9
RC/MMT-6	284.0	309.4	34.0
RC/MMT-8	242.0	306.0	28.4

Table 2. Oxygen and water permeability rate values of RC and RC/MMT nanocomposites (Reprinted from Ref. [20] Mahmoudian, S., Wahit, M.U., Ismail, A.F., Yussuf, A.A., (2012). Carbohydr. Polym. 88, 1251)

Samples	Oxygen permeability $\times 10^{-18}$ m^3 m/(m^2 s Pa)	Water absorption (%)	
		2h	24h
RC	1.12 ± 0.04	83.6 ± 1.5	90.2 ± 1.7
RC/MMT-2	1.03 ± 0.10	57.7 ± 1.1	65.8 ± 1.7
RC/MMT-4	0.84 ± 0.13	56.3 ± 1.3	62.7 ± 1.5
RC/MMT-6	0.76 ± 0.11	55.9 ± 1.2	58.7 ± 1.0
RC/MMT-8	0.82 ± 0.08	56.1 ± 1.8	61.3 ± 1.2

Gas and water barriers are also an important application of regenerated cellulose films composites. Mahmoudian et al., [20] investigated the resistance of RC/MMT nanocomposites to gas and water permeability. Oxygen (O_2) permeability values of RC film and RC/MMT nanocomposite films are presented in Table 2. Permeability values decreased gradually with increasing MMT concentration. O_2 permeability decreased by about 33% by addition of 6 wt.% MMT. The reduction in permeability was attributed to the high aspect ratio of MMT particles combined with intercalation of MMT in cellulose matrix, which resulted in an increase in tortuosity, and increasing MMT volume fraction improves this property. It was also reported that, the water absorption (%) of RC and RC/MMT nanocomposites for 2 h and

24 h (Table 2). The presence of MMT in cellulose leads to improvement in water resistance for RC/MMT nanocomposite films. The 24 h water absorption of the nanocomposite films decreased from 90.2% to 61.3% as the MMT content increased from 0 to 8 wt.%. It was deduced that the addition of MMT diminishes the water absorption of the RC films due to the presence of impermeable MMT particles in the nanocomposites which can lower the rate of water transport in the polymer matrix. This also was attributed to the low maximum water uptake of MMT compared to RC.

2.2. Regenerated Cellulose/Sepiolite

Preparation and characterization of unique regenerated cellulose/sepiolite (RC/SEP) nanocomposite films was reported by Soheilmoghaddam et al., [51] using an environmentally friendly ionic liquid, 1-butyl-3 methylimidazolium chloride (BMIMCl), through a simple, cost effective and "green" method. As can be seen in Figure 12, the XRD patterns of the regenerated cellulose films with different loadings of sepiolite represented cellulose II crystalline structure. After adding the sepiolite, the intense diffraction peak of (110) (d-spacing of 1.2 nm) was visible in the nanocomposites but some characteristic sepiolite peaks have disappeared in the nanocomposite films, especially at low sepiolite concentration. This was attributed to the homogenous dispersion of the sepiolite into the RC matrix due to its bundles being generally delaminated to fiber sticks.

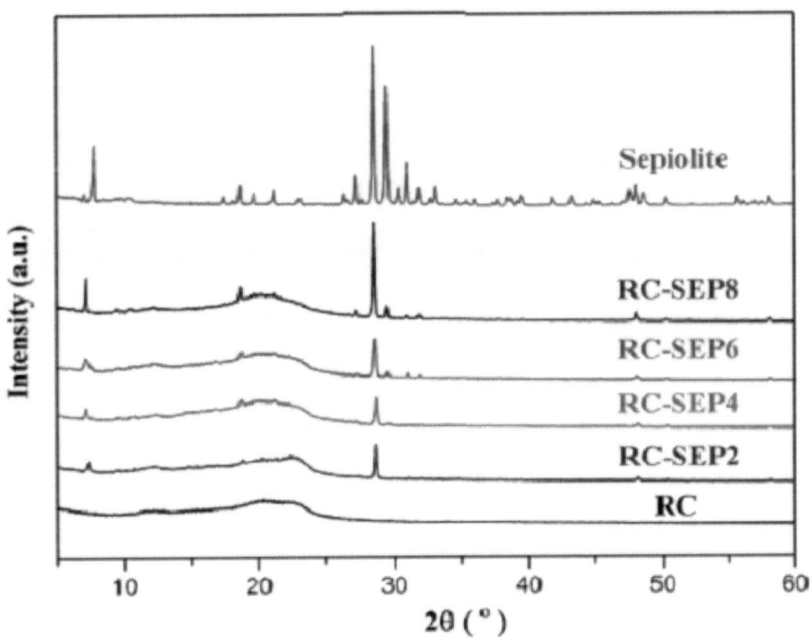

Figure 12. XRD patterns of sepiolite, RC and RC/sepiolite nanocomposite films (Reprinted from Ref. [51] Soheilmoghaddam, M., Wahit, M.U., Yussuf, A.A., Al-Saleh, M.A., Whye, W.T., (2014). Polym. Test. 33, 121).

Good dispersion of sepiolite in polymers for targeted applications is also a fundamental challenge. In general, the properties of polymer-based composites depend strongly on the spatial distribution of sepiolite dispersed in the polymer matrices. In order to have conclusive evidence regarding the dispersion of the sepiolite nanoclay in RC, Soheilmoghaddam et al., [51] studied the morphology of the nanocomposites through field emission scanning electron microscope (FESEM) and transmission electron microscope (TEM).

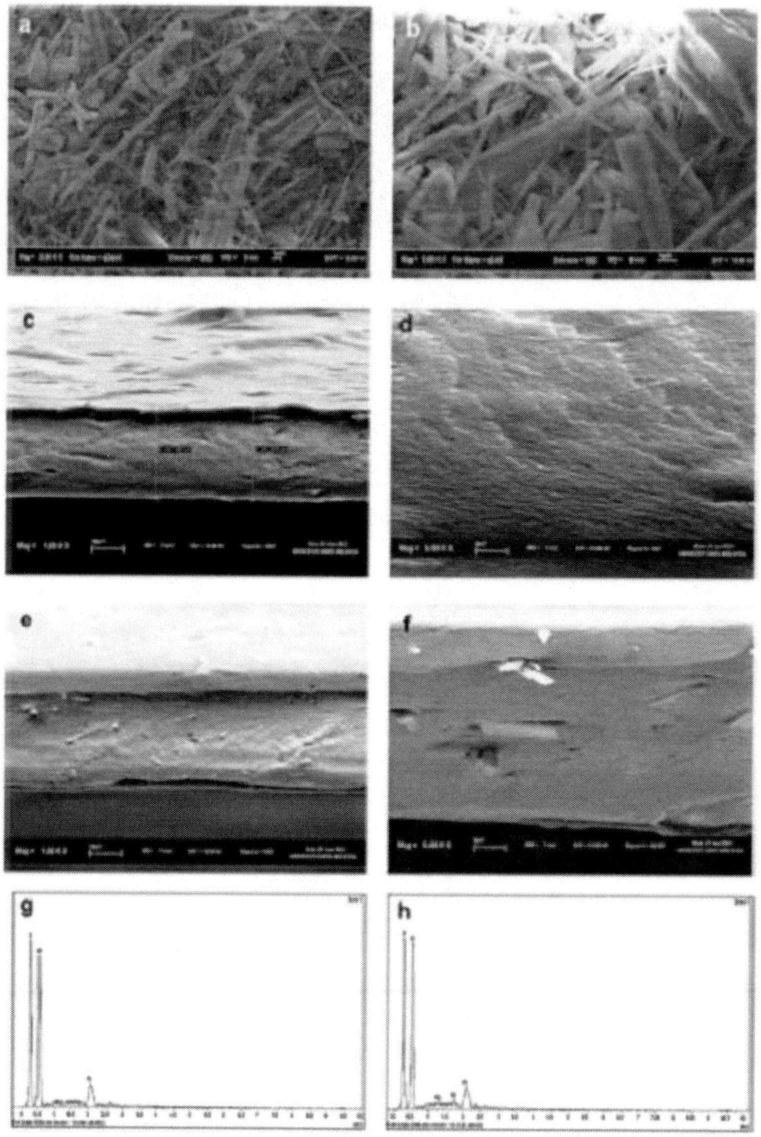

Figure 13. FESEM images of (a), (b) sepiolite, (c), (d) RC. (e), (f) RC with 6 wt.% sepiolite content. EDX spectra (g) RC and (h) RC with 6 wt.% sepiolite content (Reprinted from Ref. [51] Soheilmoghaddam, M., Wahit, M.U., Yussuf, A.A., Al-Saleh, M.A., Whye, W.T., (2014). Polym. Test. 33, 121).

Figure 13 shows the FESEM micrographs of sepiolite, RC and RC nanocomposite with 6 wt.% sepiolite content (RC/SEP-6) . It is obvious that the sepiolite has a fibrous and needle like morphology (Figure 13 (a), (b)). The cross sectional view shows a dense and nonporous structure of pure RC with no flaws on the surface. Film thickness was found to be about 30 mm (Figure 13 (c), (d)). It is clearly observed from Figure 14 (e) that sepiolite nanofillers are uniformly dispersed in the RC matrix. In addition, the RC appeared to cover the surface of sepiolite, suggesting interaction between the surface of sepiolite and RC matrix (Figure 13(f)). The result of the elemental measurement for RC and RC/sepiolite nanocomposite films indicated that there was no N in the films, suggesting that ionic liquid had been washed out (Figure 13(g), (h)).

To better verify the sepiolite structure and sepiolite dispersion in RC matrix obtained by FESEM above, TEM images of RC nanocomposite with 6 wt.% sepiolite content is shown in Figure 14. It can be observed from the TEM images that the sepiolite is a fibrous and needle like morphology (Figure 14 (a)) with nanofillers well dispersed in the RC matrix (Figure 14 (b)), and exhibiting a random orientation of the contact between them, which was consistent with the FESEM results (Figure 14(c)).

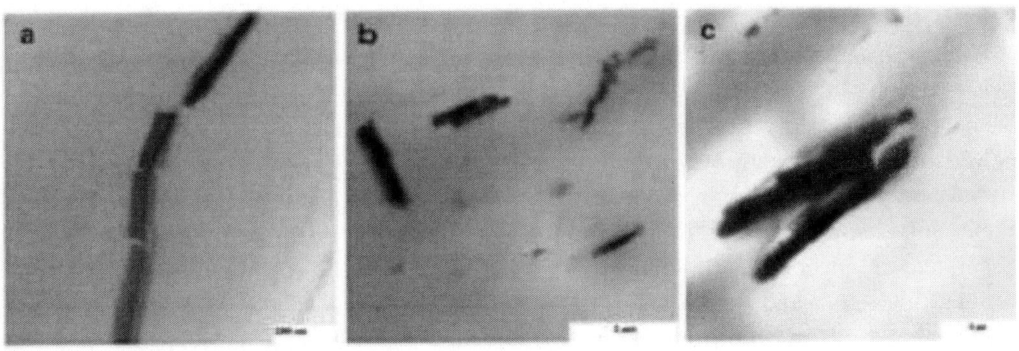

Figure 14. TEM pictures of RC-SEP6 nanocomposite film at (a)(magnification: 25K×), (b) 5K× and (c) 6.3K×(Reprinted from Ref. [51] Soheilmoghaddam, M., Wahit, M.U., Yussuf, A.A., Al-Saleh, M.A., Whye, W.T., (2014). Polym. Test. 33, 121).

Table 3. The mechanical properties of RC and RC/sepiolite nanocomposite films (Reprinted from Ref. [51] Soheilmoghaddam, M., Wahit, M.U., Yussuf, A.A., Al-Saleh, M.A., Whye, W.T., (2014). Polym. Test. 33, 121)

Samples	Young's modulus (GPa)	Tensile strength (MPa)	Elongation at break (%)
RC	1.8±1.20	35.30±1.12	5.26±0.72
RC/SEP2	2.1±0.75	37.70±0.80	5.41±0.56
RC/SEP4	3.1±0.68	46.50±0.93	6.35±1.28
RC/SEP6	3.5±0.45	61.6±0.75	7.12±1.00
RC/SEP8	4.6±0.35	59.8±1.33	7.00±1.26

Soheilmoghaddam et al., [51] also reported that the mechanical properties of regenerated cellulose/sepiolite nanocomposites were greatly influenced by the amount of sepiolite addition. Young's modulus of nanocomposite film with 8 wt.% of sepiolite increased by 166% compared to that of the RC film. It was found that the tensile strength also increased by

74% for RC/SEP-6 compared to the pure RC film, but as the sepiolite loading further increased to 8 wt.% the tensile strength slightly decreased, this result was attributed to a reduced level of dispersion (Table 3). The authors reported that the considerable enhancement in the mechanical properties of the RC/sepiolite nanocomposite films can not only be explained by uniformly dispersed sepiolite nanofibers with high aspect ratio and high intrinsic stiffness, but also can be ascribed to the strong interaction between sepiolite nanofibers and RC matrix. It is also noted that, ductility of the nanocomposite films were also improved when sepiolite was added. The incorporation of 6 wt.% sepiolite was resulted in 35 % increase in elongation at break; thereafter, the elongation at break was nearly constant (Table 3). This behavior was attributed to the sufficiently dispersed state of sepiolite nanofillers in the cellulose matrix.

FUTURE DIRECTIONS

Regenerated cellulose is one of the most promising natural polymeric resources as alternative for petroleum-based polymers. It can be used in packaging, membranes and medical products as well as in electronic devices. However, only about 2 % of cellulose renewed annually by nature has been used as raw material in manufacturing of cellulose material and goods. The main problem to the more extensive use of cellulose is the lack of suitable solvents for cellulose regeneration. To date, most of the solvents used to prepare regenerated cellulose materials have many drawbacks such as toxicity, volatility, side reactions, high cost and difficult solvent recovery. In recent years, ionic liquids (IL) have begun to be used in cellulose materials processing as environmentally friendly solvents. However, high price of IL at the current market, enable regenerated cellulose to compete with petroleum based plastics. Another challenge is the limited ductility (elongation at break) of regenerated cellulose products which has urged researchers to develop regenerated cellulose blends or nanocomposites to improve elongation of RC. RC films are now commercially used as biodegradable material. However, the major reason for the low utilization of regenerated cellulose films is their relatively higher water absorption and gas permeability than conventional thermoplastic packaging materials. Therefore, it appears necessary to improve the properties of this polymer to make it fully competitive with common thermoplastics. Despite the setback, the properties of these nanocomposites up to now confirm that the promise of a dream material for future can be fulfilled. Further research is therefore necessary to develop a proper understanding of the formulation/structure/property relationship/ toughening mechanisms and interactions involved in RC nanocomposites system.

REFERENCES

[1] Sinha Ray, S., Bousmina, M., (2005). Biodegradable polymers and their layered silicate nanocomposites: In greening the 21st century materials world. *Prog. Polym. Sci.* 50, 962.

[2] Klemm, D., Heublein, B., Fink, H-P., Bohn, A., (2005). Cellulose: fascinating biopolymer and sustainable raw material. *Angew. Chem. Int. Ed.* 44, 3358.

[3] Chivrac, F., Pollet, E., Avérous, L., (2009). Progress in nano-biocomposites based on polysaccharides and nanoclays. *Mat. Sci. Eng.: R: Reports* 67, 1.

[4] Potters, G., Goethem, V.D., Schutte, F., (2010). Promising biofuel resources: lignocellulose and algae. *Nature Education* 3,14.

[5] Langan, P., Nishiyama, Y., Chanzy, H., (1999). A revised structure and hydrogen-bonding system in cellulose ii from a neutron fiber diffraction analysis. *J. Am. Chem. Soc.* 121, 9940.

[6] Zugenmaier, P., (2001). Conformation and packing of various crystalline cellulose fibers. *Prog. Polym. Sci.* 26, 1341.

[7] Heinze, T., Liebert, T., (2001). Unconventional methods in cellulose functionalization. *Prog. Polym. Sci.* 26, 1689.

[8] Zhu, S., Wu, Y., Chen, Q., Yu, Z., Wang, C., Jin, S., Ding, Y., Wu, G., (2006). Dissolution of cellulose with ionic liquids and its application: a mini-review. *Green Chemistry* 8, 325.

[9] El Seoud, O.A., Koschella, A., Fidale, L.C., Dorn, S., Heinze, T., (2007). Applications of ionic liquids in carbohydrate chemistry: A window of opportunities. *Biomacromolecules* 8, 2629.

[10] Zhang, L., Ruan, D., Zhou, J., (2001). Structure and properties of regenerated cellulose films prepared from cotton linters in NaOH/urea aqueous solution. *Ind. Eng. Chem. Res.* 40, 5923.

[11] Li, R., Zhang, L., Xu, M., (2012). Novel regenerated cellulose films prepared by coagulating with water: Structure and properties. *Carbohydr. Polym.* 87, 95.

[12] Scherer, P.C., (1931) Solubility of cellulose in ammonia salt solutions. *J. Am. Chem. Soc.* 53, 4009.

[13] Hudson, S.M., Cuculo, J.A., Wadsworth, L.C., (1983). The solubility of cellulose in liquid ammonia/ammonium thiocyanate solution: The effect of composition and temperature on dissolution and solution properties. *J. Polym. Sci.: Polym. Chem. Edition* 21, 651.

[14] Terbojevich, M., Cosani, A., Conio, G., Ciferri, A., Bianchi, E., (1985). Mesophase formation and chain rigidity in cellulose and derivatives. 3. Aggregation of cellulose in N,N-dimethylacetamide-lithium chloride. *Macromolecules* 18, 640.

[15] Tamai, N., Tatsumi, D., Matsumoto, T., (2004). Rheological properties and molecular structure of tunicate cellulose in LiCl/1,3-Dimethyl-2-imidazolidinone. *Biomacromolecules* 5, 422.

[16] Fink, H.P., Weigel, P., Purz, H.J., Ganster, J., (2001). Structure formation of regenerated cellulose materials from NMMO-solutions. *Prog. Polym. Sci.* 26, 1473.

[17] Dogan, H., Hilmioglu, N.D., (2009). Dissolution of cellulose with NMMO by microwave heating. *Carbohydr. Polym.* 75, 90.

[18] Rosenau, T., Hofinger, A., Potthast, A., Kosma, P., (2003). On the conformation of the cellulose solvent< i> N</i>-methylmorpholine-< i> N</i>-oxide (NMMO) in solution. *Polymer* 44, 6153.

[19] Zhang, H., Wu, J., Zhang, J., He, J., (2005). 1-Allyl-3-methylimidazolium chloride room temperature ionic liquid: a new and powerful nonderivatizing solvent for cellulose. *Macromolecules* 38, 8272.

[20] Mahmoudian, S., Wahit, M.U., Ismail, A.F., Yussuf, A.A., (2012). Preparation of regenerated cellulose/montmorillonite nanocomposite films via ionic liquids. *Carbohydr. Polym.* 88, 1251.

[21] Wilkes, J.S., (2002). A short history of ionic liquids-from molten salts to neoteric solvents. *Green Chemistry* 4, 73.

[22] Koel, M., (2005). Ionic liquids in chemical analysis. *Cr. Rev. Anal. Chem.* 35, 177.

[23] Wilkes, J.S., Wasserscheid., Welton, T., (2008). Ionic liquids in synthesis. (Vol.1). New Dehli : WILLEY-VCH Verlags Gmbh & Co KgaA.

[24] Hameed, N., Guo, Q., Tay, F.H,, Kazarian, S.G., (2011). Blends of cellulose and poly(3-hydroxybutyrate-co-3-hydroxyvalerate) prepared from the ionic liquid 1-butyl-3-methylimidazolium chloride. *Carbohydr. Polym.* 86, 94.

[25] Azubuike, C., Rodríguez, H., Okhamafe, A., Rogers, R., (2012). Physicochemical properties of maize cob cellulose powders reconstituted from ionic liquid solution. *Cellulose* 19, 425.

[26] Fort, D.A., Remsing, R.C., Swatloski, R.P., Moyna, P., Moyna, G., Rogers, R.D., (2007). Can ionic liquids dissolve wood? Processing and analysis of lignocellulosic materials with 1-n-butyl-3-methylimidazolium chloride. *Green Chemistry* 9, 63.

[27] Pu, Y., Jiang, N., Ragauskas, A.J., (2007). Ionic liquid as a green solvent for lignin. *J. Wood Chem. Techn.* 27, 23.

[28] Casas, A., Palomar, J., Alonso, M.V., Oliet, M., Omar, S., Rodriguez, F., (2012). Comparison of lignin and cellulose solubilities in ionic liquids by COSMO-RS analysis and experimental validation. *Ind. Crops Prod.* 37, 155.

[29] Zhao, D., Li, H., Zhang, J., Fu, L., Liu, M., Fu, J., Ren, P., (2012). Dissolution of cellulose in phosphate-based ionic liquids. *Carbohydr. Polym.* 87, 1490.

[30] Swatloski, R.P., Spear, S.K., Holbrey, J.D., Rogers, R.D., (2002). Dissolution of cellose with ionic liquids. *J. Am. Chem. Soc.* 124, 4974.

[31] Ren, Q., Wu, J., Zhang, J., He, J., Guo, M., (2003). Synthesis of 1-allyl,3-methylimidazolium-based room-temperature ionic liquid and preliminary study of its dissolving cellulose. *Acta Polymerica Sinica* 3, 448.

[32] Feng, L., Chen, Z., (2008). Research progress on dissolution and functional modification of cellulose in ionic liquids. *Journal of Molecular Liquids* 142, 1.

[33] Zavrel, M., Bross, D., Funke, M., Büchs, J., Spiess, A.C., (2009). High-throughput screening for ionic liquids dissolving (ligno-)cellulose. *Bioresour. Technol.* 100, 2580.

[34] Zhai, R., Zhang, B., Liu, L., Xie, Y., Zhang, H., Liu, J., (2010). Immobilization of enzyme biocatalyst on natural halloysite nanotubes. *Catalysis Communications* 12, 259.

[35] Kadokawa, J-I., Murakami, M-A., Kaneko, Y., (2008). A facile preparation of gel materials from a solution of cellulose in ionic liquid. *Carbohydr. Res.* 343, 769.

[36] Alexandre, M., Dubois, P., (2000). Polymer-layered silicate nanocomposites: preparation, properties and uses of a new class of materials. *Mat. Sci. Eng.: R: Reports* 28, 1.

[37] Sinha, Ray, S., Okamoto, M., (2003). Polymer/layered silicate nanocomposites: a review from preparation to processing. *Prog. Polym. Sci.* 28, 1539.

[38] Gerhard, L., (2005). From clay mineral crystals to colloidal clay mineral dispersions. In: Coagulation and Flocculation, Second Edition. Surfactant Science. CRC Press, 519.

[39] Bordes, P., Pollet, E., Avérous, L., (2009). Nano-biocomposites: Biodegradable polyester/nanoclay systems. *Prog. Polym. Sci.* 34, 125.

[40] Merlin, F., Lombois, H., Joly, S., Lequeux, N., Halary, J-L., Van Damme, H., (2002). Cement-polymer and clay-polymer nano- and meso-composites: spotting the difference. *J. Mater. Chem.* 12, 3308.
[41] Tartaglione, G., Tabuani, D., Camino, G., Moisio, M., (2008). PP and PBT composites filled with sepiolite: Morphology and thermal behaviour. *Compos. Sci. Technol.* 68, 451.
[42] Fukushima, K., Tabuani, D., Camino, G., (2009). Nanocomposites of PLA and PCL based on montmorillonite and sepiolite. *Mat. Sci. Eng.: C* 29, 1433.
[43] Huang, D., Mu, B., Wang, A., (2012). Preparation and properties of chitosan/poly (vinyl alcohol) nanocomposite films reinforced with rod-like sepiolite. *Mater. Lett.* 86, 69.
[44] Ruiz-Hitzky, E., (2001). Molecular access to intracrystalline tunnels of sepiolite. *J. Mat. Chem.* 11, 86.
[45] Nohales, A., Solar, L., Porcar, I., Vallo, C.I., Gómez, C.M., (2006). Morphology, flexural, and thermal properties of sepiolite modified epoxy resins with different curing agents. *Eur. Polym. J.* 42, 3093.
[46] Ma, J., Bilotti, E., Peijs, T., Darr, J.A., (2007). Preparation of polypropylene/sepiolite nanocomposites using supercritical CO_2 assisted mixing. *Eur. Polym. J.* 43, 4931.
[47] Chen, H., Zheng, M., Sun, H., Jia, Q., (2007). Characterization and properties of sepiolite/polyurethane nanocomposites. *Mat. Sci. Eng.: A* 445–446, 725.
[48] García-López, D., Fernández, J.F., Merino, J.C., Santarén, J., Pastor, J.M., (2010). Effect of organic modification of sepiolite for PA 6 polymer/organoclay nanocomposites. *Compos. Sci. Technol.* 70, 1429.
[49] Yu, Y., Qi, S., Zhan, J., Wu, Z., Yang, X., Wu, D., (2011). Polyimide/sepiolite nanocomposite films: Preparation, morphology and properties. *Mat. Res. Bull.* 46, 1593.
[50] Liu, M., Pu, M., Ma, H., (2012). Preparation, structure and thermal properties of polylactide/sepiolite nanocomposites with and without organic modifiers. *Compos. Sci. Technol.* 72, 1508.
[51] Soheilmoghaddam, M., Wahit, M.U., Yussuf, A.A., Al-Saleh, M.A., Whye, W.T., (2014). Characterization of bio regenerated cellulose/sepiolite nanocomposite films prepared via ionic liquid. *Polym. Test.* 33, 121.
[52] García-Romero, E., Suárez, M., (2013). Sepiolite–palygorskite: Textural study and genetic considerations. *Appl. Clay Sci.* 86, 129.
[53] Zhang, F., Guo, Z., Gao, H., Li, Y., Ren, L., Shi, L., Wang, L., (2005). Synthesis and properties of sepiolite/poly (acrylic acid-co-acrylamide) nanocomposites. *Polym. Bull.* 55, 419.
[54] White, L.A., (2004). Preparation and thermal analysis of cotton–clay nanocomposites. *J. Appl. Polym. Sci.* 92, 2125.
[55] Cerruti, P., Ambrogi, V., Postiglione, A., Rychlý, J., Matisová-Rychlá, L., Carfagna, C., (2008). Morphological and thermal properties of cellulose−montmorillonite nanocomposites. *Biomacromolecules* 9, 3004.
[56] Lee, J., Sun, Q., Deng, Y., (2008). Nanocomposites from regenerated cellulose and nanoclay. *J. Biobased Mat. Bioenerg.* 2, 162.

In: Polymer-Matrix Composites
Editor: Rakesh Kumar

ISBN: 978-1-61122-005-6

Chapter 3

RECENT RESEARCH INTO NATURAL POLYSACCHARIDE BASED NANOCRYSTALS: PREPARATION, PROPERTIES AND APPLICATIONS IN BIONANOCOMPOSITES FIELD

Huafeng Tian[1*], Dagang Liu[2] and Aimin Xiang[1]
[1]School of Material and Mechanical Engineering,
Beijing Technology and Business University, Beijing, China
[2]Department of Chemistry,
Nanjing University of Information Science and Technology, Nanjing, China

ABSTRACT

Natural polysaccharides, including starch, cellulose, chitin, etc. are renewable and biodegradable biopolymer produced by many organisms as a source of stored energy or tissue materials. The amorphous domains of these biopolymers, especially starch, cellulose and chitin, can also be removed under certain conditions such as acid hydrolysis to produce crystallites in nanoscale, which are called nanocrystals or whiskers. These nanocrystals from natural polymers show many advantages over traditional inorganic nanoparticles such as easy availability, nontoxicity, biodegradability, low density, and easy modification. During the past decade, many works have been devoted to use them as substitutes for inorganic nanoparticles in preparing polymer nanocomposites and mimic biocomposites by blending whiskers from different sources with polymer matrices. This chapter intends to give a clear overview of preparation, properties, and applications of natural polymer based nanoparticles.

Keywords: Polysaccharide, Nanocrystals, Nanocomposites

---

* Corresponding author. Tel.: +86 10 68985531; fax: +86 10 68988056. E-mail address: tianhuafeng@th.btbu.edu.cn (H. Tian).

1. INTRODUCTION

Polysaccharides known as carbohydrates, such as cellulose, starch and chitin are naturally occurring polymers found abundantly in nature as structural building blocks or energy storage in the form of sugar [1]. In recent years, researches have attracted the tremendous academic interest not only replacing the present oil-based chemistry, but also opening new fields and applications for these biopolymers. Natural polysaccharides including cellulose, chitin and starch, consist of crystalline and amorphous regions. The amorphous regions are highly susceptible to hydrolysis and could be removed, leaving the rigid crystalline regions [2]. The obtained crystalline material occurs in the form of polysaccharide nanocrystals, most commonly rod-like cellulose nanocrystals (or cellulose whiskers, nanocrystalline cellulose), chitin whiskers and platelet-like starch nanocrystals. Generally speaking, the research on cellulose whiskers is much more than that on chitin and starch nanocrystals.

The advantages of natural polysaccharide nanocrystals over inorganic nanoparticles include the availability, biocompatibility, biodegradability, easy processability, low cost, high specific strength and modulus and relatively reactive surface (sufficient –OH groups on the surface). The outstanding properties and characteristics of polysaccharide nanocrystals lead to their wide applications in the field of materials science. The incorporation of polysaccharide nanocrystals with high specific modulus into polymeric matrices led to the improvement of properties of the resulted bionanocomposites, especially for mechanical performance. Over the past two decades, polysaccharide nanocrystals have been introduced in both natural polymeric matrices, such as natural rubber [3], starch [4] and soy protein [5], cellulose [6], as well as synthetic polymer matrices, such as polyesters [7, 8], polyvinyl alcohol [9, 10], polyolefin [11], polyurethane [12], etc.

The investigation of natural nanocrystals in the field of functional materials has also been expanded for their high aspect ratio, rigidity, highly crystalline properties, active functional groups, and chiral nematic phase. This review gives a clear overview of preparation, properties, and applications of natural polysaccharide based nanoparticles as well as their bionanocomposites.

2. PREPARATION OF NATURAL POLYSACCHARIDE BASED NANOCRYSTALS AND THEIR PROPERTIES

Different methods have been developed for the preperation of nanocrystals. For instance, acid hydrolysis [13], enzymatic hydrolysis [14], ionic liquid preparation [15, 16], sonochemical-assisted hydrolysis preparation [17], and some combined technology [18,19] have been introduced.

Due to the controllable and convenient treatment, acid hydrolysis is mostly utilized in the preparation of polysaccharide nanocrystals. The hydrolysis conditions, such as the type of acid, acid concentration, temperature, and duration, are different for different source of polysaccharide. The morphology and geometrical dimensions [length (L), diameter (D), thickness (T) and aspect ratio (L/D)] of polysaccharide nanocrystals always directly varied due to the different sources and extracting methods [2]. Both cellulose and chitin nanocrystals exhibit rod-like morphology, therefore are often called whiskers. The source of cellulose whiskers includes cotton, jute, flax, kenaf, sugar-beet pulp, bamboo and tunicin, etc. Tunicin cellulose whiskers possess the length and diameter of 1 μm and 15 nm, with the aspect ratio of 67[20], while cotton whiskers possess the length and diameter of 189 nm and 13 nm, with the aspect ratio of about 10 [21]. Chitin whiskers usually derived from crab shell, *Riftia* tubes, shrimp, squid pen, etc., with the aspect ratio ranges from 10 to several hundreds. The dimension of whiskers from squid pen, shrimp and crab shell is close to that of cotton whiskers. Whiskers from *Riftia* tubes possess the average length of 2.2 μm, and the aspect ratio of 120 [22]. As for the platelet-like starch nanocrystals, they are generally derived from pea, potato, corn and waxy maize, with a length of 20–100 nm, a width of around 25–30 nm and a thickness of 6–8 nm [23]. The hydrolysis of starch displays two well-separated time zones [24]. For lower times (t<15 days), the hydrolysis is fast and corresponds to the hydrolysis of amorphous starch domains. For higher times (t>15 days), the hydrolysis is slow and corresponds to the hydrolysis of crystalline domains. The critical time corresponding to fast/slow hydrolysis zones depends on the origin of starch, and it appears that for potato starch, a hydrolysis time of 15 days allows the removal of most of the amorphous zones without damaging the crystalline zones. Waxy maize starch nanocrystals consist of platelet-like particles with a thickness of 6-8 nm, a length of 40-60 nm, a width of 15-30 nm. Such starch nanocrystals are generally observed in the form of aggregates having an average size around 4.4 m, as measured by laser granulometry [25].

The morphology of these nanocrystals could be investigated by transmission electron microscopy (TEM) or atomic force microscopy (AFM). The crystalline structure could be studied using X-ray diffraction (XRD). The average dimensions of some polysaccharide nanocrystals are listed in Table 1.

Table 1. Geometrical characteristics of some polysaccharide nanocrystals: Average length (L), average diameter (D) of rod-like particles and average length (L), average width (W) and thickness (T) of platelet-like particles

Nature	Source	L / nm	D / nm	
Cellulose	Algal (*Valonia*) [26]	>1000	10-20	
	Bacterial[27]	1000-2000	10-20	
	Cotton[28]	189	13	
	Cottonseed linter[29]	170-490	40-60	
	MCC[30]	200-400	10	
	Sisal[31]	100-500	3-5	
	Sugar beet pulp[32]	210	5	
	Tunicin[20]	100-several 1000	10-20	
	Wheat straw[33]	150-300	5	
	Ramie[34]	50-150	5-10	
	Wood[35]	100-300	3-5	
Chitin	Crab shell[36]	80-600	8-50	
	Riftia tubes[22]	500-10000	18	
	Shrimp[37]	50-300	5-70	
	Squid pen[37]	150-800	10	
		L / nm	W / nm	T / nm
Starch	Waxy maize[38]	30	20	6
	Potato[39]	25	14	6

Some interesting shapes of the nanocrystals are also reported in the literature. Using a method of combined HNO_3–$KClO_3$ treatment and sulfuric acid hydrolysis, bamboo cellulose crystals (BCCs) were prepared [40]. BCCs were of typical cellulose I structure, and the morphology was dependent on its concentration in the suspension. BCC of 50–100 nm were assembled into leaf nervations at low concentration (i.e. 0.1 wt.% of solids), but congregated into a micro-sized "flower" geometry at high concentration (i.e. 10.0 wt.% of solids) as shown in Figure 1. BCC freeze–dried from suspension containing 0.1 wt.% solids seemed to be a leaf nervation composed of many sphere shaped nano-crystallites. The different geometries resulting from aggregation intensity of BCCs were due to high surface electrostatic energy and large surface area.

Figure 1. SEM micrograph of freeze–dried bamboo fiber (a), BCC freeze–dried from suspensions with 0.1% (b) and 10% concentrations (c) (Reprinted from Ref. [40] Liu, D., Zhong, T., Chang, P.R., Li, K., Wu, Q., (2010). Bioresour. Technol. 101, 2529).

Different from the commonly used acid hydrolysis method, preparation of nanocrystalline cellulose (NCC) from microcrystalline cellulose (MCC) via purely physical method of high-intensity ultrasonication was reported recently [41]. The resulting cellulose nanocrystals exhibited a rod-shaped structure, with diameters between 10 and 20 nm and lengths between 50 and 250 nm. The length and crystallinity of NCC decreased with increasing ultrasonication duration as shown in Figure 2. The ultrasonication effect was non-selective, which means it can remove both the amorphous and crystalline cellulose. The ultrasonic activation effect in heterogeneous systems is primarily the consequence of cavitation. During cavitation the potential energy of the expanded bubble is converted into the kinetic energy of a liquid jet that moves through the bubble's interior and penetrates the opposite bubble wall at several hundreds of meters per second, and hits the surface of MCC. These jets hit the surface with tremendous force, which can cause severe damage at the point of impact and produce newly exposed, highly reactive surfaces. In the MCC–water system, both the crystalline and amorphous regions were subjected intense collisions. In Figure 3, it can be seen that on the MCC surface, the intense physical stresses cause particle breakage and then decomposed to NCC.

For the highly crystalline properties, polysaccharide nanocrystals possess high modulus compared with metals or inorganic materials. Take cellulose for example, cellulose crystallites have been calculated to have tensile strength of 10 GPa and modulus of 150 GPa [42]. The bending strength that is nearly one-sixth of the corresponding value (63 GPa) for carbon nanotubes whose tensile strength is predicted to be as high 300 GPa at modulus of 1 TPa [43]. The moduli and the specific moduli (modulus/density) of a number of commonly used engineering materials are reported in Table 2, which show that the specific modulus of crystalline cellulose exceeds engineering materials such as steel, concrete, glass and aluminium [44], which ensure these nanocrystals excellent reinforcing agents for polymeric composites.

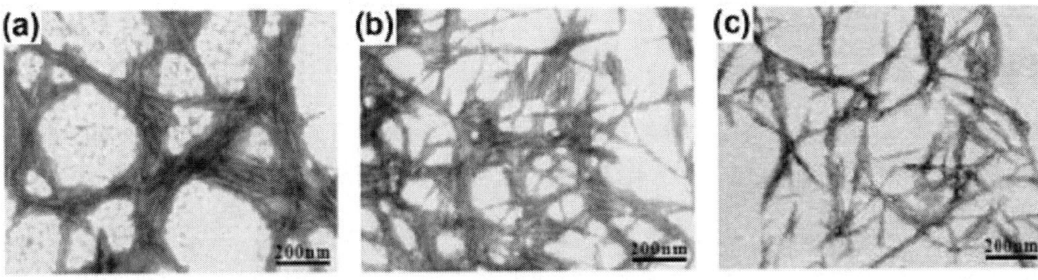

Figure 2. TEM images of NCC treated using different ultrasonication times: (a) 5 min, (b) 10 min and (c) 15 min. (Reprinted from Ref. [41] Li, W., Yue, J., Liu, S., (2012). Ultrason. Sonochem. 19, 479).

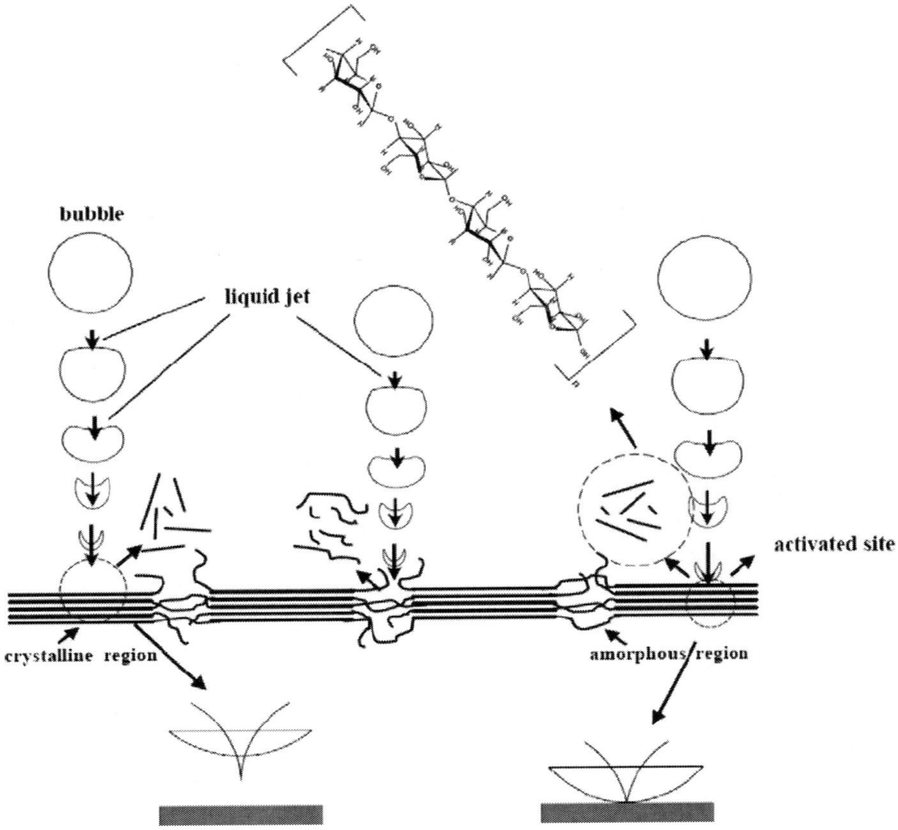

Figure 3. Schematic diagram of the MCC ultrasonication process. (Reprinted from Ref. [41] Li, W., Yue, J., Liu, S., (2012). Ultrason. Sonochem. 19, 479).

Table 2. Moduli of engineering materials compared to cellulose nanocrystals [44]

Material	Modulus (GPa)	Density (mg/m^3)	Specific Modulus ($GPa\ mg^{-1}\ m^3$)
Aluminum	69	2.7	26
Steel	200	7.8	26
Glass	69	2.5	28
Cellulose whisker	138	1.5	92

3. THE PREPARATION OF BIONANOCOMPOSITES WITH NATURAL POLYSACCHARIDE BASED NANOCRYSTALS

Currently, there is a considerable interest in preparing high performance bionanocomposites filled with natural nanocrystals. Water is the best dispersion media for polysaccharide nanocrystals. Stable suspensions could be formed by dispersing these nanocrystals in water. One drawback related to the use of these nanocrystals for polymeric nanocomposites is their inherent difficulty to disperse in non-polar medium, because of their polar surface [45]. In other words, the incorporation of polysaccharide nanocrystals as a reinforcement material has so far been mainly limited to aqueous or polar environments. Processing techniques have an important impact on the final properties of the resulting composites. The main techniques used to prepare polysaccharides nanocomposites include water or organic solvent evaporation by solvent casting, extrusion with freeze-dried nanoparticles, etc [46].

3.1. Solvent Casting Technique

The solvent casting has been the main technique to transfer polysaccharide nanocrystals from an aqueous dispersion into an organic polymer matrix [47]. Three systems can be distinguished depending on the polymer used as matrix, i.e. (i) water soluble polymers, (ii) polymer emulsions, and (iii) non hydrosoluble polymers.

3.1.1. Water Soluble Polymer Systems

Polysaccharide nanocrystals are recovered in water suspension after acid hydrolysis. Because of the high stability of aqueous nanoparticle suspensions, water is the preferred medium for preparing nanocomposite films. The preparing process of the composites consists of mixing the nanocrystal suspension with the polymer solution previously dissolved in water first, and then evaporating the water. However, it restricts the choice of the matrix to hydrosoluble polymers, such as polyvinyl alcohol (PVA) [48, 49], starch [50], cellulose [51] and poly(oxyethylene) [52], etc.

Besides solution casting, the water could also be removed by lyophilization. The freeze-drying of the mixtures of polysaccharide nanocrystal suspensions and soy protein/water dispersions resulted in the soy protein/nanocrystal nanocomposite powders. With the added plasticizers, the nanocrystals reinforced nanocomposite films could be prepared through extrusion and hot molding [53, 54].

3.1.2. Polymer Emulsion Systems

Polymer emulsion could also be processed by casting or extrusion after freeze drying as polymer solutions. Many research studies were done in order to process nanocomposites by using matrices such as copolymers of styrene and butyl-acrylate [55], poly (β-hydroxyoctanoate) (PHO) [56], polyvinylchloride (PVC) [57], waterborne epoxy [58], waterborne polyurethane [59, 60], natural rubber [61] and polyvinyl acetate [30], etc. In all these reports, the polysaccharide nanoparticles suspensions were mixed with the polymer emulsion in aqueous medium, and then nanocomposite films were obtained by casting the mixed dispersions. Some hydrophobic polymer/natural nanocrystal composites could also be fabricated through this method by preparing emulsions. Poly(caprolactone) (PCL) is a kind of hydrophobic thermoplastic polymers. A procedure was optimized to prepare a latex of poly(caprolactone), and a nanocomposite materials were obtained from a colloidal suspension of high aspect ratio β-chitin whiskers as the reinforcing phase and poly(caprolactone) as the matrix [22]. After mixing and stirring the two aqueous suspensions, solid films were obtained by either freeze-drying or hot-pressing or casting and evaporating the preparations.

3.1.3. Non-hydrosoluble Systems

Due to the high surface area and hydrophilic nature, strong inter-particle hydrogen bonding will cause the nanocrystals to aggregate. Sulfuric acid is widely used as the hydrolyzing agent. Hydrolysis treatment with sulfuric acid will result in the introduction of negatively charged surface acid groups (OSO^{3-}/H^{+}) on polysaccharide nanocrystals, leading to the electrostatic stabilization of nanocrystals in aqueous medium and a perfectly homogeneous dispersion [62]. The original nanocrystals could hardly dispersed in organic solvent and possess poor compatibility with hydrophobic polymer matrix. Two routes can be envisaged in order to obtain non-flocculated dispersions of polysaccharide nanocrystals in an appropriated organic medium [63]:

1. Coating of the surface of the nanocrystals with surfactants having polar heads and long hydrophobic tails.
2. Chemical surface modification of the nanocrystals.

Surfactant coating is the most simple modification method. The natural nanocrystals could be dispersed in toluene, cyclohexane [64] or chloroform [65] after surfactant coating. Stable suspensions were formed as observed by TEM. For the high specific surface area of nanocrystals, the coating technology usually needs several times surfactant higher than the quantity of nanocrystals to obtain highly dispersed nanocrystal suspensions [66], which would limit their applications.

Another way to disperse nanocrystals in organic solvent is chemical modification, including modification with small molecules and long chain grafting. Chemical functionalization on the surface of nanocrystals is the most effective approach to avoid irreversible agglomeration during drying, and enhance the compatibility between nanoparticles and nonpolar matrices [67]. The large amount of –OH groups provides reacting routes to chemical reaction. Stable suspensions of cellulose whiskers dispersed in organic solvents were obtained through partially silylation [68]. When the surface substitution was low, the nanocrystals kept their morphological integrity. But due to the surface silylation, they became readily dispersible in solvent of low polarity such as THF. The resulting suspensions, which did not flocculate, were stable and appeared birefringent when viewed between cross polars. With higher surface substitution, the core of the nanocrystals became silylated, leading to the loss of birefringent phenomenon. Namazi et al. [69] introduced a method for preparation of hydrophobically modified starch nanocrystals with using fatty acids. The modified starch nanoparticles substrates were synthesized with using fatty acid chlorides in a homogenous condition, resulting in compositions with chemical grafting.

Grafting of larger chains on the surfaces of nanocrystals could also change the surface properties. The surface modifications with medium-chain-length stearic acid chloride and poly(ethylene glycol) methyl ether had a layer thickness on the order of magnitude of the thickness of starch nanoparticles [70]. However, the length of the grafted chains was not sufficient to allow the formation of a continuous film by hot-pressing the modified nanoparticles. Monocrystalline starch nanoparticles were successfully grafted with poly(tetrahydrofuran), poly(caprolactone), and poly(ethylene glycol) monobutyl ether chains using toluene 2,4-diisocyanate as a linking agent [71]. The initial crystalline structure of the nanoparticles was preserved. The surface chemical modification with isocyanate or anhydride functions allowed enhancement of the nonpolar nature of original starch nanocrystals. Chemically modified chitin nanocrystals were synthesized by grafting poly(3-hydroxybutyrate-co-3-hydroxyvalerate) (PHBV) onto chitin backbone via chlorination [72]. Acetyl amino group was maintained in the reaction. Contact angle measurements showed that the lypophilicity of chitin nanocrystals was improved.

With the well dispersed nanocrystals in organic solvent, polysaccharide nanocrystals reinforced polymeric composites could be fabricated through solution casting. The surface chemical modification of cellulose nanocrystals with a long chain isocyanate (n-octadecyl isocyanate) allows dispersion of the nanoparticles in organic solvents and may allow processing of nanocomposite films from a casting/evaporation technique for a broad range of polymeric matrices [73]. Poly(ε-caprolactone) was grafted to the surface of starch nanocrystals via microwave-assisted ring-opening polymerization. The resultant nanoparticles were then incorporated into a poly(lactic acid) matrix to produce fully-biodegradable nanocomposites through solution casting in dichloromethane [74]. The resultant nanocomposites showed great improvement in their performance, and especially the elongation and strength were simultaneously enhanced for the nanocomposites containing a loading-level of 5 wt.-% fillers. Such starch nanocrystal-graft-polycaprolactone fillers also exhibited a simultaneous enhancement of tensile strength and elongation at break at a loading-level of 5 wt % in PCL-based waterborne polyurethane [75]. Surface acetylated chitin nanocrystals by acetic anhydride exhibited improved dispersion in CH_2Cl_2, as shown in Figure 4 [76]. The acetylated nanocrystals was incorporated into a PLA matrix and when 4 wt.% nanocrystals was introduced, the tensile strength and Young's modulus of the nanocomposites reached the maximum values, a 45 and 37% increase over the neat PLA film, respectively. This simultaneous enhancement could be attributed to the well dispersion of acetylated nanocrystals in the PLA matrix and strong interfacial adhesion between acetylated nanocrystals and PLA.

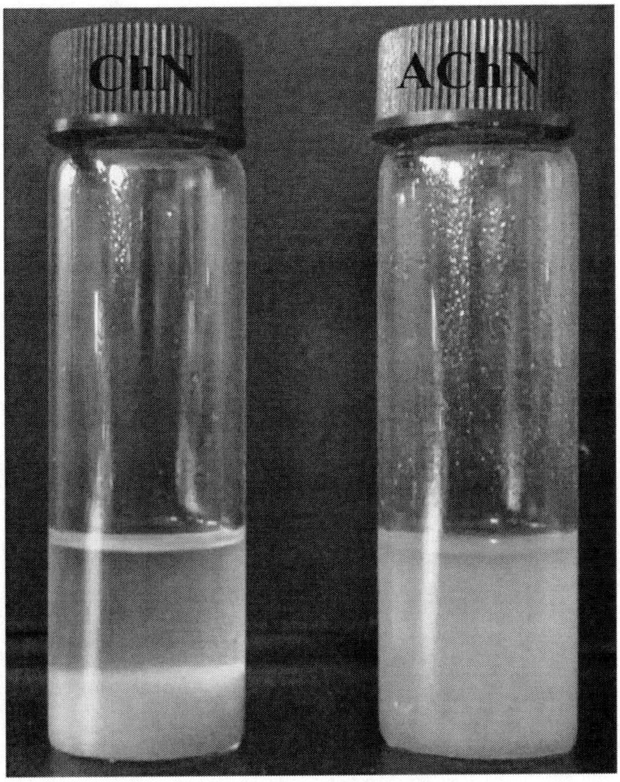

Figure 4. Dispersibility of the unmodified and acetylated chitin whisker in dichloromethane. (Reprinted from Ref. [76] Zhang, Q., Wei, S., Huang, J., Feng, J., Chang, P., (2014). J. Appl. Polym. Sci. 131, 10.1002/APP.39809).

After being acetylated, PHBV/modified chitin nanocrystals films could be prepared via a solution casting method in chloroform [77]. The hydrophilic hydroxyl groups were partly replaced by hydrophobic acetyl groups on the surface of chitin nanocrystals, leading to the improved hydrophobicity of chitin nanocrystals. Since acetylated modification reduced the intermolecular hydrogen bonds between PHBV matrix and chitin nanocrystals, acetylated chitin nanocrystals assisted the crystallization behavior and reinforced the mechanical properties comparing to the original chitin nanocrystals in the weak polar PHBV matrix. The tensile strength and Young's modulus of PHBV/acetylated chitin nanocrystals composites were improved by 44% and 67%, comparing to the improvement of 24% and 43% for PHBV/chitin nanocrystals composites with the addition of 5.0 wt.% nanocrystals into PHBV.

Liu et al. introduced a novel method to disperse cellulose nanocrystals into organic solvent without surface modification [78]. Nano-sized cellulose crystals (CNs) were fabricated from microcrystalline cellulose (MCC) using combined sulfuric acid hydrolysis and high-pressure homogenization techniques. Cellulose nanocrystals with an average length of 124 nm and an average diameter of 8–10 nm were obtained by acid-hydrolysis first. The crystals were further shortened to about 62 nm through five high-pressure homogenization treatment passes (Figure 5). High pressure treatment effectively dispersed the aggregated crystals by strong mechanical shearing forces. Spindle-shaped cellulose nanocrystals presented a relative uniform size. The homogenized samples showed a decreased crystal length and with increased treatment cycles. The crystals were then utilized to prepare polymethylmethacrylate (PMMA) nanocomposites by the solution casting in DMF. The composite sheets retained good transparency due to the size effect and dispersion of the cellulose nanocrystals. Samples with increased nanocrystal loading levels became increased opalescent due to nanocrystal agglomeration. The thermogravimetric analysis indicated retained thermal stability of the composites. The storage modulus of the nanocomposite sheets from dynamic mechanical analysis showed significantly enhanced property in comparison with that of the pure PMMA sheets.

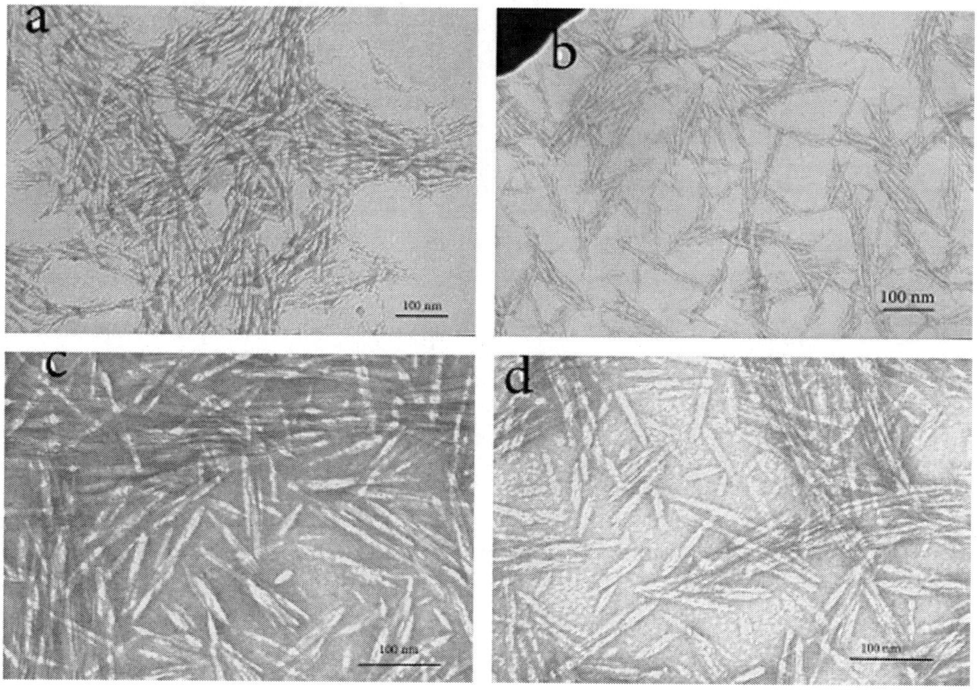

Figure 5. TEM micrographs of MCC (a) and CNs (b, c, and d representing CNs with 1, 3, and 5 homogenizing cycles, respectively). (Reprinted from Ref. [78] Liu, H., Liu, D., Yao, F., Wu, Q., (2010). Bioresour. Technol. 101, 5685).

3.2. Melt Extrusion Technique

The second way to obtain natural nanoparticles reinforced nanocomposites is melting extrusion technique. For solution casting technique is energy costing and may bring out pollution, it is only possible to be used in the laboratory work. More and more studies have been carried out to use melt extrusion technique to prepare nanocomposites which is more realistic to industrial scale [79]. Extrusion is a high volume manufacturing process in which polymeric solid material is transported by a screw and melted to form a continuous profile by passing through a die. However, this conventional processing technique is infrequently employed for the preparation of polysaccharide nanocrystal reinforced polymer nanocomposites. This is ascribed to the issues of low thermal stability of nanocrystals and inherent incompatibility between nanocrystal and most synthetic polymers [80], as well as the irreversible agglomeration during drying. Indeed, as soon as these nanoparticles are dried, strong hydrogen bonds establish and most of the time aggregates form, resulting in the limiting of nanosized reinforcement. It was demonstrated that nanocomposites prepared by casting and evaporating a mixture of cellulose [81, 82] or chintin [83] whiskers and polymer latex presented higher mechanical properties than nanocomposites of the same mixture prepared by freeze-drying and hot-pressing. Similar conclusions were obtained from nanocomposites prepared from poly(oxyethylene) (PEO) as the matrix and cellulose whiskers as the reinforcing phase [84]. Microscopic observations show some whiskers aggregations and a small decrease of the whiskers aspect ratio for extruded sample, although for both processes employed, films display homogeneous surfaces. The rheological behavior for cast/evaporated films shows that viscoelastic and creep measurements have a solid-like behavior, exhibiting a spectacular reinforcement after melting temperature, ascribed to the formation of a rigid cellulosic network. For the extruded composites, the rheological behavior through the viscoelastic and creep measurements shows a liquid-like behavior. This weak mechanical reinforcement after the PEO melting temperature leads to the conclusion that the extrusion process prevents the formation of a strong whiskers network. One possibility to improve this reinforcement, taking into account the aggregation and the decrease of the aspect ratio of whiskers during the extrusion process, is to increase the filler content.

Moreover, because of the introduction of sulfate groups resulting from common acid hydrolysis procedure involving sulfuric acid, polysaccharide nanocrystals present inferior thermal stability with degradation temperature lower than 150 °C, which is lower than the most melt processing temperature (about 200 °C or above) [85]. Higher thermal stability for nanocrystals is requires for them to process through extrusion.

Azouz et al. introduced a novel method for melt processing of nanocomposites with a hydrophobic polymeric matrix applicable at the industrial scale [86]. A dispersion of cellulose nanocrystals (CNCs) in an aqueous solution of PEO was prepared. Upon adding CNC, the viscosity of the suspension first decreases and then increases. Adsorption of PEO chains on the surface of the nanoparticles has been suspected. Freeze drying of this PEO-adsorbed CNC dispersion was performed, and the ensuing lyophilizate was extruded with low density polyethylene (LDPE). Compared to neat CNC-based nanocomposites, both improved dispersibility and thermal stability were observed. Figure 6 shows the appearance of extruded LDPE-based nanocomposites. The neat LDPE film is obviously translucent as any low thickness polymeric film with a relatively low degree of crystallinity induced by ramifications. When adding 3 wt % CNC, the film becomes homogeneously dark. This dark coloration of the film after extrusion is an indication of the degradation of the filler. When using PEO-adsorbed CNC, the appearance of the film becomes similar to the one of the unfilled film, revealing probably a much more homogeneous material. This is a strong indication that the adsorbed polymeric layer probably plays the role of compatibilizer because of the hydrophobic moieties of the monomer unit of PEO. Moreover, the dark color observed for uncompatibilized samples disappears. It could be related to the protection of sulfate groups induced by the coating polymeric layer.

However, the weak combination between cellulose crystals and PEO allows the free movement of surface polymer, which can induce the self-aggregation of CNC and microphase separation in composites especially during melt processing. Using carboxylation-amidation reaction, short chains poly(ethylene glycol) (PEG) can be grafted on the surface of the cellulose nanocrystals, and immobilize long PEO chains on modified nanocrystals through physical adsorption and entanglement [87]. Better PEO adsorption for PEG-grafted nanocrystals than pristine CNC was observed. Two polymeric layers should further improve the thermal stability of CNCs, and surface polymeric chains should provide significant dispersibility for extruded nanocomposites. Results from AFM and SEM revealed homogeneous dispersion and good compatibility of modified nanocrystals in polystyrene matrix. The chemical and physical compatibilization imparted by PEG and PEO layers promoted the interfacial interactions between cellulosic nanoparticles and apolar matrix. The surface modifications of nanocrystals did not block totally the interactions between the nanoparticles, and kept at least partially the three-dimensional network architecture, which is beneficial for the improvement of mechanical and barrier properties of extruded nanocomposites.

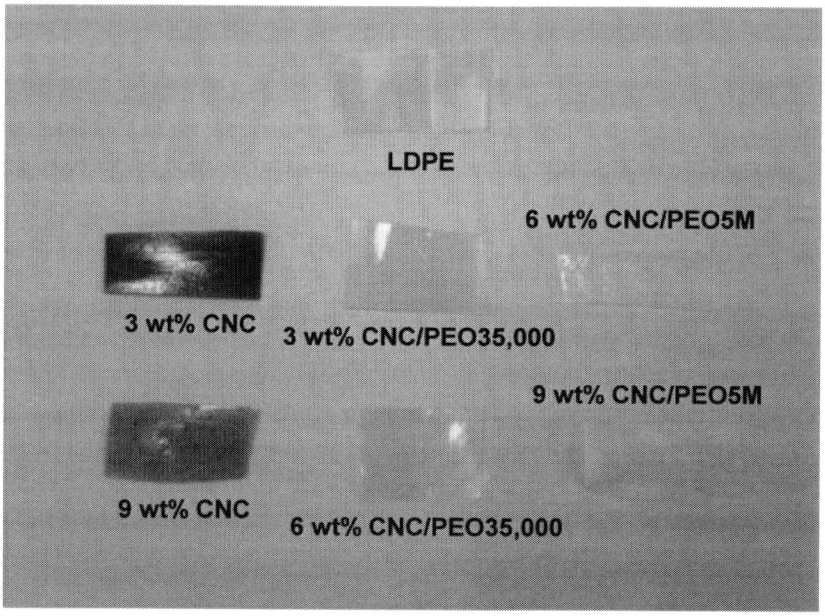

Figure 6. Pictures of the extruded films: unfilled LDPE matrix and LDPE reinforced with neat CNC and PEO-adsorbed CNC. (Reprinted from Ref. [86] Azouz, K., Ramires, E., Fonteyne, W., Kissi, N., Dufresne, A., (2012). ACS Macro Lett. 1, 236).

Goffin et al. introduced the "grafting from" approach to graft PCL to cellulose nanocrystals by ring-opening polymerization (ROP) of ε-caprolactone (CL) initiated by hydroxyl groups available onto nanocrystal surface catalyzed with tin octoate ($Sn(Oct)_2$) [88]. Then ramie cellulose nanocrystals grafted with PCL (obtained by this "grafting from" method) were melt-blended with the corresponding commercial PCL matrix using extrusion technique [89]. The polyester grafting allowed largely improving the mechanical performances of the PCL matrix. Nanocrystals extracted from ramie present an excellent dispersion within the hydrophobic matrix. Interestingly, a solid-like behavior was observed by rheological analyses most likely due to the formation of a polymer physical network. The entanglement of the surface-grafted polyester chains with the PCL matrix contributes to increased thermomechanical properties and decreased chain relaxation phenomenon. This significant improvement of the mechanical properties can be ascribed to the unique combination of a fine nanofiller dispersion and physical chain network formation.

4. Properties of Composites with Natural Polysaccharide Based Nanocrystals

Polysaccharide nanocrystals are used as the loading-bearing constituent in developing new and inexpensive biodegradable materials due to a high aspect ratio, a high bending strength of about 10 GPa, and a high Young's modulus of approximately 150 GPa [90]. As compared to other inorganic reinforcing fillers, these natural polymer based nanocrystals have many additional advantages, including a positive ecological aspects, wide variety of fillers available throughout the world, low density, low-energy consumption in manufacturing, ease for recycling by combustion, high sound attenuation, and comparatively easy processability due to their nonabrasive nature, which allows high filling levels, in turn resulting in significant cost savings [91, 92]. The properties of the resulting nanocomposites should take into account the intrinsic properties of nanocrystals, the nature of polymer matrix, and the dispersibility of the nanocrystals in the matrix as well as the interaction (compatibility) between the filler and matrix. Good dispersibility of the nanocrystals in polymer matrix is the prerequisite to prepare high-performance nanocomposites.

4.1. Mechanical Properties

Polysaccharide nanocrystals with different aspect ratios were evaluated as a reinforcing phase in nanocomposites using different polymer matrices. The improvement in mechanical properties, especially tensile strength and modulus, has been achieved in many polymer/polysaccharide nanocrystals systems. There are several factors that affect the reinforcing effect, including the glass transition temperature of polymer matrix, the aspect ratio (L/D ratio) and loading of nanocrystals, the compatibility between the filler and matrix, and the processing techniques of nanocomposite. In most cases, the mechanical properties could be substantially improved depending on the amount and homogeneity of filler dispersion. The strong reinforcing effect of the nanocrystals is generally attributed to the formation of a percolating network structure above the percolation threshold resulting from hydrogen bonding between nanoparticles [93]. The percolation mechanism, determined by three-dimensional networks formed by the interconnection of polysaccharide nanocrystals and stabilized by hydrogen bonding, is the predominant contributor to reinforcement of the resultant nanocomposites, as it enhances tensile strength and Young's modulus, gives a higher storage modulus, a thermally stable rubbery plateau, and affects the diffusion coefficient [94]. Dynamical mechanical analysis (DMA) is presented as a powerful tool to investigate the linear mechanical behavior of nanocomposites in different temperature/frequency ranges. Classical tensile or compressive tests can be used to evaluate the nonlinear mechanical properties.

Nanocrystals usually brought great reinforcing effect especially in the rubbery plateau region, that is, at temperatures higher than T_g of the matrix. The plot of the logarithm of the storage modulus and damping factor versus temperature for NR films filled with starch nanocrystals are displayed in Figure 7 [95]. Starch nanocrystals displayed a significant reinforcing effect, in particular on the rubbery plateau of the nonvulcanized NR polymer. This reinforcing effect was attributed to the formation of a percolating starch nanocrystals network through hydrogen linkages between starch nanoparticles clusters. The addition of starch nanocrystals led to not only the decrease in the magnitude of this peak but also to the shift of its maximum toward lower temperatures, in particular for filler contents higher than 10 wt %.

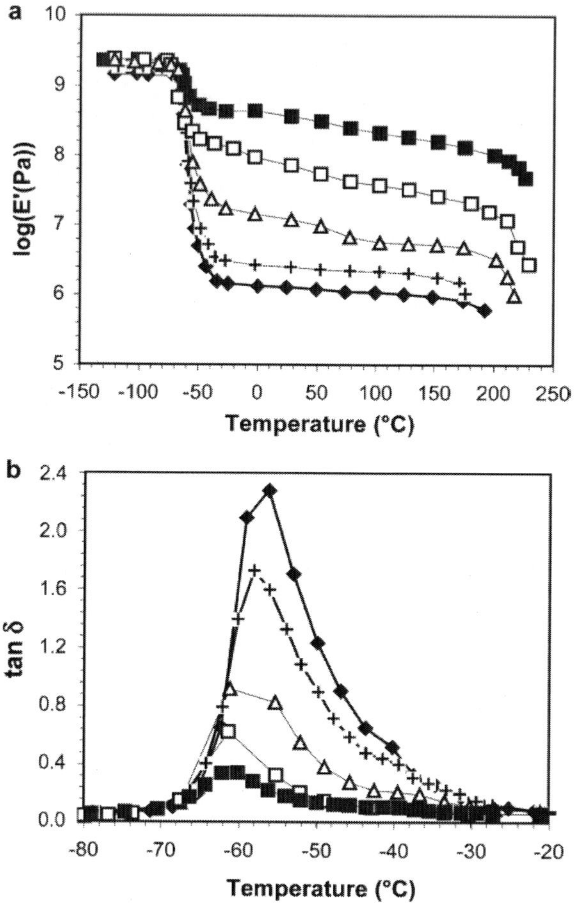

Figure 7. (a) Logarithm of the storage tensile modulus E' and (b) tangent of the loss tangent angle tan δ versus temperature at 1Hz for NR filled with different waxy maize starch nanocrystals contents: (◆) 0, (+) 5, (△) 10, (□) 20, and (■) 30 wt %. (Reprinted from Ref. [95] Mélé, P., Angellier-Coussy, H., Molina-Boisseau, S., Dufresne, A., (2011). Biomacromolecules 12, 1487).

For starch nanocrystal reinforced waterborne polyurethane (WPU) [96] significant enhancements in strength, elongation and Young's modulus was achieved, and the key role of starch nanocrystals in simultaneous reinforcing and toughening was disclosed, namely active surface and rigidity facilitated forming the interface of transferring stress and contributed to enduring stress, respectively. Figure 8 shows the stress-strain curves of the nanocomposite films prepared from a suspension of cellulose nanocrystals as the filler and polycaprolactone-based WPU as the matrix [97]. No evidence of necking phenomenon at the stress-strain curves confirms the good dispersion of the nanocrystals in the matrix and homogeneous morphology of the nanocomposites. The films showed a significant increase in Young's modulus and tensile strength from 0.51 to 344 MPa and 4.27 to 14.86 MPa, respectively, with increasing filler content from 0 to 30 wt %. The synergistic interaction between fillers and between the filler and WPU matrix played an important role in reinforcing the nanocomposites.

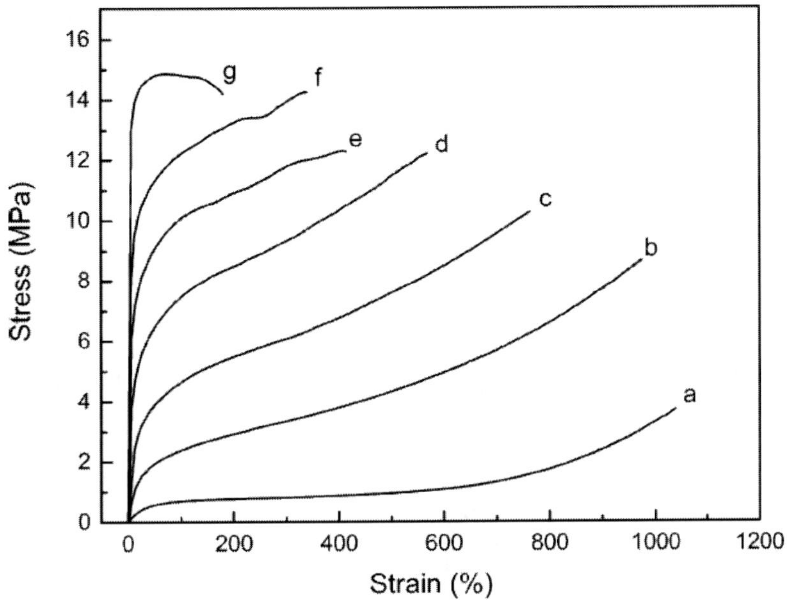

Figure 8. Stress-strain curves of WPU/flax cellulose nanocomposite films with filler content: (a) 0 wt %; (b) 5 wt %; (c) 10 wt %; (d) 15 wt %; (e) 20 wt %, (f) 25 wt %; (g) 30 wt %. (Reprinted from Ref. [97] Cao, X., Dong, H., Li, C., (2007). Biomacromolecules 8, 899).

Generally speaking, increase of the tensile strength and Young's modulus was achieved at the cost of ductility of the composites [98]. Whereas a few exceptions of composites with improved strength and flexibility were reported. Wang et al. [99] prepared high-strength elastic nanocomposite materials from waxy maize starch nanocrystals as the reinforcing phase and waterborne polyurethane as the matrix by casting and evaporating. The films having 1 wt% starch nanocrystal exhibited a significant increase from 0.6 to 3.2 MPa for Young's modulus and from 10.4 to 24.1 MPa for tensile strength. Interestingly, they showed high elongation at break, and remained basically in the range from 1148 to 1136%. The improvement in the mechanical properties of the WPU/starch nanocrystal composites could be mainly ascribed to network structure formed through the strong hydrogen-bonding interactions between filler and filler and between filler and matrix, and it is affected by the sizes of the aggregates of starch nanocrystals.

A more dramatic reinforcing effect would be observed with highly oriented nanocrystals dispersed in a polymer matrix [100]. The macroscopically homogeneous polyvinyl alcohol-cotton derived nanocrystals suspensions were extruded into cold methanol to form gel fibers followed by a hot drawing. Compared to the neat PVA fiber, the as-spun fiber containing a small amount of nanocrystal (5 wt % of solid PVA) showed higher drawability, leading to an extremely high orientation of nanocrystals with the matrix PVA. The drawn fibers exhibited extremely high orientation of nanocrystals and excellent mechanical properties. The stress on the incorporated cellulose nanocrystals was monitored by applying an in situ nondestructive load to the composite fibers. The applied stress to the whole sample was found to be effectively transferred to the nanocrystals inside the composites, suggesting strong interfacial bonding between the filler and the matrix. Effective stress transfer to the oriented whiskers resulted in outstanding enhancement in mechanical properties of the nanocomposites.

A great synergistic reinforcing effect of waxy maize starch nanocrystals (SN) and cellulose whiskers (CW) from cotton linter pulp has been observed on WPU based nanocomposite films [101]. WPU/1% SN/0.4% CW system exhibited a much better reinforcing effect, compared to all of WPU/SN and WPU/CW systems (Figure 9). All SN were under 100 nm, so the interactions between SN and WPU existed and SN became the crosslinking points of the networks, leading to the reinforcement of WPU matrix. In the meantime, CW was rod-like nanocrystal and can insert into the WPU/SN networks as observed in Figure 10. CW located in SN platelets regions and connected with SN to form a new type of SN–CW networks. In the ternary system, the formation of much jammed network consisted of nanocrystals and whiskers with different geometrical characteristics from polysaccharides played an important role in the enhancement of the crosslink networks. Strong hydrogen bonding interactions existed both between the nanofillers and between the nanofillers and the hard segments of WPU matrix, leading to the improvement of the mechanical and thermal properties. This work provided important information that utilizing nanocrystals and whiskers together can increase the density of the networks, so as to improve the performance of the materials.

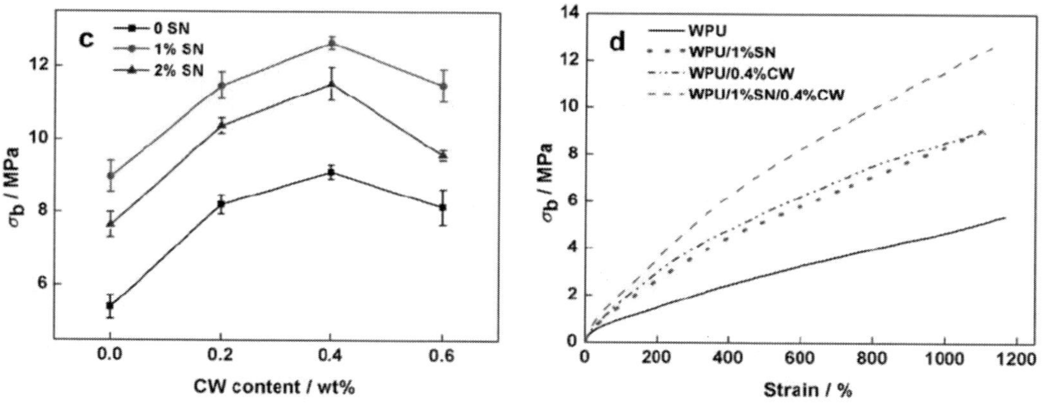

Figure 9. Dependence of tensile strength of WPU/SN/CW nanocomposite films on CW content and stress–strain curves of pure WPU, WPU/1% SN, WPU/0.4% CW, and WPU/ 1% SN/0.4% CW nanocomposite films. (Reprinted from Ref. [101] Wang, Y., Tian, H., Zhang, L., (2010). Carbohyd. Polym. 80, 665).

Recently, Khoshkava et al. reported that dispersion quality and polymer-filler interaction are important factors in determining the final properties of polymer nanocomposites [102]. The dispersion of nanoparticles in polymer melts is strongly influenced by the surface energy characteristics at the processing temperature of both the particles and the polymer, which determine the interfacial interactions between the particles and the melt. Similarly, the interfacial energy characteristics at ambient temperature have a strong influence on the mechanical properties of the final composite. The dispersion factor has been defined to indicate the balance of adhesive and cohesive energy forces, which governs the effectiveness of applied shear or extensional forces/stresses in achieving good dispersion in the polymer matrix during processing. Surface energy of nanocrystalline cellulose (NCC) and some polymers (polypropylene (PP) and polylactic acid (PLA)) was measured at room and high temperatures. NCC had higher polarity and surface energy than PP and PLA at room temperature but had a lower surface energy at higher temperatures. Total surface energy of NCC was lowered after alkenyl succinic anhydride (ASA) surface modification. Thermodynamic work of adhesion for PP/NCC and PLA/NCC was lowered by NCC surface modification. A thermodynamic analysis is proposed to estimate the dispersion energy, based on surface energy measurements at room and high temperatures. Also, a dispersion factor is defined to provide a quantitative indication of the dispersibility of nanoparticles in a polymer matrix under various conditions. The required dispersion energy was reduced by lowering the interfacial tension. On the other hand, it increased as the quality of NCC dispersion (i.e. the nanoparticle surface area) in the system was improved. Surface modification of NCC with ASA had a negative effect on the compatibility between NCC and PLA, whereas it had a positive influence on compatibility between PP and NCC.

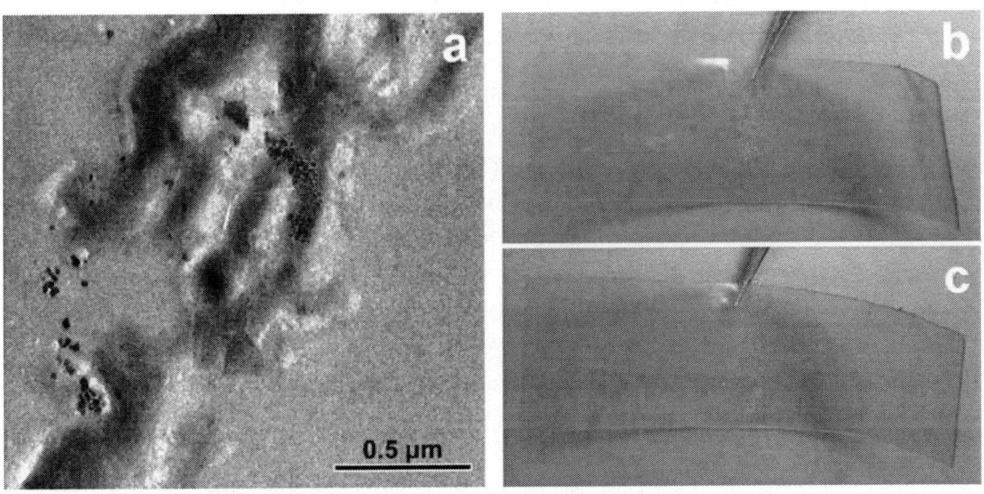

Figure 10. TEM image of ultra-thin section of WPU/1% SN/0.4% CW nanocomposite films (a), and photographs of WPU (b) and WPU/1% SN/0.4% CW (c) (Reprinted from Ref. [101] Wang, Y., Tian, H., Zhang, L., (2010). Carbohyd. Polym. 80, 665).

4.2. Barrier Properties

Because of the possibility of producing films with high transparency, and improved mechanical and barrier properties, nanocomposites have obtained interest in packaging area. Barrier properties increase shelf life duration by protecting the inside product from deteriorations such as oxidation, humidity, and bacteria. Natural nanocrystals exhibit novel significant potential in coatings, membranes and food agrobased packaging applications. The ability of polysaccharide nanoparticles to form a dense percolating network held together by strong, inter-particle bonds suggests their use to improve both mechanical and barrier properties. Resistance to water vapor permeation (WVP) is an essential requirement in composites used in specific applications in packaging. The use of natural nanocrystal as fillers that can reduce water vapor permeation have reported for chitosan films [103], poly(vinyl alcohol) [104], polyvinyl acetate [30], carboxymethyl cellulose (CMC) [105], etc. Both the water vapor and oxygen permeability of waxy maize starch filled natural rubber (NR) films were found to decrease [23]. Transport properties of poly(vinyl alcohol) films cross-linked with poly(acrylic acid) (PAA) and reinforced with cellulose whiskers were studied [106]. Reduced moisture transport was reported upon filler addition and improved performance was obtained for surface carboxylated nanocrystals due to improved dispersion. Polylactic acid nanocomposites using cellulose nanowhiskers as a reinforcing element possessed the reduced water permeability by up to 82% and the oxygen permeability by up to 90% with only 3 wt% of nanofiller content [107]. The presence of highly crystalline cellulose nanoshields, PLA crystallinity development (e.g. transcrystallinity) and sorbed moisture filling the free volume were put forward as the most likely factors behind this behavior. Contrarily, the mechanical performance was seen lower than that of neat PLA and than expected by typical modelling work. This observation was ascribed to both the filler ranged screened being below the percolation threshold and most importantly to filler-induced plasticization by sorbed moisture.

Natural rubber based nanocomposite films with Na-montmorillonite (MMT) and cellulose whiskers (W) was prepared, keeping a total filler content equal to 5wt.-% [108]. In the binary NR/MMT system, small stacks of intercalated montmorillonites were homogeneously dispersed within the polymer matrix whereas they were clearly lying in the vicinity of cellulose whiskers in the ternary NR/MMT/W blends. The simultaneously introduced montmorillonite and cellulose whiskers into NR could greatly improve the rubbery tensile modulus. The improvement of the gas barrier properties observed upon filler addition was explained by a tortuosity effect. The calculated tortuosity values indicated that the simultaneous use of MMT and W could greatly slow down the gas diffusion rate in NR. Formation of MMT-W subassembly should be responsible for this synergism effect.

LeCorre et al. investigated a kind of all starch nanocomposite coating for barrier material [109]. Starch nanocrystals were introduced in a starch-based coating. Starch nanocrystals compensate some of the water induced loss of mechanical properties of the base paper and decrease the water vapor permeability (WVP) up to 40% compared with the base paper.

4.3. Thermal Stability

Thermogravimetric analysis (TGA) experiments were performed to investigate the thermal stability of natural nanocrystal nanocomposites. For cellulose nanocrystals reinforced poly(oxyethylene), no significant influence of the cellulosic filler on the degradation temperature of the POE matrix was reported [110]. For polymethylmethacrylate/ cellulose nanocrystals composites [76], the degradation of the nanocomposites can be divided into two processes, which are ascribed to the decomposition of cellulose nanocrystals (150–250 ^{o}C) and PMMA (300–450 ^{o}C), respectively, according to the DTG curves. The nanocomposites began to degrade at a slightly lower temperature than that of pure PMMA. However, the weight loss of the nanocomposites in the temperature range up to 300 ^{o}C was slightly higher (around 5 wt.%) than that of pure PMMA. The maximum decomposition temperature was increased 5–15 ^{o}C with the variation of the content of cellulose nanocrystals in the nanocomposites. The nanocomposites with outstanding thermal stability up to 210-230 ^{o}C was observed for polyurethane elastomer with low volume fraction of cellulose nanocrystals, the temperature at which cellulose starts to degrade [111]. The reason for such thermal stability enhancement was attributed to the formation of a rigid network by the cellulose whiskers in the host polymer matrix, which was governed by the percolation mechanism.

4.4. Biodegradability

Biodegradable nanocomposites are important products in the area of packaging materials. Natural polysaccharide nanocrystals present fine biodegradability in natural environment. The incorporation of cellulose whiskers could significantly enhanced biodegradation of rubber in soil [112]. Neat NR lost about 19% of its weight after being buried in soil for 4 weeks while NR containing 7.5 and 12.5 wt% of cellulose whiskers lost about 62 and 71%, respectively after the same period buried in soil. Because cellulose biodegradation is faster than rubber, cellulose component in the nanocomposites films is consumed by the microorganisms faster than rubber leading to increased porosity, void formation, and the loss of the integrity of the rubber matrix. The rubber matrix will be broken down into smaller particles. Therefore overall faster disintegration of nanocomposites films containing cellulose whiskers than that of neat rubber film is observed.

Conclusion and Outlook

The present chapter reports the preparation of natural polysaccharide based nanocrystals, as well as the fabrication and properties of the nanocomposite. During the last two decades, an increasing number of researches in the area of natural polysaccharide nanocrystal involving nanocomposites have been developed. The kind of nanocrystals could be extracted from different sources and by different methods. The processing techniques and the properties of the resulting bionanocomposites were also reported. The wide availability, renewable and biodegradable features, simple hydrolysis process, and high intrinsic strength and modulus, high aspect ratio and reactivity make these nanocrystals superior to other traditional nanofillers. The dispersion difficulty due to hydrophilicity of whiskers has been overcome by either physical or chemical methods. These natural nanocrystal filled nanocomposites would create new and specific applications such as composites, foams, fibers, barrier films, and so on.

Acknowledgments

This work was supported by the National Natural Science Foundation of China (51373004 and 51203004), Beijing Natural Science Foundation (2122014), the Scientific and Technological Development Project of Beijing Municipal Commission of Education (KM201310011001) and Youth Elite Project of Beijing Colleges (YETP1453).

REFERENCES

[1] Nair, K.G., Dufresne, A., (2003). Crab shell chitin whisker reinforced natural rubber nanocomposites. 2. Mechanical behavior. *Biomacromolecules* 4, 666.

[2] Lin, N., Huang J., Dufresne, A., (2012) Preparation, properties and applications of polysaccharide nanocrystals in advanced functional nanomaterials: a review. *Nanoscale* 4, 3274.

[3] Angellier, H., Molina-Boisseau S., Dufresne, A., (2005). Mechanical properties of waxy maize starch nanocrystal reinforced natural rubber. *Macromolecules* 38, 9161.

[4] Angles, M.N., Dufresne, A., (2001). Plasticized starch/tunicin whiskers nanocomposite materials. 2. Mechanical behavior. *Macromolecules* 34, 2921.

[5] Wang, Y., Cao, X., Zhang, L., (2006). Effects of cellulose whiskers on properties of soy protein thermoplastics. *Macromol. Biosci.* 6, 524.

[6] Qi, H., Cai, J., Zhang, L., Kuga, S., (2009). Properties of films composed of cellulose nanowhiskers and a cellulose matrix regenerated from alkali/urea solution. *Biomacromolecules* 10, 1597.

[7] Zoppe, J.O., Peresin, M.S., Habibi, Y., Venditti R.A., Rojas, O.J., (2009). Reinforcing Poly(caprolactone) nanofibers with cellulose nanocrystals. *ACS Appl. Mater. Interfaces* 1, 1996.

[8] Xu, W., Qin, Z., Yu, H., Liu, Y., Liu, N., Zhou, Z., Chen, L., (2013). Cellulose nanocrystals as organic nanofillers for transparent polycarbonate films. *J. Nanopart. Res.* 15:1562.

[9] Peresin, M.S., Habibi, Y., Zoppe, J.O., Pawlak J.J., Rojas, O.J., (2010). Nanofiber composites of polyvinyl alcohol and cellulose nanocrystals: manufacture and characterization. *Biomacromolecules* 11, 674.

[10] Zhang, W., He, X., Li, C., Zhang, X., Lu, C., Zhang, X., Deng, Y., (2014). High performance poly (vinyl alcohol)/cellulose nanocrystals nanocomposites manufactured by injection molding. *Cellulose* 21, 485.

[11] Ljungberg, N., Bonini, C., Bortolussi, F., Boisson, C., Heux, L., Cavaille, J.Y., (2005). New nanocomposite materials reinforced with cellulose whiskers in atactic polypropylene: effect of surface and dispersion characteristics. *Biomacromolecules* 6, 2732.

[12] Marcovich, N.E., Auad, M.L., Bellesi, N.E., Nutt, S.R., Aranguren, M.I., (2006). Cellulose micro/nanocrystals reinforced polyurethane. *J. Mater. Res.* 21, 870.

[13] Habibi, Y., Lucia, L.A., Rojas O.J., (2010). Cellulose nanocrystals: Chemistry, self-assembly, and applications. *Chem. Rev.* 110, 3479.

[14] Campos, A., Correa, A.C., Cannella., D., Teixeira, E., Marconcini, J.M., Dufresne, A., Mattoso, L.H.C., Cassland, P., Sanadi, A.R., (2013). Obtaining nanofibers from curauá and sugarcane bagasse fibers using enzymatic hydrolysis followed by sonication. *Cellulose* 20, 1491.

[15] Kadokawa, J., Takegawaa, A., Mine, S., Prasad, K., (2011). Preparation of chitin nanowhiskers using an ionic liquid and their composite materials with poly(vinyl alcohol). *Carbohyd. Polym.* 84, 1408.

[16] Man, Z., Muhammad, N., Sarwono, A., Bustam, M., Kumar, M. V., Rafiq, S., (2011). Preparation of cellulose nanocrystals using anionic liquid. *J. Polym. Environ.* 19, 726.

[17] Filson, P.B., Dawson-Andoh, B.E., (2009). Sono-chemical preparation of cellulose nanocrystals from lignocellulose derived materials. *Bioresour. Technol.* 100, 2259.
[18] Siqueira, G., Tapin-Lingua, S., Bras, J., Perez, D., Dufresne, A., (2010). Morphological investigation of nanoparticles obtained from combined mechanical shearing, and enzymatic and acid hydrolysis of sisal fibers. *Cellulose* 17, 1147.
[19] Siqueira, G., Tapin-Lingua, S., Bras, J., Perez, D., Dufresne, A., (2011). Mechanical properties of natural rubber nanocomposites reinforced with cellulosic nanoparticles obtained from combined mechanical shearing, and enzymatic and acid hydrolysis of sisal fibers. *Cellulose* 18, 57.
[20] Angles, M.N., Dufresne, A., (2000). Plasticized starch/tunicin whiskers nanocomposites. 1. Structure analysis. *Macromolecules* 33, 8344.
[21] Kalashnikova, I., Bizot, H., Bertoncini, P., Cathala B., Capron, I., (2013). Cellulosic nanorods of various aspect ratios for oil in water Pickering emulsions. *Soft Matter* 9, 952.
[22] Morin, A., Dufresne, A., (2002). Nanocomposites of chitin whiskers from Riftia tubes and poly(caprolactone). *Macromolecules* 35, 2190.
[23] Angellier, H., Molina-Boisseau, S., Lebrun, L., Dufresne, A., (2005). Processing and structural properties of waxy maize starch nanocrystals reinforced natural rubber. *Macromolecules* 38, 3783.
[24] Dufresne, A., Cavaille, J., (1998). Clustering and percolation effects in microcrystalline starch-reinforced thermoplastic. *J. Polym. Sci. Polym. Phy.* 36, 2211.
[25] Angellier, H., Choisnard, L., Molina-Boisseau, S., Ozil, P., Dufresne, A. (2004). Optimization of the preparation of aqueous suspensions of waxy maize starch nanocrystals using a response surface methodology. *Biomacromolecules* 5, 1545.
[26] Revol, J.F., (1982). On the cross-sectional shape of cellulose crystallites in Valonia ventricosa. *Carbohydr. Polym.* 2, 123.
[27] Araki, J, Kuga, S., (2001). Effect of trace electrolyte on liquid crystal type of cellulose microcrystals. *Langmuir* 17, 4493.
[28] Lu, Y., Weng, L., Cao, X., (2005). Biocomposites of plasticized starch reinforced with cellulose crystallites from cottonseed linter. *Macromol. Biosci.* 5, 1101.
[29] Petersson, L., Kvien, I., Oksman, K., (2007). Structure and thermal properties of poly(lactic acid)/cellulose whiskers nanocomposite materials. *Comp. Sci. Technol.* 67, 2535.
[30] Garcia de Rodriguez, N. L., Thielemans, W., Dufresne, A., (2006). Sisal cellulose whiskers reinforced polyvinyl acetate nanocomposites. *Cellulose* 13, 261.
[31] Azizi Samir, M.A.S., Alloin, F., Paillet, M., Dufresne, A., (2004). Tangling effect in fibrillated cellulose reinforced nanocomposites. *Macromolecules* 37, 4313.
[32] Helbert, W., Cavaille, J. Y., Dufresne, A., (1996). Thermoplastic nanocomposites filled with wheat straw cellulose whiskers. Part Ⅰ: Processing and mechanical behavior. *Polym. Compos.* 17, 604.
[33] Junior, M.A., Siqueira, G., Curvelo, A.A.S., Dufresne, A., (2009). Extrusion and characterization of functionalized cellulose whiskers reinforced polyethylene nanocomposites. *Polymer* 50, 4552.
[34] Beck-Candanedo, S., Roman, M., Gray, D. G., (2005). Effect of reaction conditions on the properties and behavior of wood cellulose nanocrystal suspensions. *Biomacromolecules* 6, 1048.

[35] Gopalan Nair, K., Dufresne, A., (2003). Crab shell chitin whisker reinforced natural rubber nanocomposites. 1. Processing and swelling behavior. *Biomacromolecules* 4, 657.
[36] Sriupayo, J., Supaphol, P., Blackwell, J., Rujiravanit, R., (2005). Preparation and characterization of -chitin whisker-reinforced poly(vinyl alcohol) nanocomposite films with or without heat treatmen. *Polymer* 46, 5637.
[37] Paillet, M., Dufresne, A., (2001). Chitin whisker reinforced thermoplastic nanocomposites. *Macromolecules* 34, 6527.
[38] Angellier, H., Putaux, J., Molina-Biosseau, S., Dupeyre, D., Dufresne, A., (2005). Starch nanocrystal fillers in an acrylic polymer matrix. *Macromol. Symp.* 221, 95.
[39] Dufresne, A., Cavaille, J.Y., Helbert, W., (1996). New nanocomposite materials: Microcrystalline starch reinforced thermoplastic. *Macromolecules* 29, 7624.
[40] Liu, D., Zhong, T., Chang, P.R., Li, K., Wu, Q., (2010). Starch composites reinforced by bamboo cellulosic crystals. *Bioresour. Technol.* 101, 2529.
[41] Li, W., Yue, J., Liu, S., (2012). Preparation of nanocrystalline cellulose via ultrasound and its reinforcement capability for poly(vinyl alcohol) composites. *Ultrason. Sonochem.* 19, 479.
[42] Hamad, W., (2006). On the development and applications of cellulosic nanofibrillar and nanocrystalline materials. *Can. J. Chem. Eng.* 84, 513.
[43] M., Yu, O, Lourie, M., Dyer, K, Moloni, TF, Kelly, RS., Ruoff, (2000). Strength and breaking mechanism of multiwalled carbon nanotubes under tensile load. *Science* 287, 637.
[44] Saxena, A., (2013). Nanocomposites based on whiskers, Doctor thesis, Georgia Institute of Technology.
[45] Oksman, K., Mathew, A.P., Bondeson, D., Kvien, I., (2006). Manufacturing process of cellulose whiskers/polylactic acid nanocomposites. *Compos. Sci. Technol.* 66, 2776.
[46] Dufresne, A., Polymer nanocomposites from Biological Sources. In Encyclopedia of Nanoscience and Nanotechnology, 2nd ed.; Nalwa, H.S., Ed.; American Scientific Publisher: Valencia, CA, USA.
[47] Habibi, Y., Lucia, L.A., Rojas, O., (2010). Cellulose nanocrystals: Chemistry, self-assembly, and applications. *Chem. Rev.* 110, 3479.
[48] Kvien, I., Oksman, K., (2007). Orientation of cellulose nanowhiskers in polyvinyl alcohol. *Appl. Phys. A-Mat. Sci. Process.* 87, 641.
[49] Roohani, M., Habibi, Y., Belgacem, N.M., Ebrahim, G., Karimi, A.N., Dufresne, A., (2008). Cellulose whiskers reinforced polyvinyl alcohol copolymers nanocomposites. *Europ. Polym. J.* 44, 2489.
[50] Mathew, A. P., Dufresne, A., (2002). Morphological investigation of nanocomposites from sorbitol plasticized starch and tunicin whiskers. *Biomacromolecules* 3, 609.
[51] Huang, Y., Zhang, L., Yang, J., Zhang, X., Xu, M., (2013). Structure and properties of cellulose films reinforced by chitin whiskers. *Macromol. Mater. Eng.* 298, 303.
[52] Azizi Samir, M.A.S., Alloin, F., Sanchez, J.Y., Dufresne, A., (2004). Cellulose nanocrystals reinforced poly(oxyethylene). *Polymer* 45, 4149
[53] Lu, Y., Weng, L., Zhang, L., (2004). Morphology and properties of soy protein isolate thermoplastics reinforced with chitin whiskers. *Biomacromolecules* 5, 1046.
[54] Zheng, H., Ai, F., Chang, P.R. Huang, J., Dufresne, A., (2009). Structure and properties of starch nanocrystal-reinforced soy protein plastics. *Polym. Compos.* 30, 474.

[55] Favier, V., Chanzy, H., Cavaille, J.Y., (1995). Polymer nanocomposites reinforced by cellulose whiskers. *Macromolecules* 28, 6365.

[56] Dubief, D., Samain, E., Dufresne, A., (1999). Polysaccharide microcrystals reinforced amorphous poly(beta-hydroxyoctanoate) nanocomposite materials. *Macromolecules* 32, 5765.

[57] Chazeau, L., Cavaille, J.Y., Canova, G., Dendievel, R., Boutherin, B., (1999). Viscoelastic properties of plasticized PVC reinforced with cellulose whiskers. *J. Appl. Polym. Sci.* 71, 1797.

[58] Ruiz, M.M., Cavaille, J.Y., Dufresne, A., Graillat, C., Gerard, J.F., (2001). New waterborne epoxy coatings based on cellulose nanofillers. *Macromol. Symp.* 169, 211.

[59] Zou, J., Zhang, F., Huang, J., Chang, P., Su, Z., Yu, J., (2011). Effects of starch nanocrystals on structure and properties of waterborne polyurethane-based composites. *Carbohydr. Polym.* 85 824.

[60] Liu, H., Cui, S., Shang, S., Wang, D., Song, J., (2013). Properties of rosin-based waterborne polyurethanes/cellulose nanocrystals composites. *Carbohydr. Polym.* 96, 510.

[61] Bendahou, A., Habibi, Y., Kaddami, H., Dufresne, A., (2009). Physico-chemical characterization of palm from Phoenix Dactylifera-L, preparation of cellulose whiskers and natural rubber-based na nocomposites. *J. Biobased Mat. Bioenerg.* 3, 81.

[62] Roman, M., Winter, W. T., (2004). Effect of sulfate groups from sulfuric acid hydrolysis on the thermal degradation behavior of bacterial cellulose. *Biomacromolecules* 5, 1671.

[63] Kamel, S., (2007). Nanotechnology and its application in lignocellulosics composites, a mini review. *Exp. Polym. Lett.* 1, 546.

[64] Chazeau, L., Cavaille, J.Y., Canova, G.R., Dendievel, R., Boutherin, B., (1999). Viscoelastic properties of plasticized PVC reinforced with cellulose whiskers. *J. Appl. Polym. Sci.* 71, 1797.

[65] Kvien, I., Bjorn, S. T., Oksman, K., (2005). Characterization of cellulose whiskers and their nanocomposites by atomic force and electron microscopy. *Biomacromolecules* 6, 3160.

[66] Siqueira, G., Bras J., Dufresne, A., (2010). Cellulosic bionanocomposites: A review of preparation, properties and applications. *Polymers* 2, 728.

[67] Lin, N., Chen, G., Huang, J., Dufresne, A., Chang, P. R., (2009). Effects of polymer-grafted natural nanocrystals on the structure and mechanical properties of poly(lactic acid): A case of cellulose whisker-graft-polycaprolactone. *J. Appl. Polym. Sci.* 113, 3417.

[68] Gousse, C., Chanzy, H., Exoffier, G., Soubeyrand, L., Fleury, E., (2002). Stable suspensions of partially silylated cellulose whiskers dispersed in organic solvents. *Polymer* 43, 2645.

[69] Namazi, H., Dadkhah, A., (2010). Convenient method for preparation of hydrophobically modified starch nanocrystals with using fatty acids. *Carbohyd. Polym.* 79, 731.

[70] Thielemans, W., Belgacem, M.N., Dufresne, A., (2006). Starch nanocrystals with large chain surface modifications. *Langmuir* 22, 4804.

[71] Labet, M., Thielemans, W., Dufresne, A., (2007). Polymer grafting onto starch nanocrystals. *Biomacromolecules* 8, 2916.

[72] Wang, J., Wang, Z., Li, J., Wang, B., Liu, J., Chen, P., Miao, M., Gu, Q., (2012). Chitin nanocrystals grafted with poly(3-hydroxybutyrate-co-3-hydroxyvalerate) and their effects on thermal behavior of PHBV. *Carbohyd. Polym.* 87, 784.

[73] Siqueira, G., Bras, J., Dufresne, A., (2010) New process of chemical grafting of cellulose nanoparticles with a long chain isocyanate, *Langmuir* 26, 402.

[74] Yu, J., Ai, F., Dufresne, A., Gao, S., Huang, J., Chang, P.R., (2008). Structure and mechanical properties of poly(lactic acid) filled with (starch nanocrystal) –graft –poly (ε-caprolactone). *Macromol. Mater. Eng.* 293, 763.

[75] Chang, P.R., Ai, F., Chen, Y., Dufresne, A., Huang, J., (2009). Effects of starch nanocrystal-graft-polycaprolactone on mechanical properties of waterborne polyurethane- based nanocomposites. *J. Appl. Polym. Sci.* 111, 619.

[76] Zhang, Q., Wei, S., Huang, J., Feng, J., Chang, P., (2014). Effect of surface acetylated-chitin nanocrystals on structure and mechanical properties of poly(lactic acid). *J. Appl. Polym. Sci.* 131, 10.1002/APP.39809.

[77] Wang, B., Li, J., Zhang, J., Li, H., Chen, P., Gu, Q., Wang, Z., (2013). Thermo-mechanical properties of the composite made of poly (3-hydroxybutyrate-co-3-hydroxyvalerate) and acetylated chitin nanocrystals. *Carbohyd. Polym.* 95, 100.

[78] Liu, H., Liu, D., Yao, F., Wu, Q., (2010). Fabrication and properties of transparent polymethylmethacrylate/cellulose nanocrystals composites. *Bioresour. Technol.* 101, 5685.

[79] Dufresne, A., (2008). Polysaccharide nano crystal reinforced nanocomposites. *Can. J. Chem.* 86, 484.

[80] Dufresne, A., (2012). Nanocellulose. From nature to high performance tailored materials, Walter de Gruyter GmbH: Berlin and Boston, MA, 252–260.

[81] Helbert, W., Cavaille, J.Y., Dufresne, A., (1996). Thermoplastic nanocomposites filled with wheat straw cellulose whiskers. Part I. Processing and mechanical behavior. *Polym. Compos.* 17, 604.

[82] Dufresne, A., Cavaille, J.Y., Helbert, W., (1997). Thermoplastic nanocomposites filled with wheat straw cellulose whiskers. Part II. Effect of processing and modeling. *Polym. Compos.* 18, 198.

[83] Nair, K. G., Dufresne, A., (2003) Crab shell chitin whiskers reinforced natural rubber nanocomposites. 3. Effect of chemical modification of chitin whiskers, *Biomacromolecules* 4, 1835.

[84] Alloin, F., Aprea, A., Dufresne A., Kissi N., Bossard, F., (2011). Poly(oxyethylene) and ramie whiskers based nanocomposites: influence of processing: extrusion and casting/evaporation. *Cellulose* 18, 957.

[85] Corrêa, A., Teixeira, E., Carmona, V., Teodoro, K., Ribeiro, C., Mattoso, L., Marconcini, J., (2014). Obtaining nanocomposites of polyamide 6 and cellulose whiskers via extrusion and injection molding. *Cellulose* 21, 311.

[86] Azouz, K., Ramires, E., Fonteyne, W., Kissi, N., Dufresne, A., (2012). Simple method for the melt extrusion of a cellulose nanocrystal reinforced hydrophobic polymer. *ACS Macro Lett.* 1, 236.

[87] Lin, N., Dufresne, A., (2013). Physical and/or chemical compatibilization of extruded cellulose nanocrystal reinforced polystyrene nanocomposites. *Macromolecules* 46, 5570.

[88] Habibi, Y., Goffin, A-L., Schiltz, N., Duquesne, E., Dubois, P., Dufresne, A., (2008). Bionanocomposites based on poly(ε-caprolactone)-grafted cellulose nanocrystals by ring-opening polymerization. *J. Mater. Chem.* 18, 5002.
[89] Goffin, A.-L., Raquez, J.-M., Duquesne, E., Siqueira, G., Habibi, Y., Dufresne, A., Dubois, Ph., (2011). Poly(e-caprolactone) based nanocomposites reinforced by surface-grafted cellulose nanowhiskers via extrusion processing: Morphology, rheology, and thermo-mechanical properties. *Polymer* 52, 1532.
[90] Sturcova, A., Davies, G.R., Eichhorn, S.J., (2005). Elastic modulus and stress-transfer properties of tunicate cellulose whiskers. *Biomacromolecules* 6, 1055.
[91] Podsiadlo, P., Choi, S.Y., Shim, B., Lee, J., Cuddihy, M., Kotov, N.A., (2005). Molecularly engineered nanocomposites: Layer-by-layer assembly of cellulose nanocrystals. *Biomacromolecules* 6, 2914.
[92] Azizi Samir, M.A.S., Alloin, F., Dufresne, A., (2005). Review of recent research into cellulosic whiskers, their properties and their application in nanocomposite field. *Biomacromolecules* 6, 612.
[93] Favier, V., Canova, GR., Cavaillé, JY., Chanzy, H., Dufresne, A., Gauthier, C., (1995). Nanocomposite materials from latex and cellulose whiskers. *Polym. Adv. Technol.* 6, 351.
[94] Lin, N., Huang, J., Chang, P.R., Anderson, D.P., Yu, J., (2011). Preparation, modification, and application of starch nanocrystals in nanomaterials: A review. *J. Nanomater.* Article ID 573687, 1-13.
[95] Mélé, P., Angellier-Coussy, H., Molina-Boisseau, S., Dufresne, A., (2011). Reinforcing mechanisms of starch nanocrystals in a nonvulcanized natural rubber matrix. *Biomacromolecules* 12, 1487.
[96] Chen, G., Wei, M., Chen, J., Huang, J., Dufresne, A., Chang, P.R., (2008) Simultaneous reinforcing and toughening: New nanocomposites of waterborne polyurethane filled with low loading level of starch nanocrystals. *Polymer* 49, 1860.
[97] Cao, X., Dong, H., Li, C., (2007). New nanocomposite materials reinforced with flax cellulose nanocrystals in waterborne polyurethane. *Biomacromolecules* 8, 899.
[98] Zeng, J., He, Y., Li, S., Wang, Y., (2012). Chitin whiskers: An overview. *Biomacromolecules* 13, 1.
[99] Wang, Y., Zhang, L., (2008). High-strength waterborne polyurethane reinforced with waxy maize starch nanocrystals. *J. Nanosci. Nanotechnol.* 8, 5831.
[100] Uddin, A., Araki, J., Gotoh, Y., (2011). Toward "strong" green nanocomposites: Polyvinyl alcohol reinforced with extremely oriented cellulose whiskers. *Biomacromolecules* 12, 617.
[101] Wang, Y., Tian, H., Zhang, L., (2010). Role of starch nanocrystals and cellulose whiskers in synergistic reinforcement of waterborne polyurethane. *Carbohyd. Polym.* 80, 665.
[102] Khoshkava, V., Kamal, M. R., (2013). Effect of surface energy on dispersion and mechanical properties of polymer/nanocrystalline cellulose nanocomposites. *Biomacromolecules* 14, 3155.
[103] Khan, A., Khan, R.A., Salmieri, S., Tien, C.L., Riedl, B., Bouchard, J., Chauve, G., Tan, V., Kamal, M.R., Lacroix, M., (2012). Mechanical and barrier properties of nanocrystalline cellulose reinforced chitosan based nanocomposite films. *Carbohyd. Polym.* 90, 1601.

[104] Silvério, H., Neto, W., Pasquini, D., (2013). Effect of incorporating cellulose nanocrystals from Corncob on the tensile, thermal and barrier properties of poly(vinyl alcohol) nanocomposites. *J. Nanomater.* Article ID 289641, 1-9.

[105] Choi, Y., Simonsen, J., (2006). Cellulose nanocrystal-filled carboxymethyl cellulose nanocomposites. *J. Nanosci. Nanotechnol.* 6, 633.

[106] Paralikar, S.A., Simonsen, J., Lombardi, J. (2008). Poly(vinyl alcohol)/cellulose nanocrystal barrier membranes. *J. Membr. Sci.* 320, 248.

[107] Sanchez-Garcia, M.D., Lagaron, J.M., (2010). On the use of plant cellulose nanowhiskers to enhance the barrier properties of polylactic acid. *Cellulose* 17, 987.

[108] Bendahou, A., Kaddami, H., Espuche, E., Gouanve´, F., Dufresne, A., (2011). Synergism effect of montmorillonite and cellulose whiskers on the mechanical and barrier properties of natural rubber composites. *Macromol. Mater. Eng.* 296, 760.

[109] LeCorre, D., Dufresne, A., Rueff, M., Khelifi, B., Bras, J., (2014). All starch nanocomposite coating for barrier material. *J. Appl. Polym. Sci.* 10.1002/app.39826.

[110] Samir, M.A.S.A., Alloin, F., Gorecki, W., Sanchez, J.Y., Dufresne, A., (2004). Nanocomposite polymer electrolytes based on poly(oxyethylene) and cellulose nanocrystals. *J. Phys. Chem. B* 108, 10845.

[111] Pei, A., Malho, J., Ruokolainen, J., Zhou, Q., Berglund, L.A., (2011). Strong nanocomposite reinforcement effects in polyurethane elastomer with low volume fraction of cellulose nanocrystals. *Macromolecules* 44, 4422.

[112] Bras, J., Hassan, M.L., Bruzesse, C., Hassan, E.A., El-Wakil, N.A., Dufresne, A., (2010). Mechanical, barrier, and biodegradability properties of bagasse cellulose whiskers reinforced natural rubber nanocomposites. *Ind. Crops Prod.* 32, 627.

In: Polymer-Matrix Composites
Editor: Rakesh Kumar

ISBN: 978-1-61122-005-6

Chapter 4

ARYLATED AND NONARYLATED SOY PROTEIN ISOLATE BASED NANOCOMPOSITES

Rakesh Kumar*

Birla Institute of Technology, Mesra, Patna Campus, Patna, India

ABSTRACT

In recent years, nanoparticles/nanomaterials have attracted great industrial and academic interests because of their high aspect ratio and better reinforcement ability when compared to microparticles. The incorporation of these nanoparticles/nanomaterials in the polymer-matrix results in the fabrication of nanocomposites. Soy protein isolate (SPI) is one of the few natural polymers that can be processed as films and composites. They are explored as packaging films and commodity plastics. Nanomaterials such as cellulosic whiskers, chitin whiskers, nanoclay, and carbon nanotube have been incorporated into SPI to improve the properties of soy protein isolate based nanocomposites. These nanocomposites have good mechanical properties but low water resistance limit their applications. In the year 2008, it has been reported that 2,2-diphenyl-2-hydroxyethanoic acid (DPHEAc) can be used to prepare arylated SPI. These arylated SPI films exhibited improved thermal stability and high water resistance. Researchers have also incorporated cellulosic whiskers into arylated SPI to fabricate nanocomposites with significant improvement in water resistance, mechanical properties, and thermal stability. Thus arylated SPI is emerging as a new area of research and it is expected that the arylation of protein materials could serve as a general motif to fabricate water resistant protein based composites/nanocomposites. In this chapter, discussion on composites from arylated SPI and fabrication of SPI based nanocomposites with focus on mechanical, thermal and water uptake properties have been presented.

Keywords: SPI, arylated SPI, nanoparticles, nanocomposites

* Corresponding author: Birla Institute of Technology, Mesra, Patna Campus, Patna, India; Tel.: +91-9709474061. Email address: krrakesh72@gmail.com (Rakesh Kumar).

1. INTRODUCTION

Soybeans (*Glycine max*) serve as one of the most valuable crops in the world, not only as oil, but also as a good source of protein. Soy proteins are commercially available as soy flour (SF), soy protein concentrate (SPC), and soy protein isolate (SPI), all differing in protein content. Figure 1 shows the presence of major components in soy protein. The SF contains about 50% protein content. The SF is converted to SPC that contains about 60–70% of protein, by removing most of the water-soluble components. After sugar and other water-soluble materials as well as fiber have been removed, the defatted soy flour is processed into SPI that contains about 90% protein. The SPI is more expensive than SF due to complex processing of SPI [1].

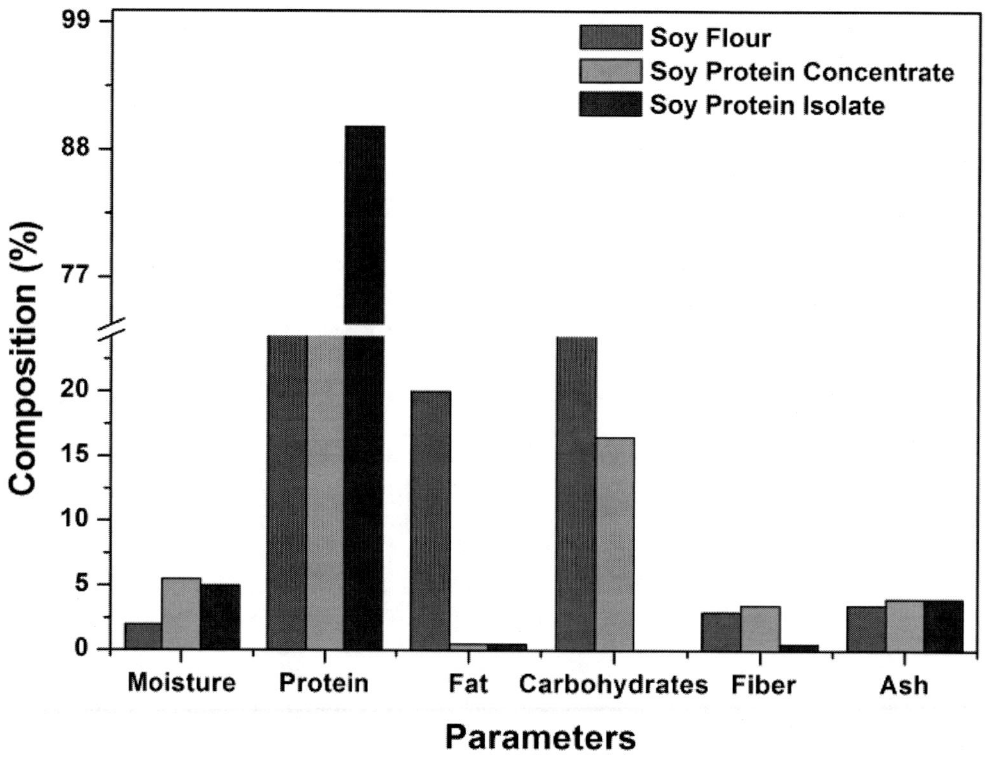

Figure 1. Different compositions of components present in soy flour (SF), soy protein concentrate (SPC) and soy protein isolate (SPI).

Soy protein consists of two major protein fractions referred to as the 7S (conglycinin, 35%) and 11S (glycinin, 52%) fraction. Both 7S and 11S contain cysteine residues leading to disulphide bridge formation.

For the readers, it is important to know that the world soybean production increased by 4.6% annually from 1961 to 2007 and reached average annual production of 217.6 million tons in 2005-07. World production of soybeans is predicted to increase by 2.2% annually to 371.3 million tons by 2030. Most of the soybean seeds are processed into edible oil. India is the fourth largest oilseed producing country in the world, next only to USA, China and Brazil.

Since 1995, Indian share in world production of oilseeds has been around 10%. However, in this book chapter, we are concerned about SPI based nanocomposites and arylated SPI composites.

SPI films are prepared extensively in various research laboratories all over the world. There are two common technologies i.e., "wet (or solvent) processes" and "dry processes" to prepare the SPI films. "Wet process," also is called solution casting which is based on the dispersion or solubilisation of proteins in a solvent medium method. "Dry processes" often includes hot-pressing or compression-molding as well as melt and extrusion techniques to prepare protein films.

Protein plastics prepared without any plasticizers often have brittle properties. Hence, plasticizers are needed to improve processability and flowability of the proteins. The processability and flexibility of soy proteins have been improved by the addition of plasticizers such as glycerol, ethylene glycol, triethylene glycol, poly(ethylene glycol), butane diols [2, 3] urea, [4] and acetamide [5]. Glycerol has been found to be a good plasticizer among polyols. Soy protein films are also prepared by using different plasticizers such as 1,2 propanediol [6], trimethylol propane [6], thiodiglycol [6, 7], and formamide [6]. In general, the solution casted films present higher percentage elongation, whereas hot-pressed films are more resistant to rupture and exhibited higher stress values than cast films [8]. The effects of molding condition and glycerol content on hot pressed films indicate that tensile strengths are higher at lower glycerol content and higher molding temperature. However, Foulk et al., [9] found that acetylated soy protein could be compression-molded into films without any plasticizers.

2. Matrix

2.1. Soy Protein Isolate (SPI)

As mentioned in the Introduction section, soy protein is an abundant side-product of an oil industry. Edible/biodegradable films, adhesives, plastics and foams can be prepared from SPI. In general, soy protein films, similar to films from other proteins, provide limited resistance to water vapour transfer which is attributed to the inherent hydrophilicity of proteins and the notable amounts of hydrophilic plasticizers incorporated into protein films. The mechanical properties of soy protein plastics are also influenced by the amount of plasticizer and the water content in the samples. With increasing plasticizer content and moisture content, the behavior of the SPI plastics changes from hard and brittle to soft and weak.

At higher relative humidity (RH) or water content, absorbed water in the protein films also acts as a plasticizer in addition to the added plasticizer and thereby the tensile strength and modulus decreases. Figure 2 shows the stress–strain curves of the SPI films with thiodiglycol (TDG) as a plasticizer at 57% RH to verify the above facts [7]. As the content of TDG increases from 20 to 40%, the tensile strength decreases. The modulus also exhibits a decrease with increasing content of TDG.

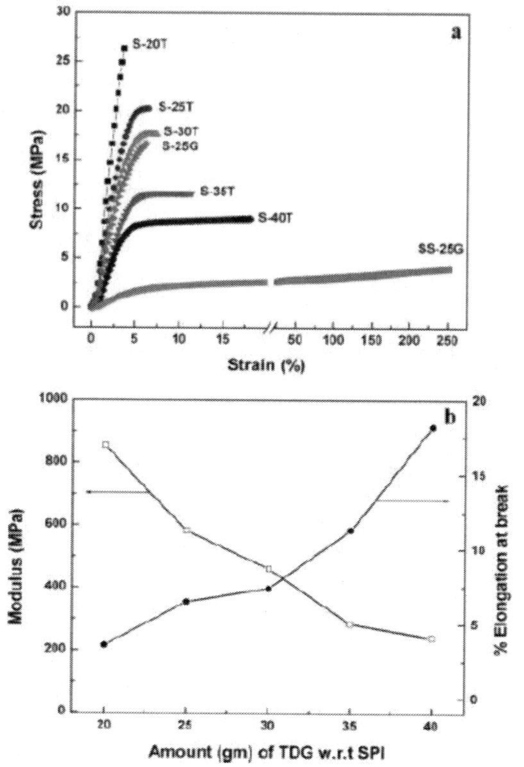

Figure 2. (a) Stress–strain curves and (b) modulus and elongation at break of the SPI films with different contents of thiodiglycol plasticizer at 57% RH. (Reprinted from Ref. [7] Kumar, R., Wang, L., Zhang, L., (2009). J. Appl. Polym. Sci. 111, 970).

Figure 3 shows the stress–strain curves of the SPI films with TDG as a plasticizer at 97% RH [7]. At higher water content there is decrease in the mechanical properties.

The main problem for protein-based plastic is hydrophilicity. Few authors have tried to prepare the blends of SPI and synthetic polymers. The strength and elongation of the blend decreases at higher content of SPI. By compatibilization, blends of SPI/polycaprolactone have been reported with improved toughness [10]. The blends from soy protein and polylactide have also been prepared [11]. However, it is not easy to prevent the soy protein from degradation when blending protein with polylactide. The preparation of injection molded biodegradable articles from soy protein in presence of glycerol plasticizer has been reported [12]. All the articles from SPI can be exploited for packaging containers or packaging films. These films or containers are transparent and yellowish-brown with low product cost [13].

Not only blends but SPI composite films have also been prepared with natural fibers such as ramie [14, 15], Indian grass fibers [16], flax [17, 18] and jute [19, 20]. The mechanical properties of natural fiber reinforced SPI biocomposites increase but the problems of low water resistance are still there because of the use of natural fibers as reinforcing agents.

Antimicrobial based SPI films can be also be prepared. SPI and octenyl succinate (OSA) modified starch have been used as paper coating with the inclusion of antimicrobial compounds such as cinnamaldehyde and carvacrol [21]. Cross-linking reaction by dialdehyde starch (DAS) in SPI imparts better mechanical properties in the antimicrobial based SPI films [22].

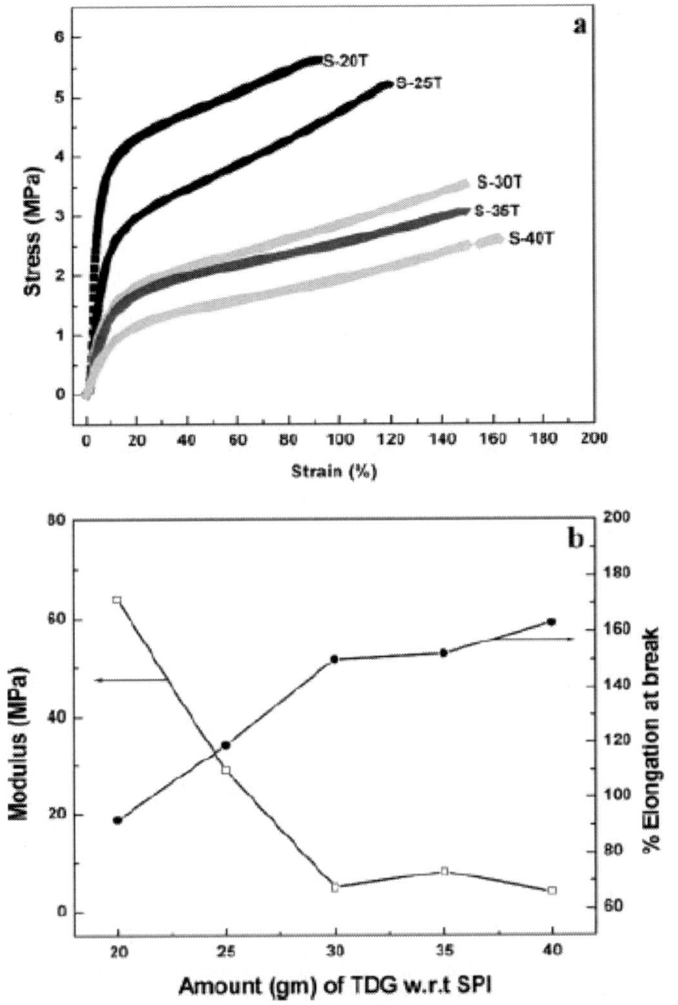

Figure 3. (a) Stress–strain curves and (b) modulus and elongation at break of the SPI films with different contents of thiodiglycol plasticizer at 97% RH. (Reprinted from Ref. [7] Kumar, R., Wang, L., Zhang, L., (2009). J. Appl. Polym. Sci. 111, 970).

2.2. Arylated Soy Protein Isolate

Arylated soy protein films can be prepared by two methods i.e., dry as well as wet method. First, the dry method will be disucced. Plasticizers either glycerol or thiodiglycol are mixed with SPI powder, separately, in an electronic mixer for about 15 min to prepare plasticized soy protein powder. The resulting mixtures are further mixed with 2,2-diphenyl-2-hydroxyethanoic acid (DPHEAc) and subjected to hot press at 155 °C for 10 min under 15 MPa pressure. The resulting protein films are immersed in distilled water at room temperature for 26 h. Subsequently, the films are removed from water and dried in an air oven to obtain the arylated soy protein films [23]. The scheme for this type of reaction is given in Figure 4.

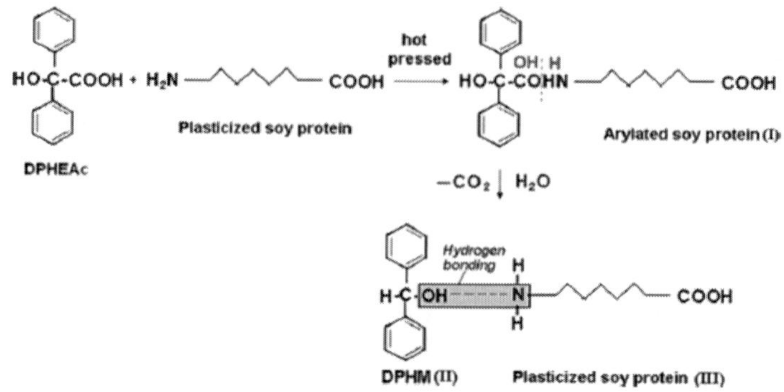

Figure 4. Reactions to prepare arylated soy protein. (Reprinted from Ref. [23] Kumar, R., Zhang, L., (2008). Biomacromolecules 9, 2430).

Figure 5 shows the photograph of native SPI films (a) and arylated SPI films (b) as well as the scanning electron microscope (SEM) pictures of the surface and cross section of films. In the SEM images, the native SPI film displayed a smooth structure on the surface, whereas the surface and the interior of the arylated SPI film exhibited nanospheres having a diameter of 30 - 40 nm [23].

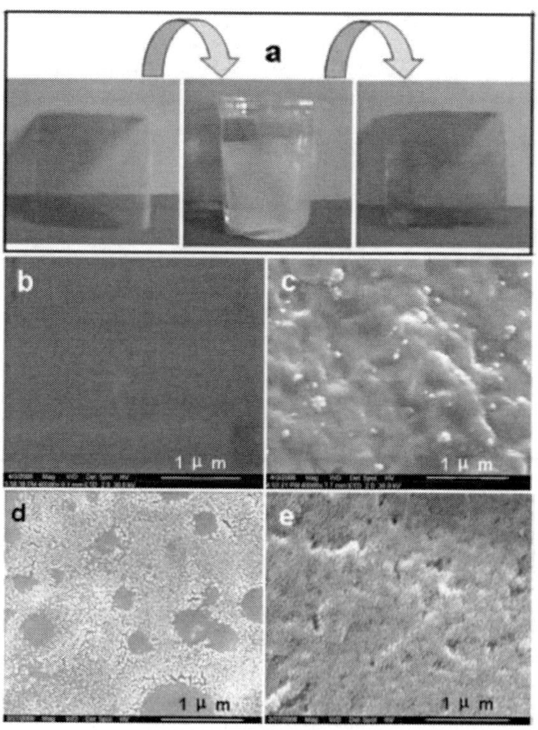

Figure 5. Photographs of the native (left) and arylated (right) soy protein films. SEM images of native SPI (b,c) and arylated SPI (d,e) films. Surface morphology (left) and cross-section (right) (Reprinted from Ref. [23] Kumar, R., Zhang, L., (2008). Biomacromolecules 9, 2430).

In the wet method, plasticizers are used to prepare soy protein films by compression molding at 140 °C and 20MPa. Then the native SPI films are immersed in 0.5% (w/v) DPHEAc solution for 26h to get arylated soy protein films with the evolution of CO_2, leading to the formation of diphenylhydroxymethane (DPHM) on the surface [6]. Figure 6 shows the increase in average contact angle of arylated SPI films when compared to native SPI films in the presence of different plasticizers. This shows that there is increase in the hydrophobicity of the arylated SPI films.

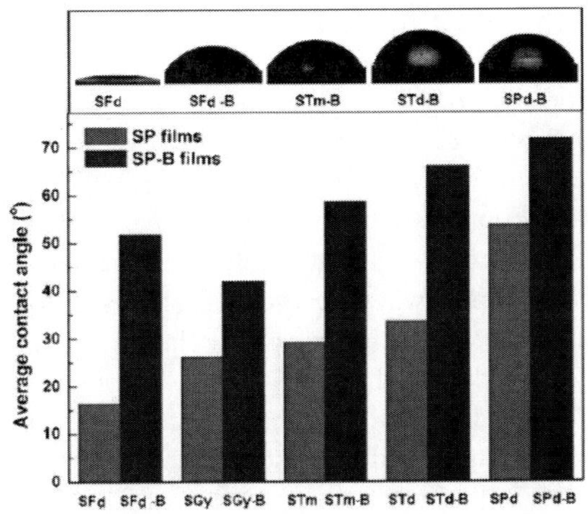

Figure 6. Contact angle measurements of native SPI (SP) and arylated SPI (SP-B) films in the presence of different plasticizers. Here FD, Gy, Tm, Td and Pd represent formamide, glycerol, trimethylol propane, thiodiglycol and propanediol, respectively. (Reprinted from Ref. [6] Kumar, R., Zhang, L., (2009). Ind. Crops Prod. 29, 485).

Not only the hydrophobicity but also the mechanical properties of the arylated films are higher than native SPI films and this can be easily observed in the mechanical properties factor as shown in Figure 7.

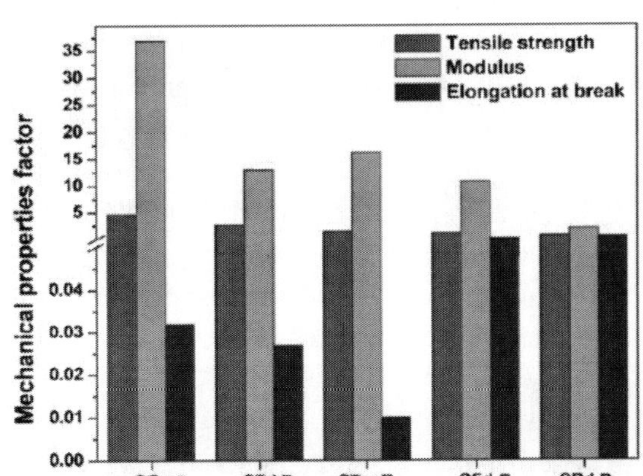

Figure 7. Mechanical properties factor of arylated soy protein films in the presence of different plasticizers. Here FD, Gy, Tm, Td and Pd represent formamide, glycerol, trimethylol propane, thiodiglycol and propanediol, respectively. (Reprinted from Ref. [6] Kumar, R., Zhang, L., (2009). Ind. Crops Prod. 29, 485).

3. REINFORCEMENT BY NANOMATERIALS

3.1. Organic and Inorganic Nanoparticles

In recent years, montmorillonite (MMT) has attracted great industrial and academic interests because of its high aspect ratio of silicate nanolayers and high surface area [24, 25]. The nanocomposites may be formed by incorporating MMT into polymers. The polymer/MMT nanocomposites thus fabricated can be divided into intercalated and exfoliated nanocomposites [26]. To give better properties, the dispersion of nanoclay in the polymer matrix is important [27]. Highly exfoliated polymer/clay nanocomposites constitute fundamental and application research [28-34]. Solution intercalation, melt intercalation, and in situ intercalative polymerization can result in exfoliated nanocomposites [35-39].

Like MMT, as discussed above, carbon nanotubes (CNTs) are considered ideal reinforcing fillers for polymer matrices to achieve high performance and special functions [40, 41]. However, CNTs exhibit extraordinary mechanical strength as well as high electrical and thermal conductivity [42]. Multi-walled carbon nanotubes (MWCNTs) or Single-walled SWCNTs with various diameters can be used to reinforce SPI. Thus the blending of nanofillers such as chitin [43], industrial lignin [44] and, layered silicate [45, 46] and carbon nanotube [47] and so on, showed obvious reinforcing effects in soy protein polymers. The reinforcement effect were not only limited to soy protein but are also extended to other biodegradable polymers [48-50]. Bionanocomposites, or econanocomposites, are novel materials based on renewable resources combined with the nanoscale effects by in situ assembling nanophase [50] or blending organic and inorganic nanoparticles [51]. It is just the prominent reinforcing effects and low loading of nanofiller that draw attentions to developing SPI or other biodegradable polymers based bionanocomposite.

3.2. Cellulosic Whiskers

Additionally, whiskers derived from natural fibers have also become one of the most promising means of dramatically increasing the physical properties of the composites [51-54]. Whiskers are considered as very promising reinforcing materials for composites because of their high stiffness and strength [54]. They are nearly free of internal defects because of their small diameter; thereby they yield strengths close to the maximum theoretical value predicted by the theory of elasticity. The extent of their reinforcement has been found to depend on several factors such as the nature of the matrix, the generation of a strong fiber matrix interface through physicochemical bonding, and the dispersion of the whiskers in the matrix [55-58].

4. PROPERTIES OF SPI NANOCOMPOSITES

We have discussed in earlier section about the incorporation of several nanoparticles in SPI. In this section, we are discussing the effect of these nanoparticles on the properties of SPI sheets.

4.1. Mechanical

The values of the Young's modulus (*E*) increase from 180.2 to 587.6 MPa with an increase of the MMT content from 0 to 20 wt % [45]. When the MMT content reaches 20 wt %, the value of tensile strength decreases to 14.48 MPa. Figure 8 shows the mechanical properties of SPI/MMT plastic sheets [45]. As evident from the Figure 8, the elongation at break keeps decreasing with the increase of the MMT addition. These results confirm that the strong electrostatic and hydrogen bonding interactions between the soy protein and the highly dispersed MMT layers are significant to restrict the segmental motion of the soy protein, and this restriction leads to the improvement of the modulus and tensile strength of the soy protein plastics.

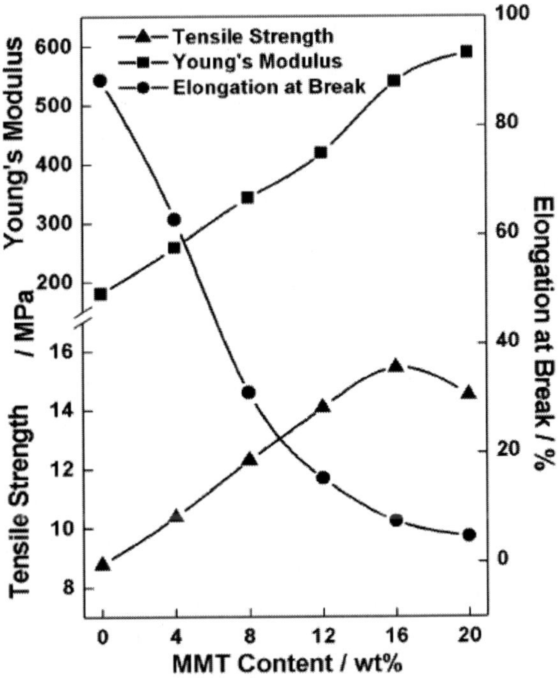

Figure 8. Effects of MMT contents on tensile strength, Young's modulus and elongation at break for SPI/MMT plastics. (Reprinted from Ref. [45] Chen, P., Zhang, L., (2006). Biomacromolecules 7, 1700).

SPI/hemp whiskers based composites are also prepared and optimum wt.% of the whiskers has been found to be 30 and 20 for the composites prepared by solution cast and hot press methods, respectively [59]. Tensile strength and modulus of compression molded composites with 20% hemp whisker are 9.9 MPa and 464.8 MPa, respectively for solution-casted sample. Figure 9 shows the effect of MWNT diameter (<10, 10–15, 20–40, 40–60, and >60 nm) on mechanical properties [60]. As the diameter of MWNT increases up to 10–15 nm, tensile strength and Young's modulus increase and reach maximum values of 11.7 MPa and 258.6 MPa respectively. Thereafter, the mechanical properties decrease with increasing MWNT diameter up to 40–60 nm, followed by an increase for nanocomposites based on >60 nm MWNT.

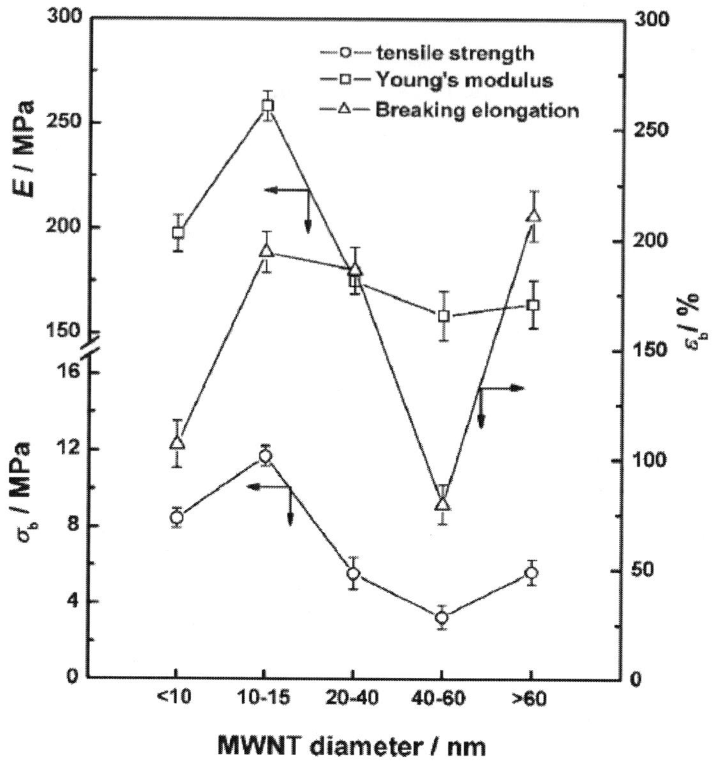

Figure 9. Effects of MWNT diameters on tensile strength, Young's modulus and elongation at break for SPI/MWNT plastics. (Reprinted from Ref. [60] Zheng, H., Ai, F., Wei, M., Huang, J., Chang, P. R., (2007). Macromol. Mater. Eng. 292, 780).

Figure 10 shows the effects of nano-SiO_2 contents on the mechanical properties of nano silica incorporated soy protein nanocomposite sheet (S-nSi) [61]. The effects of nano-SiO_2 content on tensile strength and modulus are similar at 4 wt % nano-SiO_2 followed by a sharp decrease at 8 wt % nano-SiO_2. Except for expected strength increase, the elongation is enhanced for the sheets with nano-SiO_2 content lower than 8 wt %. It is worth noting from Figure 10 that the maximum strength of 11.3 MPa happened at the addition of 4wt% nano-SiO_2 with an enhancement of elongation, and that adding 8 wt % nano-SiO_2 results in optimal elongation as well as a reinforcing effect.

4.2. Thermal

The thermal stability of SPI/MMT composites had been investigated and reported in Figure 11 [45]. Initial weight loss between 0 and 120 °C is attributed to the water absorbed in SPI/MMT plastics.

The weight loss in the temperature range of 120-250 °C is mainly related to the evaporation of glycerol which had been used as a plasticizer (Figure 11). The residue weight during the whole thermo-degradation period of soy protein is higher for nanocomposite plastics than that of neat SPI plastics. These facts evidence the improvement of thermo-stability in the protein-degradation range of 250-800 °C for the SPI/MMT plastics.

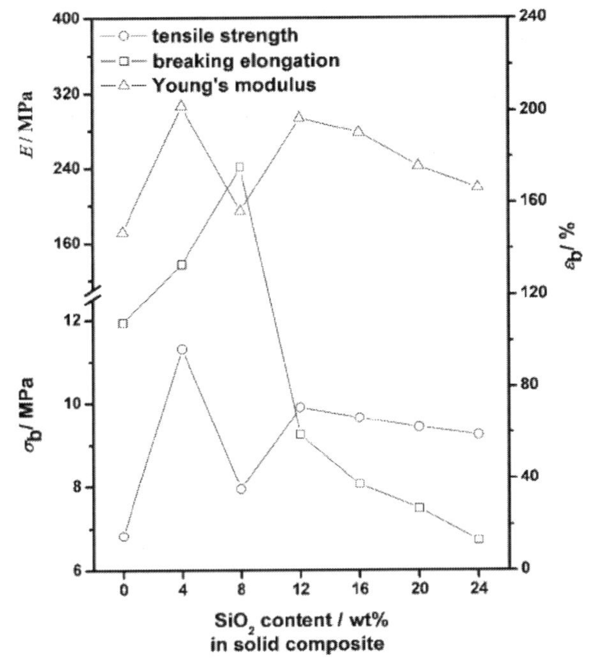

Figure 10. Effects of nano SiO_2 contents on tensile strength, Young's modulus and elongation at break for SPI/nanoSiO_2 plastics. (Reprinted from Ref. [61] Ai, F., Zheng, H., Wei, M., Huang, J., Chang, P. R., (2007). J. Appl. Polym. Sci. 105, 1597).

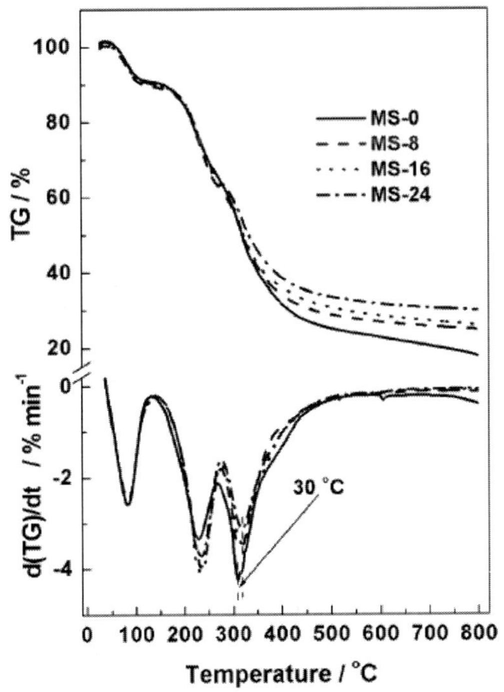

Figure 11. Effects of MMT contents (from 8 to 24) on the thermal properties for SPI/MMT plastics. Here MS-0 represents neat SPI plastics. (Reprinted from Ref. [45] Chen, P., Zhang, L., (2006). Biomacromolecules 7, 1700).

TGA and DTG curve of the SPI in absence of hemp whisker (SC0) and in presence of 20% hemp whisker (SC20-W) are evaluated [59]. Three T_{max} values designated as T_{max1}, T_{max2}, and T_{max3} corresponding to temperature range of room temperature (RT) to 200°C, 200–300°C, and 300–400°C, respectively, are observed for the samples. The first step mass in SC20-W, attributed to the loss of adsorbed moisture, is lower than SC0. The second step mass loss from 200–300°C is attributed to the loss of thiodiglycol as plasticizer. The T_{max2} for SC20-W increases from 247°C to 255°C, but the mass loss in this temperature range is lower than that of SC0. Third mass loss is attributed to the degradation of soy protein. T_{max3} values also increases from 308 to 337°C for SC20-W sample. There is an increase in the thermal stability of the composites because of strong interactions between hemp whisker and soy protein.

4.3. Water Uptake

The water uptake up to equilibrium for SPI/MWNT nanocomposite sheets under a relative humidity (RH) of 98% is plotted as a function of time (t) in Figure 12 [60]. Two well separated zones were observed as Zone I (t<75 h) for rapidly increasing water uptake and Zone II (t>75 h) for approaching the equilibrium of water uptake, respectively. The decrease in water uptake of the nanocomposites can be attributed to the restriction of the relatively rigid MWNTs and their aggregates to SPI motion associated with the wrapping of SPI chains on MWNTs.

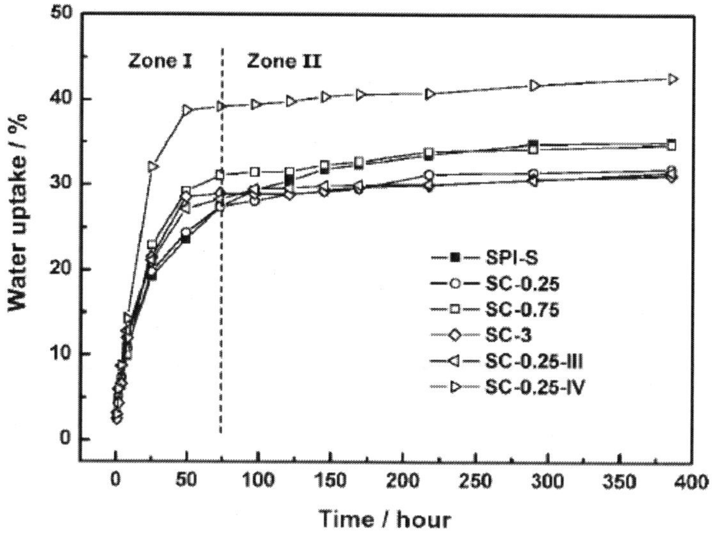

Figure 12. Effects of MWNT contents on the water uptake properties for SPI/MWNT plastics. (Reprinted from Ref. [60] Zheng, H., Ai, F., Wei, M., Huang, J., Chang, P. R., (2007). Macromol. Mater. Eng. 292, 780).

The water uptake for SPI/hemp whisker based composites with 20% of hemp whisker (SC20-W) prepared by the solution casting and hot pressing is 154% and 58%, respectively [59]. This may be attributed to less compact structure of composites prepared by solution casting method.

4.4. Antimicrobial

Sun et al., prepared bactericidal films from SPI based film-forming dispersions (FFDs) for use in the food and medical fields [62]. The FFD and films were prepared after the incorporation of different concentrations of $AgNO_3$ as a bactericidal agent. The transparency, tensile strength, and antimicrobial features are evaluated and reported. The opacity of these FFD decreases after the incorporation of $AgNO_3$. The minimum inhibitory concentration of $AgNO_3$ had been found to be 336 μg/mL FFD for both *Escherichia coli* ATCC 25923 and *Staphylococcus aureus* ATCC 25922.

An antimicrobial silver nanoparticles (AgNPs) embedded SPI film is prepared by Zhao et al. [63]. AgNPs were *in situ* synthesized from SPI/$AgNO_3$ solution, taking the advantage of the reducibility of tyrosine residue in SPI. The whole reaction process is highly energy-efficient and eco-friendly. The SPI/AgNPs films are prepared by blending *in situ* synthesized AgNPs in SPI solution and additional pristine SPI solution, which shows an effective antimicrobial activity against both Gram-positive and Gram-negative bacteria.

Zhang et al., used solution blowing to form soy protein-containing nanofibers which are decorated with silver nanoparticles [64]. These nanofibers demonstrate significant antibacterial activity against *E. coli* colonies without exposure to UV light (Figure 13).

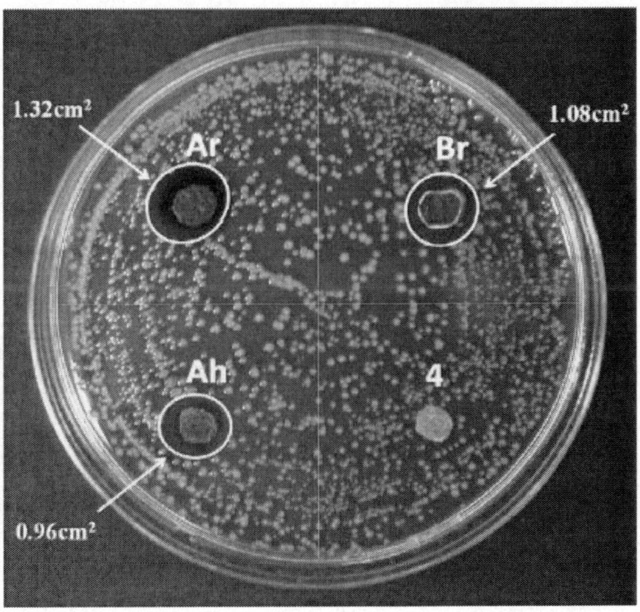

Figure 13. Inhibition of *E. coli* by silver-decorated soy protein nanofiber mats. (Reprinted from Ref. [64] Zhang, Y., Lee, M.W., An, S., Ray, S.S., Khansari, S., Joshi, B., Hong, S., Hong, J-H., Kim, J-J., Pourdeyhim, B., Yoon, S.S., Yarin, A.L., (2013). Catalysis Communications 34, 35).

4.5. Morphology

The microstructure of SPI/MMT plastics has been visualized by using TEM, and the images are shown in Figure 14 [45]. At lower content of MMT, the dimensions of the silicate layers are diminished to about 30 nm in length and 1 nm in thickness. It indicated the layered

MMT is highly exfoliated by the soy protein molecules (Figure 14a). At higher contents of MMT in SPI/MMT plastic plastic sheet (Figure 14b), most of the layered MMT tactoids are intercalated with a d-spacing of about 6 nm.

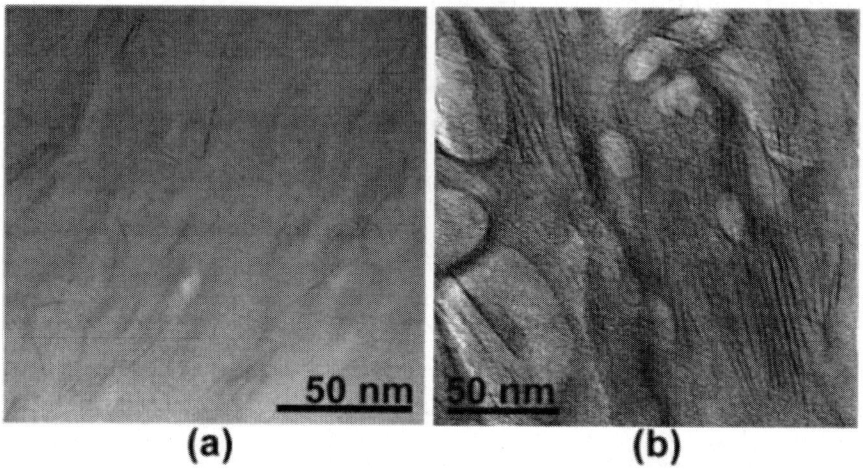

Figure 14. TEM images of SPI/MMT plastics, (a) MS-8 and (b) MS-16. (Reprinted from Ref. [45] Chen, P., Zhang, L., (2006). Biomacromolecules 7, 1700).

The SEM images of fractured surfaces (cross sections) for the nanocomposite sheets and SPI without MWNTs are shown in Figure 15 [60]. The neat SPI sheet shows morphology of fractured lamellae (Figure 15(A)). After introducing MWNTs, such morphology is replaced by a surface with donut-like and raised blob-like objects, shown in Figure 15(B)–15(G). With an increase in MWNT content, the number of donut-like objects increase [Figure 15(B)–15(D)].

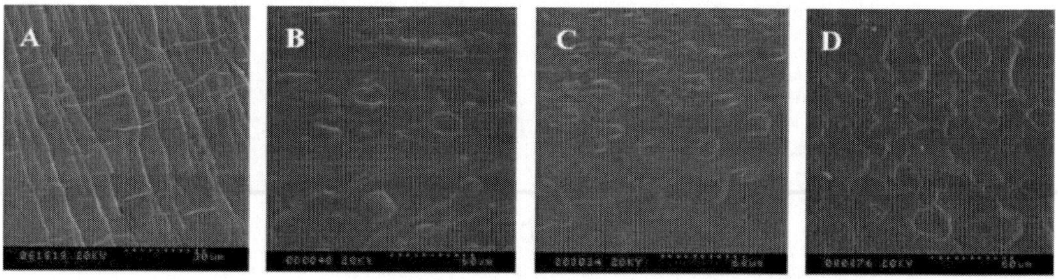

Figure 15. SEM images of cross-sections of the SC sheets containing various contents of MWNTs with different sizes, as well as the SPI-S sheet without MWNTs (A: SPI-S; B: SC-0.25; C: SC-0.75; D: SC-1). (Reprinted from Ref. [60] Zheng, H., Ai, F., Wei, M., Huang, J., Chang, P. R., (2007). Macromol. Mater. Eng. 292, 780).

The SEM images of fractured surface for nanocomposite sheets with and without adding nano-SiO_2 have been reported [61]. Similar to previous images, SPI showed a heterogeneous morphology originated from the complex component and structure in soy protein. Fractured morphology did not obviously changed after adding 4 wt % nano silica. However, the obvious change of fractured surface after adding 8 wt % nano silica can be observed at high concentration.

5. Properties of Arylated SPI Composites / Nanocomposites

We have discussed in earlier section about the preparation of arylated SPI. In this section, we will discuss about the incorporation of natural fibers such as ramie, hemp and flax as well as cellulosic whiskers in arylated SPI. The effect of these fibers and cellulosic whiskers on the properties of arylated SPI sheets are discussed in details.

5.1. Mechanical

To enhance the strength of the composites prepared from neat SPI as well as arylated SPI and ramie fiber, the fibers are aligned in vertical (V), horizontal (H) as well as both vertical and horizontal (X) directions [65]. The composites thus prepared are coded as VSC, HSC and XSC. Figure 16a shows the mechanical properties of the VSC composites with the different content of the ramie fibers. The optimum content of the ramie fibers in the SPI composites is found to be 30 wt%, which lead to a significant improvement in both tensile strength and modulus. There is a decrease in tensile strength and improper wetting of fibers with SPI resins above 30 wt% ramie fiber. Mechanical properties of the composites aligned in different directions, having 30 wt% of the fibers, are also evaluated, and the results are shown in Figure 16b. The VSC composites have higher tensile properties in vertical directions of ramie fibers than others. The HSC composites prepared by taking horizontal aligned fibers show the lowest mechanical properties. The mechanical properties of the arylated SPI/ramie fiber composites are also found to be strongly dependent on the alignment of the fibers.

Kumar et al., has also reported the mechanical properties of arylated and non arylated SPI and ramie whiskers based composites [66]. The optimum wt.% of the ramie whiskers for the reinforcement in the SPI composites is found to be 15 wt.%. Above this wt.%, there is no significant increase in the tensile strength and modulus. Interestingly, there is a ~10–15 times increase in the modulus for the arylated SPI/whisker composites, whereas increase in the tensile strength is around 1.5–2.0 times. However, both arylated and non arylated SPI/ramie whisker based composites show a decrease in the mechanical properties above optimum wt.% of the fibers.

From the nonarylated SPI and flax fabric, one can prepare damage-insensitive composites [67]. The tensile stress and modulus increase with increasing weight fraction up to 30% of the flax fabric. After that, there is decrease in the tensile stress for composites containing 40% flax fabric with increasing tensile modulus. There is decrease in the percentage elongation at break because of the flax fabric.

On the other hand, from the arylated SPI and flax fabric one can prepare damage-sensitive composites [67]. Figure 17 shows the mechanical properties of arylated soy protein composites. For 20% (S-2F-A) and 30 % (S-3F-A) of flax fabric in arylated SPI, the maximum load on the composites is the same as that of the matrix alone; the composites continue to carry the decreasing load after the peak. The postpeak resistance is primarily provided by the breaking of fibers from the cracked surface. With incorporation of 40 % (S-4F-A) flax fabric in arylated SPI, even after the cracking of the matrix, the composite continue to bear a tensile load; the peak stress is greater than that of the matrix alone. During the inelastic range in S-4F-A, multiple cracking of the matrix, fiber debonding, and stick–slip occurred (68-70). Thus, in the arylated composites the stick-slip mechanism dominates and it prevents easy deformation of the composite film.

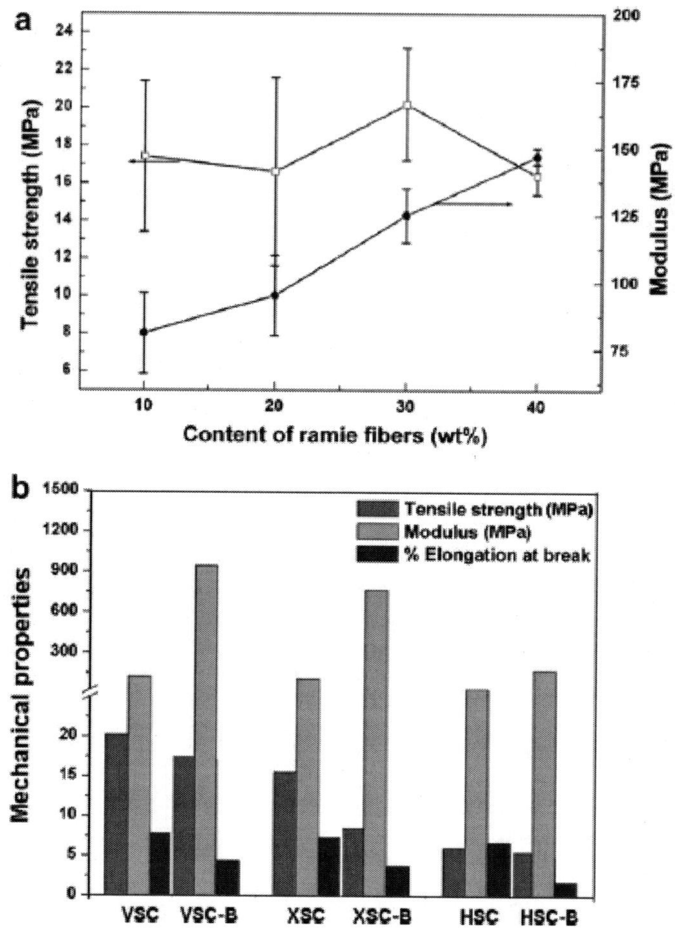

Figure 16. Effects of content of ramie fiber on the mechanical properties of SC composites (a). Mechanical properties of VSC, XSC, HSC, VSC-B, XSC-B, and HSC-B, having 30 wt% of the fibers (b). (Reprinted from Ref. [65] Kumar, R.; Zhang, L., (2009). Compos. Sci. Techn. 69, 555).

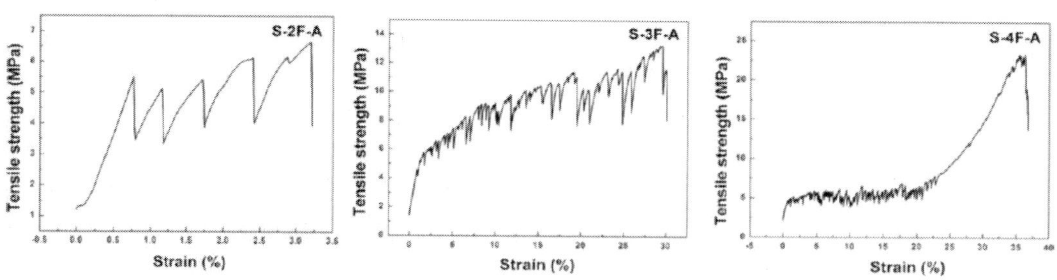

Figure 17. Tensile stress–strain curves for arylated flax-reinforced soy protein composites. (Reprinted from Ref. [67] Kumar, R., Anandjiwala, R. D., (2012). J. Appl. Polym. Sci. 124, 3132).

5.2. Thermal

Figure 18 shows the thermal patterns of SPI/ramie fiber (SC) and arylated SPI/ramie fiber (SC-B) composites [65]. The SC composites show three stages degradation with three T_{max} as 190, 259, and 312 °C in the temperature range of 133–219, 225–275, and 280–376 °C, respectively. The first stage degradation is attributed to the loss of absorbed moisture, whereas second and third stage degradation is attributed to the loss of plasticizer and the degradation of the SPI and fibers, respectively. Interestingly, only single stage degradation with higher degradation temperature (T_{onset}) is observed for SC-B samples. Merger of second and third stage mass loss in the SC-B samples confirm the higher thermal stability of the SC-B composite films than that of SC.

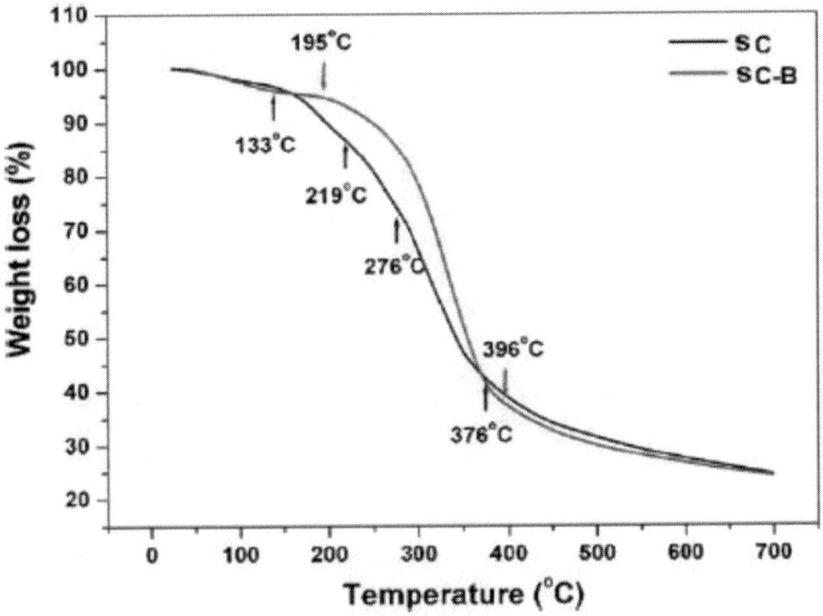

Figure 18. TGA curves for SC and SC-B composites. (Reprinted from Ref. [65] Kumar, R.; Zhang, L., (2009). Compos. Sci. Techn. 69, 555).

High thermal stability is also observed for arylated SPI/ramie whisker composites [66]. The onset temperature (T_{onset}) and final temperature (T_{final}) for degradation of arylated SPI/ramie whisker composites increase from 133°C to 195°C and from 376°C to 396°C, respectively. As mentioned above, arylated SPI/ramie whisker composites show single stage degradation while SPI/ramie whisker show three stages of degradation. The increase in the thermal stability indicates the effect of arylation on soy protein composites. Similar type of conclusion can be drawn from the study of the thermal stability of the arylated and nonarylated SPI–flax fiber composites [67]. The thermal degradation of all of the nonarylated samples, in the presence of thiodiglycol as a plasticizer, show three stage mass loss. On the other hand, the arylated SPI–flax soy protein composites show a one-stage mass loss.

5.3. Water Uptake

Water resistance of the SPI/ramie fiber (SC) and arylated SPI/ramie fiber (SC-B) composites are evaluated by water uptake experiments [65]. The water resistance of the arylated SPI/ramie fiber composites is greater. Higher contact angles (70 ± 7°) of the arylated SPI/ramie fiber composites than non-arylated ones (33 ± 5°) after 30 seconds are observed. This further proves that the SC-B composites possess hydrophobic surfaces. As the contents of ramie whisker increases in SPI/ramie whisker composites, the water uptake decreases indicating the effect of whiskers on soy protein composites [66]. The water uptake of the arylated SPI/ramie whisker composites is significantly lower than that of the non-arylated SPI/ramie whisker composites. Similar type of results is obtained from the water uptake study of the arylated and nonarylated SPI–flax fiber composites [67]. The arylated composites show a low water uptake.

5.4. Morphology

The SEM photograph of the SPI/ramie whisker composites exhibit uniform surfaces, whereas the microparticles of diphenylhydroxymethane (DPHM) are observed on the surface of the arylated SPI/ramie whisker composites [66]. A relatively uniform distribution of whiskers in the SPI matrix with a compact structure is observed for the cross-section of arylated SPI/ramie whisker composites. However, the cross-section of SPI/ramie whisker composites show agglomerates of the whiskers. Figure 19 shows the morphology of the arylated and nonarylated soy protein films after they were subjected to tensile testing [67]. The surface and cross-sectional morphologies of the arylated soy protein composites show the presence of DPHM microparticles. The surface morphology of the nonarylated composites after tensile testing shows a homogeneous surface as shown in Figure 19(a).

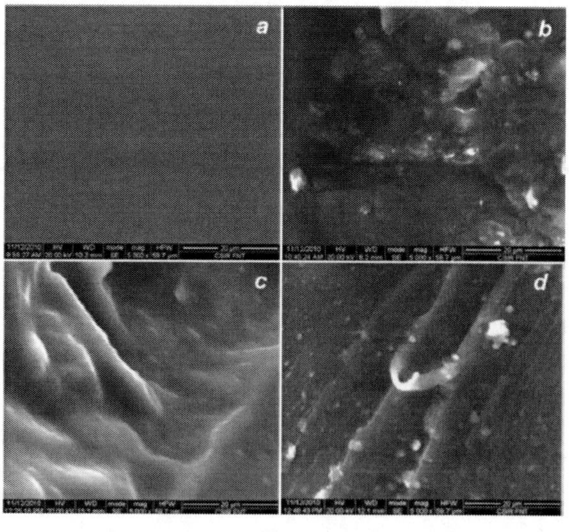

Figure 19. (a,b) Surface and (c,d) cross section morphologies of the non-arylated and arylated flax-reinforced soy protein. (Reprinted from Ref. [67] Kumar, R., Anandjiwala, R. D., (2012). J. Appl. Polym. Sci. 124, 3132).

6. Future Directions

Production of nanoparticles under ecofriendly conditions is of significance to address growing concerns on the overall toxicity of nanoparticles for medical and biotechnological applications. This can be achieved by either using natural nanoparticles or using natural polymers to generate nanoparticles. We know that natural nanoparticles are renewable and environmentally friendly. The synthesis of nanoparticles from soy protein, one of the most abundant and widely utilized plant proteins, for nutraceutical and drug encapsulation can be focussed to develop natural nanoparticles [71]. The preparation of soy protein nanoparticles mentioned in this literature consisted of dispersion, desolvation, drug incorporation, cross-linking, and evaporation.

One investigation demonstrates the production of silver nanoparticles capabilities of a miracle bean soybean *Glycine max* [72]. The researchers found that a single protein of soybean with a molecular weight of 51 kDa stabilizes the newly formed silver nanoparticles. Similarly, completely green synthesis of Ag nanoparticles stabilized by SPI under UV irradiation has been discussed [73]. In both the cases, natural polymers were used to produce nanoparticles. Transmission electronic microscopy (TEM) results indicate that the Ag nanoparticles produced in presence of natural polymers are stable and well dispersed with an average diameter about 13 nm.

SPI nanoparticles (28–179 nm) are also prepared by employing a cold gelation method [74]. The obtained nanoparticles exhibit uniform size distribution and spherical shape with a unique honeycomb-like core structure. Nanoparticle characteristics are dependent on size, surface charge and hydrophobicity that could be adjustable by changing calcium concentration and environmental pH. Soy protein nanoparticles are also produced with a high shear microfluidizer [75]. Recently use of soy protein nanoparticle aggregates as pickering stabilizers for oil-in-water emulsions have been reported [76].

In the current development, soy protein nanoparticles will have potential uses in the reinforcement of rubber materials and controlled-release of medicine, and will be beneficial to soybean farmers by creating new markets for soy products. However, control of the nanoparticle size and the size distribution has several important implications for controlled-release in food and drug delivery systems. For example, there is typically an ideal sphere size that provides the optimal release rate and route of administration.

Summary and Conclusion

In this chapter, we have discussed about the preparation of nanocomposites by incorporating nanoparticles. The maximum tensile strength and tensile modulus what we get by incorporating these nanoparticles are not more than 15.4 MPa and 587.6 MPa, respectively (Table 1).

There is obvious increase in the thermal stabilities of the SPI nanocomposites but they showed three stages of mass loss which means that SPI based nanocomposites followed almost same degradation pattern as that of SPI. However, elongation at break was around 194% at maximum tensile strength and tensile modulus.

Table 1. Properties of nanocomposites and arylated composites/nanocomposites

Parameters	SPI Based Nanocomposites with				Arylated SPI Based Bio/Nanocomposites with		
					Fibers		Whiskers
	CNT	*MMT*	*Nano silica*	*Hemp whiskers*	*Ramie*	*Flax*	*Ramie*
Tensile Strength (MPa)	11.7	15.4	11.3	9.91	17.4	18.6	11.5
Tensile Modulus (MPa)	258.6	587.6	300	464.4	942	118.5	1697
Elongation (%)	194	10	175	12.31	3	2	NR*
Water Uptake (%)	30	NR*	NR*	58	20	30	25

*NR - Not reported.

On the other hand, we have discussed about the preparation of arylated biocomposites/nanocomposites by incorporating 2,2-diphenyl-2-hydroxyethanoic acid. The maximum tensile strength and tensile modulus what we get from arylated SPI based composites are 18.6 MPa and 1697 MPa, respectively. In arylated composites also there is obvious increase in the thermal stabilities of the arylated biocomposites/nanocomposites but they showed only one stage of mass loss which means that arylated SPI based biocomposites/nanocomposites followed different degradation pattern as from neat SPI. However, elongation at break was around 3% at maximum tensile strength and tensile modulus.

Figure 20 shows the summary of the preparation of nanocomposites and arylated composites/nanocomposites as discussed in this chapter.

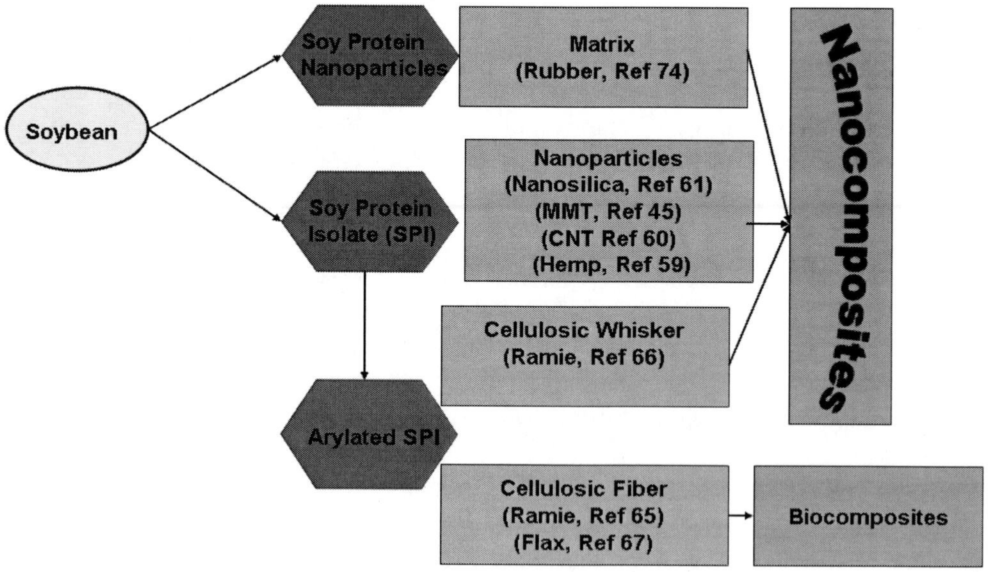

Figure 20. Summary of the prepared nanocomposites and arylated composites.

REFERENCES

[1] Endres, J.G., (2001). "Soy Protein Products Characteristics, Nutritional Aspects, and Utilization," pp. 4–8. AOCS Press, Illinois.

[2] Wang, S., Zhang, S., Jane, J.L., Sue, H., (1995). Effect of polyols on mechanical properties of soy protein. *J. Polym. Mater. Sci. Eng.* 72, 88.

[3] Wang, S., Sue, H.J., Jane, J.L., (1996). Mechanical properties of soy protein plastics. *J. Macromol. Sci. Pure Appl. Chem.* 33, 557.

[4] Mo, X., Sun, X., Wang, Y., (2001). Thermal and mechanical properties of plastics molded from urea-modified soy protein isolates. *J. Am. Oil Chem. Soc.* 78, 867.

[5] Liu, D., Zhang, L., (2006). Structure and properties of soy protein plastics plasticized with acetamide. *Macromol. Mater. Eng.* 291, 820.

[6] Kumar, R., Zhang, L., (2009). Soy protein films with the hydrophobic surface created through non-covalent interactions. *Ind. Crops Prod.* 29, 485.

[7] Kumar, R., Wang, L., Zhang, L., (2009). Structure and mechanical properties of soy protein materials plasticized by thiodiglycol. *J. Appl. Polym. Sci.* 111, 970.

[8] Mangavel, C., Rossignol, N., Perronnet, A., Barbot, J., Popineau, Y., Gue'guen, J., (2004). Properties and microstructure of thermo-pressed wheat gluten films: a comparison with cast films. *Biomacromolecules* 5, 1596.

[9] Foulk, J.A., Bunn, J.M., (2001). Properties of compression-molded, acetylated soy protein films. *Ind. Crops Prod.* 14, 11.

[10] Zhong, Z. K., Sun, X. Z., (2001). Properties of soy protein isolate/polycaprolactone blends compatibilized by methylene diphenyl diisocyanate. *Polymer* 42, 6961.

[11] Zhang, J.W., Mungara, P., Jane, J.L., (2004). Biodegradable plant protein composites and related methods. *US Patent 6806353.*

[12] Jane, J.L., Wang, S., (1996). Soy protein-based thermoplastic composition for preparing molded articles. *US Patent 5523293.*

[13] Naga, M., Kirihara, S., Tokugawa, Y., Tsuda, F., Saito, T., Hirotsuka, M., (1996). Process for producing edible proteinaceous film. *US Patent 5569482.*

[14] Lodha, P., Netravali, A.N., (2002). Characterization of interfacial and mechanical properties of ''green" composites with soy protein isolate and ramie fiber *J. Mater. Sci.* 37, 3657.

[15] Lodha, P., Netravali, A.N., (2005). Characterization of stearic acid modified soy protein isolate resin and ramie fiber reinforced 'green' composites. *Compos. Sci. Technol.* 65, 1211.

[16] Liu. W., Mohanty, A.K., Askeland, P., Drzal, L.T., Misra, M., (2004). Influence of fiber surface treatment on properties of Indian grass fiber reinforced soy protein based biocomposites. *Polymer* 45, 7589.

[17] Chabba, S., Netravali, A.N., (2005). 'Green' composites Part 1: characterization of flax fabric and glutaraldehyde modified soy protein concentrate composites. *J. Mater. Sci.* 40, 6263.

[18] Huang, X., Netravali, A.N., (2007). Characterization of flax fiber reinforced soy protein resin based green composites modified with nano-clay particles. *Compos. Sci. Technol.* 67, 2005.

[19] Behera, A.K., Avancha, S., Basak, R.K., Sen, R., Adhikari, B., (2012). Fabrication and characterization of jute reinforced soy based green composites. *Carbohydrate polym.* 88, 329.

[20] Avancha, S., Behera, A.K., Sen, R., Adhikari, B., (2013). Physical and mechanical characterization of jute reinforced soy composites. *J. Reinf. Plast. Compos.* doi:10.1177/0731684413485979.

[21]] Arfa, A.B., Belloy, L.P., Chalier, P., Gontard, N., (2007). Antimicrobial paper based on a soy protein isolate or modified starch coating including carvacrol and cinnamaldehyde. *J. Agric. Food Chem.* 55, 2155.

[22] Rhim, J.W., Gennadios, A., Weller, C.L., Cezeirat, C., Hanna, M. A., (1998). Soy protein isolate–dialdehyde starch films. *Ind. Crops Prod.* 8, 195.

[23] Kumar, R., Zhang, L., (2008). Effect of water on the hydrophobicity of soy protein materials containing 2,2-diphenyl 2-hydroxyethanoic acid. *Biomacromolecules* 9, 2430.

[24] Lepoittevin, B., Pantoustier, N., Devalckenaere, M., Alexandre, M., Kubies, D., Calberg, C., Je´ro^me, R., Dubois, P., (2002). Poly(ε-caprolactone)/Clay nanocomposites by in-situ intercalative polymerization catalyzed by dibutyltin dimethoxide. *Macromolecules* 35, 8385.

[25] Morlat-Therias, S., Mailhot, B., Gonzalez, G., Gardette, J., (2005). Photooxidation of Polypropylene/Montmorillonite Nanocomposites. 2. Interactions with Antioxidants *Chem. Mater.* 17, 1072.

[26] Giannelis, E.P., (1996). Polymer Layered Silicate Nanocomposites. *Adv. Mater.* 8, 29.

[27] Viville, P., Lazzaroni, R., Pollet, E., Alexandre, M., Dubois, P., (2004). Controlled polymer grafting of sigle clay nanoplateletes. *J.Am. Chem. Soc.*126, 9007.

[28] Plummer, C.J.G., Garamszegi, L., Leterrier, Y., Rodlert, M., Månson, J.E., (2002). Hyperbranched polymer layered silicate nanocomposites. *Chem. Mater.* 14, 486.

[29] Choi, Y.S., Choi, M.H., Wang, K.H., Kim, S.O., Kim, Y.K., Chung, I.J., (2001). Synthesis of exfoliated PMMA/NA-MMT composites via soap free emulsion polymerization. *Macromolecules* 34, 8978.

[30] Viville, P., Lazzaroni, R., Pollet, E., Alexandre, M., Dubois, P., Borcia, G., Pireaux, J., (2003). Surface characterization of poly(ε caprolactone)-based nanocomposites. *Langmuir* 19, 9425.

[31] Park, J.H., Jana, S.C., (2003). Mechanism of exfoliation of nanoclay particles in epoxy−clay composites. *Macromolecules* 36, 2758.

[32] Mariott, W.R., Chen, E.Y.X., (2003). Stereochemically controlled pmma-exfoliated silicate nanocomposites using intergallery-anchored metallocenium cations. *J. Am. Chem. Soc.* 125, 15726.

[33] Robello, D.R., Yamaguchi, N., Blanton, T., Barnes, C., (2004). Spontaneous formation of an exfoliated polystyrene−clay nanocomposite using a star-shaped polymer. *J. Am. Chem.Soc.* 126, 8118.

[34] Zhang, Z., Zhang, L., Li, Y., Xu, H., (2005). New fabricate of styrene–butadiene rubber/montmorillonite nanocomposites by anionic polymerization. *Polymer* 46, 129.

[35] Ray, S.S., Okamoto, M., (2003). Polymer/layered silicate nanocomposite: A Review from preparation to processing. *Prog. Polym. Sci.* 28, 1539.

[36] Burnside, S.D., Giannelis, E.P., (1995). Synthesis and properties of new poly(dimethylsiloxane) nanocomposites. *Chem. Mater.* 7, 1597.

[37] Strawhecher, K.E., Manias, E., (2000). Structure and properties of poly(vinyl alcohol)/Na^+ montmorillonite nanocomposites. *Chem. Mater.* 12, 2943.

[38] Chou, C., Lin, J., (2005). One-step exfoliation of montmorillonite via phase inversion of amphiphilic copolymer emulsion. *Macromolecules* 38, 230.

[39] Wang, K., Wang, L., Wu, J., Chen, L., He, C., (2005). Preparation of highly exfoliated epoxy/clay nanocomposites by "slurry compounding": process and mechanisms. *Langmuir* 21, 3613.

[40] Moniruzzaman, M., Winey, K. I., (2006). Polymer nanocomposites containing carbon nanotubes. *Macromolecules* 39, 5194.

[41] Coleman, J.N., Khan, U., Gun'ko, Y. K., (2006). Mechanical reinforcement of polymers using carbon nanotubes. *Adv. Mater.* 18, 689.

[42] Saito, R., Dresselhaus, G., Dresselhaus, M.S., (1998). *''Physical Properties of Carbon Nanotubes''*, Imperial College Press, London.

[43] Zheng, H., Tan, Z., Zhan, Y., Huang, J., (2003). Morphology and properties of soy protein plastics modified with chitin. *J. Appl. Polym. Sci.* 90, 3676.

[44] Huang, J., Zhang, L., Chen, P., (2003). Effects of lignin as a filler on properties of soy protein plastics. II. Alkaline lignin. *J. Appl. Polym. Sci.* 88, 3291.

[45] Chen, P., Zhang, L., (2006). Interaction and properties of highly exfoliated soy protein/montmorillonite nanocomposites. *Biomacromolecules* 7, 1700.

[46] Yu, J., Cui, G., Wei, M., Huang, J., (2007). Facile exfoliation of rectorite nanoplatelets in soy protein matrix and reinforced bionanocomposites thereof. *J. Appl. Polym. Sci.* 104, 3367.

[47] Zhang, J., Jiang, L., Zhu, L., Jane, J., Mungara, P., (2006). Morphology and properties of soy protein and polylactide blends. *Biomacromolecules* 7, 1551.

[48] Zhang, J., Mungara, P., Jane, J., (2001). Mechanical and thermal properties of extruded soy protein sheets. *Polymer* 42, 2569.

[49] Huang, J., Zhang, L., Chen, F., (2003). Effects of lignin as a filler on properties of soy protein plastics. I. Lignosulfonate *J. Appl. Polym. Sci.* 88, 3284.

[50] Wei, M., Fan, L., Huang, J., Chen, Y., (2006). Role of star-like hydroxylpropyl lignin in soy-protein plastics. *Macromol. Mater. Eng.* 291, 524.

[51] Angellier, H., Molina-Boisseau, S., Dole, P., Dufresne, A., (2006). Thermoplastic starch−waxy maize starch nanocrystals nanocomposites. *Biomacromolecules* 7, 531.

[52] Lu, Y., Weng, L., Zhang, L., (2004). Morphology and properties of soy protein isolate thermoplastics reinforced with chitin whiskers. *Biomacromolecules* 5, 1046.

[53] Wang, Y., Cao, X., Zhang, L., (2006). Effects of cellulose whiskers on properties of soy protein thermoplastics.*Macromol. Biosci.* 6, 524.

[54] Tjong, S.C., Meng, Y.Z., (1999). Mechanical and thermal properties of polycarbonate composites reinforced with potassium titanate whiskers. *J. Appl. Polym. Sci.* 72, 501.

[55] Chazeau, L., Cavaill, J.Y., Perez, J., (2000). Plasticized PVC reinforced with cellulose whiskers. II. Plastic behavior. *J. Polym. Sci. Part B: Polym. Phys.* 38, 383.

[56] Nair, K.G., Dufresne, A., (2003). Crab shell chitin whisker reinforced natural rubber nanocomposites. 1. processing and swelling behavior. *Biomacromolecules* 4, 657.

[57] Nair, K.G., Dufresne, A., (2003). Crab shell chitin whisker reinforced natural rubber nanocomposites. 2. Mechanical behavior. *Biomacromolecules* 4, 666.

[58] Malkapuram, R., Kumar, V., Negi, Y.S., (2009). Recent Development in Natural Fiber Reinforced Polypropylene Composites. *J. Reinf. Plast. Compos.* 28, 1169.

[59] Wang, L., Kumar, R., Zhang, L., (2009). Investigation into hemp fibers and whiskers reinforced soy protein composites. *Front. Chem. China* 4, 313.

[60] Zheng, H., Ai, F., Wei, M., Huang, J., Chang, P. R., (2007). Thermoplastic soy protein nanocomposites reinforced by carbon nanotubes. *Macromol. Mater. Eng.* 292, 780.

[61] Ai, F., Zheng, H., Wei, M., Huang, J., Chang, P. R., (2007). Soy protein plastics reinforced and toughened by SiO_2 nanoparticles. *J. Appl. Polym. Sci.* 105, 1597.

[62] Sun, Q., Li, X., Wang, P., Du, Y., Han, D., Wang, F., Liu, X., Li, P., Fu, H., (2011). Characterization and evaluation of the Ag^+-loaded soy protein isolate-based bactericidal film-forming dispersion and films. *J. Food Sc.* 76, E438.

[63] Zhao, S., Yao, J., Fei, X., Shao, Z., Chen, X., (2013). An antimicrobial film by embedding *in situ* synthesized silver nanoparticles in soy protein isolate. *Materials Letters* 95, 142.

[64] Zhang, Y., Lee, M.W., An, S., Ray, S.S., Khansari, S., Joshi, B., Hong, S., Hong, J-H., Kim, J-J., Pourdeyhim, B., Yoon, S.S., Yarin, A.L., (2013). Antibacterial activity of photocatalytic electrospun titania nanofiber mats and solution-blown soy protein nanofiber mats decorated with silver nanoparticles. *Catalysis Communications* 34, 35.

[65] Kumar, R., Zhang, L., (2009). Aligned ramie fiber reinforced arylated soy protein composites with improved properties. *Compos. Sci. Techn.* 69, 555.

[66] Kumar, R., Zhang, L., (2010). Investigation into ramie whiskers reinforced arylated soy protein composites. *Frontiers of Chemistry in China* 5, 104.

[67] Kumar, R., Anandjiwala, R. D., (2012). Flax fabric reinforced arylated soy protein composites: a brittle-matrix behaviour. *J. Appl. Polym. Sci.* 124, 3132.

[68] Evans, A.G., Zok, F.W., (1994). Review: the physics and mechanics of fiberreinforced brittle-matrix composites. *J. Mater. Sci.* 29, 3857.

[69] Marshall, D.B., Cox, B.N., Evans, A.G., (1985). The mechanics of matrix cracking in brittle-matrix fiber composites. *Acta Met.* 33, 2013.

[70] Cao, H.C., Bischoff, E., Sbaizero, O., Ru¨ hle, M., Evans, A.G., Marshall, D.B., (1990). Effect of interfaces on the properties of fiber-reinforced ceramics. *J. Am. Ceram. Soc.* 73, 1691.

[71] Teng, Z., Luo, Y., Wang, Q., (2012). Nanoparticles synthesized from soy protein: preparation, characterization, and application for nutraceutical encapsulation. *J. Agric. Food Chem.* 60, 2712.

[72] Sasikala, D., Govindaraju, K., Tamilselvan, S., Singaravelu, G., (2012). Soybean protein: A natural source for the production of green silver nanoparticles. *Biotechnology and Bioprocess Engineering* 17, 1176.

[73] Liu, R., Liu, S., Zhou, H., Yang, C., Liu, X., (2012). Completely green synthesis of Ag nanoparticles stabilized by soy protein isolate under UV irradiation. *Journal of Wuhan University of Technology-Mater. Sci. Ed.* 27, 852.

[74] Zhang, J., Liang, L., Tian, Z., Chen, L., Subirade, M., (2012). Preparation and *in vitro* evaluation of calcium-induced soy protein isolate nanoparticles and their formation mechanism study. *Food Chemistry* 133, 390.

[75] Jong, L., (2013). Characterization of soy protein nanoparticles prepared by high shear microfluidization. *Journal of Dispersion Science and Technology* 34, 469.

[76] Liu, F., Tang, C-H., (2013). Soy protein nanoparticle aggregates as pickering stabilizers for oil-in-water emulsions. *J. Agric. Food Chem.* 61, 8888.

In: Polymer-Matrix Composites
Editor: Rakesh Kumar

ISBN: 978-1-61122-005-6

Chapter 5

PERFORMANCES OF COMPOSITES BASED ON ALFA FIBER

Dalila Hammiche and Amar Boukerrou
Laboratoire des Matériaux Polymères Avancés, Département de Génie des Procédés, Université Abderrahmane MIRA, Route de Targa-Ouzemmour, Bejaia, Algérie

ABSTRACT

Composite materials are made with a polymer matrix reinforced by high-strength fibers, such as glass and carbon. The combination of natural fibers of both non renewal (petroleum based) and renewal resources used to produce polymer composites that are competitive with synthetic composites is gaining attention over the last decade. However, for the sake of sustainable development and environmental safety, there is renewed interest in natural materials and remarkable increase in introduction of new materials and products. Natural fiber use as reinforcement materials for polymeric composites has led to the development of the next generations of materials, products and processes. A great number of natural fibers such as bast fibers (Flax, Jute, Hemp, and Kenaf), leaf fibers (Sisal, Pineapple, and Henequen), seed fibers (Cotton), cereal fibers (Rice Straw, Corn, and Wheat), fruit fiber (Coconut), and Grass fibers (Stalk, Bagasse, Bamboo and Alfa) are being used for this intention. The Alfa fiber "Stippa Tenacissima" is built from strong, stiff and light cellulosic materials. The incorporation of Alfa fiber into composites improves the properties of the composites.

This chapter addresses the relevant background of Alfa fiber. The lack of an official set of specifications concerning the characterization of Alfa fiber has motive us to conduct the present work. To the best of our knowledge, this is the first attempt to assess the properties and applications of Alfa fiber. To this aim, chemical constituents, morphological, mechanical and thermal properties were reviewed. In addition, the preparation and some performances of composites based on Alfa fiber are discussed. We also report that the incompatibility between Alfa fiber and polymer matrix can be solved by chemical treatments of the surface fiber or by the use of compatibilizer. Other characteristics will be exposed as well as the extraction of nanowhiskers from this fiber and its use as nanoreinforcement for polymeric nanocomposites.

Keywords: Composites, Alfa fiber, Mechanical properties

1. Introduction

Farouk et al., [1], Summerscales et al., [2], Loh et al., [3], Akil et al., [4] have reviewed the characteristics of natural fiber. Bledzki et al., [5], Rezaur et al., [6], Velasica et al., [7], and Zainudin et al., [8] reviewed the use of natural fiber as reinforcements for polymer materials. The use of natural fibers to reinforce polymer is a well-established practice.

The development of high-performance materials made from natural resources is increasing worldwide. The greatest challenge in working with natural fiber reinforced plastic composites is their large variation in properties and characteristics.

The combination of natural fibers with polymer matrices from both non renewal (petroleum based) and renewal resources used to produce polymer composites that are competitive with synthetic composites are gaining attention over the last decade [9].

The exploitation of natural fibers as reinforcement for plastics, replacing glass fibers, and other synthetic materials occurs not only due to environmental concerns but also for providing a unique combination of high performance, great versatility, and processing advantages at favorable cost [10]. Natural fibers are abundant, renewable, cheap, recyclable and biodegradable. They exhibit low density and the extent of environmental pollution caused is less compared to synthetic fibers. They are naturally designed to be incredibly tough [11].

The properties of composites containing fibers are dependent on a number of parameters, which include the properties of the matrix, the size and aspect ratio of the fibers, dispersion of the fibers and the interface. The problem associated with the use of natural fibers in composites is the thermal stability limitation of fiber and the difficulties in obtaining strong interfacial adhesion [12, 13]. These issues can be addressed; surface of fibers must be treated with coupling or compatibilizer agents to improve the interface between the fiber and the matrix. However, the use of both matrix resin and fiber surface treatment has received considerable attention due to their effectiveness in modifying the interface by forming a link between the components [14, 15]. Another parameter has to be considered, the properties of composites also depends on the compounding process, because an effective mixing is crucial for achieving an optimal dispersion of fiber and for optimization of composite properties [16].

This chapter addresses the relevant background of Alfa fiber. The lack of an official set of specifications concerning the characterization of Alfa fiber has motive us to conduct the present work. To the best of our knowledge, this is the first attempt to assess the properties and applications of Alfa fiber. To this aim, chemical constituents, morphological, mechanical and thermal properties were reviewed. In addition, the preparation and some performances of composites based on Alfa fiber are discussed. We also report that the incompatibility between Alfa fiber and polymer matrix can be solved by chemical treatments of the surface fiber or by the use of compatibilizer. Other characteristics will be exposed as well as the extraction of nanowhiskers from this fiber and its use as nanoreinforcement for polymeric nanocomposites.

2. Alfa Fiber

Natural fibers of vegetable origin are referred to as "plant fibers", "vegetable fibers", or "lignocellulosic fibers." Natural fibers derived from plants are grouped based on the origin of the fibers as follows: Bast fibers (Flax, Jute, Hemp, and Kenaf), Leaf fibers (Sisal, Pineapple, and Henequen), Seed fibers (Cotton), Cereal fibers (Rice straw, Corn, and Wheat), Fruit fiber (Coconut), and Grass fibers (Stalk, Bagasse, Bamboo and Alfa).

The Alfa fiber "esparto grass or Stippa Tenacissima plant" is a hardy perennial grass of the family of the grasses onesis. It is widely cultivated in the dry region of North Africa (hot Mediterranean region) and southern Spain (Table 1) [17].

Table 1. The territorial distribution of the Alfa fiber

Country	Territory (ha)
Algeria	4.000.000
Morocco	3.186.000
Tunisia	600.000
Libya	350.000
Spain	300.000
Portugal	little

In Algeria, it is in the western south steppes of the country that one meets the vast and greatest esparto expanses. Alfa is mostly used in the production of high quality papers. It provides better sheet formation and good bulk. In Algeria, three mills (Celpap Company) are producing bleached pulp using Alfa grass fiber as raw material. It doesn't need insecticides nor pesticides and only a small amount of water and it needs to be grown on light clay grounds [18]. Because of those advantages the Esparto grass culture is thus much ecologically friendly. It is constituted of stems with a cylindrical shape (Figure 1) which have a maximum height of about 1 to 2 m. The plant has deep roots with many secondary roots. The Alfa stem is built of strong, stiff and light cellulosic materials [19].

Figure 1. Representation of Alfa fiber.

2.1. Morphological Studies of Alfa Fiber

Scanning electron micrographs of longitudinal and transversal cross-section of Alfa fibers are illustrated in Figure 2 [20]. The fiber consists of a number of cells associated with filaments called cellulose microfibrils. These cells are arranged in a matrix consisting of lignin [20, 21]. According to Hammiche et al., and Bessadok et al., the vessels are wrapped in a bundle of fibers. The width of the vessel could reach 200 μm. The diameter of the fibers varies from 5 to 10 μm [20, 22].

There are also exists a central hole called lacuna in each fiber to induct water and nutrients. Each cell in turn has a different size and shape arranged in some fashion with a central hole called lumen.

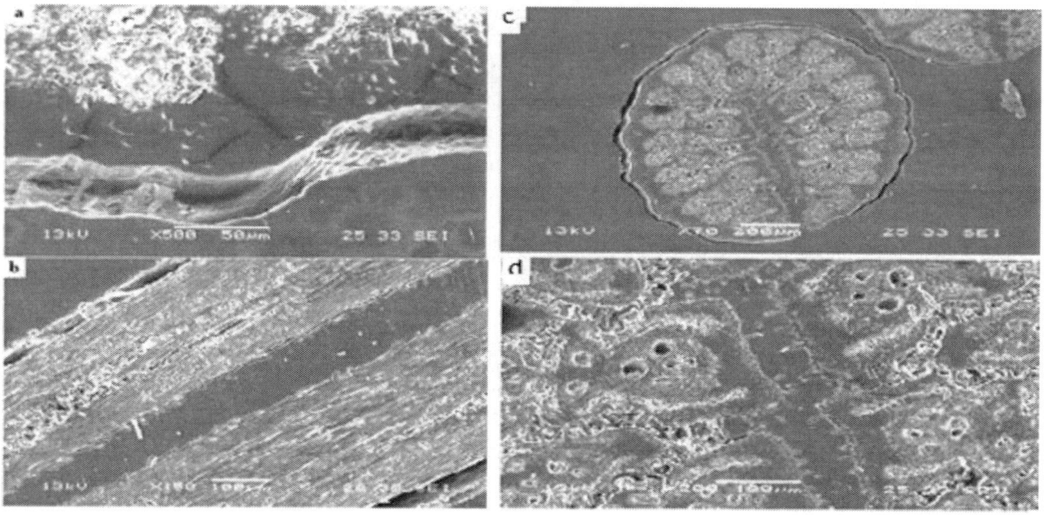

Figure 2. Scanning electron micrographs showing the longitudinal view (a, b) and cross-section of (c, d) Alfa fibers.

2.2. Chemical Composition of Alfa Fiber

The constituents and chemical composition of Alfa fibers are given in Table 2.

Alfa fiber is composed of cellulose, hemicelluloses, lignin and waxes. Cellulose is the main component of Alfa fiber. The polysaccharides content of Alfa fibers is similar to the one obtained for the rachis of leaves of date palm tree [25-27]. However the lignin content is quite higher. It is also high compared to other annual plants such as jute or kenaf for which the lignin content is lower than 15 wt%. The lignin and cellulose rates are very nearby to those of hardwood, for example the eucalyptus [28].

Table 2. Chemical composition of Alfa fiber

Content of constituent (wt.%)			References
Cellulose	hemicellulose	lignin	
45	20.9	28.9 4	[20]
46	24	20	[22]
33-38	27-32	17-19	[23]
48	-	-	[24]

Physical properties of natural fibers are basically influenced by the chemical structure such as cellulose content, degree of polymerization, orientation and crystallinity, which are affected by conditions during growth of plants as well as extraction methods used. Indeed, for natural fibers, flammability is in part due to differences in chemical composition. Higher cellulose content results in higher flammability while higher lignin content results in greater char formation [29-30]. Percentage of cellulose in a fiber affects the ultimate fiber's properties. On the other hand, the structure and chemical composition of fiber are strongly influenced by area of growth, its climate and the age of the plant.

2.3. Thermal and Mechanical Properties of Alfa Fiber

It is important to identify the type of information required in order to choose the suitable test methods. Thermogravimetric analysis (TGA) is widely used by most researchers to study on the thermal degradation of natural fiber composites. Early decomposition, observed by onset of lower degradation temperature, implies less thermal stability and this can be used to compare thermal performance of various composites.

TGA and derivative thermogravimetric (DTG) curves are used to determine the weight loss and to identify the decomposition of material at a certain temperature, respectively. A typical thermogravimetric graph for the decomposition process of Alfa fiber is shown in Figure 3. According to Sen et al [31], the degradation process of natural fibers includes dehydration combined with emission of volatile components initiating at a temperature of about 260 °C, and rapid weight loss due to oxidative decomposition corresponding to the formation of char as the temperature is increased. Early decomposition shows less thermal stability [32]. Three main stages of weight loss of natural fibers during fire can be observed.

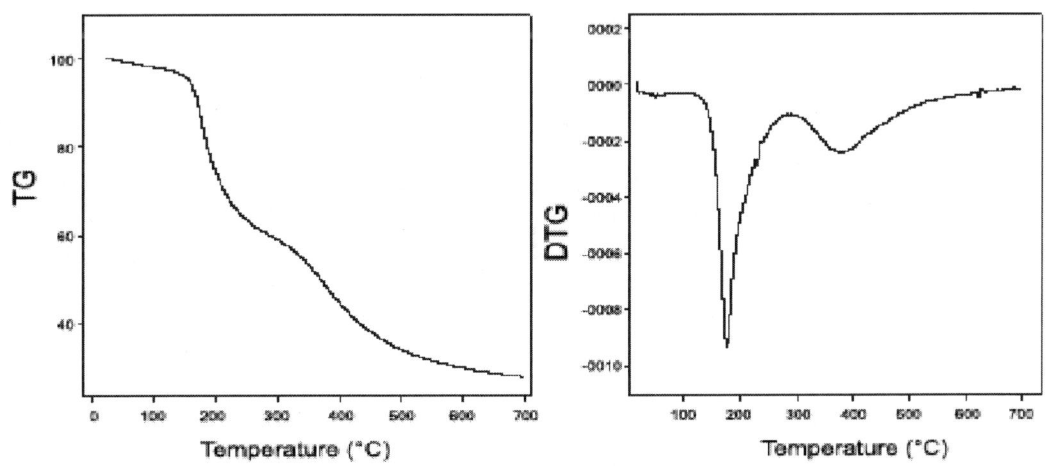

Figure 3. Thermogravimetric decomposition process of Alfa fiber.

The first stage of weight loss around 100 °C is attributed to the evaporation of moisture or to water loss in the Alfa fiber. Approximately 60% of the thermal decomposition of Alfa fiber occurred within a temperature range between 240 and 320 °C due to decomposition of hemicelluloses [20, 33]. Another stage of degradation, around 320-440 °C, corresponds to the thermal degradation of cellulose. Near 440 °C, weight loss was attributed to lignin decomposition. Lignin peak is wider and appears superimposed on the other two peaks.

The mechanical properties of Alfa fiber were also investigated and are presented in Table 3 as highlighted by a few researchers.

Table 3. Characteristics values for mechanical properties of Alfa fiber

Characteristics			References
Tensile strength (MPa)	Young's modulus (GPa)	Elongation at break (%)	
145	21	1,8	[20]
-	22	5.8	[22]
147	22	2	[34]

The mean values of tensile modulus (E), strength (σ) and strain(ε) at the breaking point of the tested fibers samples were approximately 22 GPa, 145 MPa and 2% respectively. According to literature [35-38], the mechanical performances of esparto grass were close to those of sisal, jute and bamboo. From the available properties, it is evident that the Alfa fiber is one among the future alternative of any natural fibers.

3. Composites Based on Alfa Fiber

Compared with conventional and mineral filler reinforced thermoplastic products, composites based on natural fiber have many advantages such as high specific strength and modulus, low cost, low density, and low friction during compounding.

The Alfa fibers have the potential as a reinforcing fiber in polymeric composites, because of its superior toughness and high aspect ratio in comparison to other fibers.

From mechanical point of view, natural fibers may enhance mechanical properties of polymers with some considerations and improvement to the surface characteristics natural fiber. There are several factors related to the natural fibers which influence the performance of the composites such as the interfacial adhesion, the orientation, the strength, physical properties, etc. The mechanical efficiency of the fiber-reinforced polymer composites depends on the fiber-matrix interface and the ability to transfer stress from the matrix to fiber as reported by many researchers, [39- 50].

Boukerrou et al., [51] investigated the changes in mechanical properties and the water absorption capacity of poly (vinyl chloride)/Alfa fiber composites as a function of Alfa fiber content (10, 20 and 30 wt %) and its particle size. The study showed that composites samples exhibited higher tensile modulus compared with the virgin PVC, whereas elongation at break and tensile strength were observed to decline. Moreover, the amount of absorbed water depends on the amount of fiber in the composite. The comparison of the results obtained from the samples with different sizes 160, 200, 250, and 315 mm called F160, F200, F250 and F315 and the virgin PVC indicated that the presence of small particle sizes exhibited a beneficial effect on the water absorption by improving the quality of adhesion between polymer and fiber.

Belhassen et al., [52] studied the potential of Alfa fiber as a reinforcing fiber in a starch-based biopolymer matrix. The fibers and the matrix were first mixed in the melted state under mechanical shearing using a plastograph and the obtained composites were molded by injection process. The tensile mechanical analysis showed a linear increase of the composite tensile modulus upon increasing the fiber content, together with a sharp decrease of the elongation at break. The fibers incorporation into the biopolymer matrix brings about an enhancement in the mechanical strength and the impact strength of the composite. Dynamic mechanical thermal analysis (DMTA) investigation showed two relaxations occurring at about -30 and 35°C. The addition of Alfa fibers enhanced the storage modulus E' before and after T_{α}, which is consistent with the reinforcing effect of Alfa cellulose fibers. The moisture uptake of the composite showed only a modest reduction in the fiber incorporation evolving from 2.62% for the pristine matrix to 2.37% for composite with 30 wt% fiber content. The relatively low water uptake by both the matrix and the composite was ascribed to the presence of the polymer additive imparting a hydrophobic character that contributes to decrease in the water susceptibility of the starch-based matrix.

Several studies have been carried out using Alfa fiber in different polymeric matrix, it is reported that the main disadvantages are the poor compatibility between fiber and matrix. This situation calls for the development of strategies for the surface modification of fiber surfaces, thereby an effective control over the fiber/polymer interface [53]. Indeed, the main approaches to enhancement of the interaction between the fiber and the matrix are surface modification of the fiber, application of coupling agents to the fiber surface and/or use of compatibilizers in the matrix.

3.1. Effect of Surface Treatment

Fiber-matrix interfacial adhesion can be improved with many chemical modifications of the fiber. Chemical treatment of fibers will activate hydroxyls groups or introduce new moieties that can effectively interlock with the matrix.

Generally, chemical coupling agents are molecules possessing two functions. The first function is to react with hydroxyl groups of cellulose and the second is to react with functional groups of the matrix [54].

Several works have been attempted to study the influence of the type and concentration of chemical solution on the Alfa fiber characteristics and their interfacial adhesion with various matrices.

Bessadok et al., [22] carried out various treatments on Alfa fiber. These different treatments involve acetylation (Ac), with the help of chemicals such as styrene (S), acrylic acid (AA) and maleic anhydride (MA) (Figure 4). As expected, it was found that the all chemical treatments reduced the overall water uptake of Alfa fiber. In particular styrene treatment allows increasing significantly moisture resistance of this fiber.

One of the familiar and effective modifications applied to Alfa fiber, is an alkaline treatment based on sodium hydroxide (NaOH). For example, Boukerrou et al., [55] investigated the effect of the incorporation of virgin and treated fiber in the poly (vinyl chloride) matrix (PVC) on mechanical, thermal stability, viscoelastic properties and morphology of composites. The results indicated that the chemical treatment of Alfa fiber with NaOH improved the thermal stability of Alfa fiber. The mechanical properties of treated composites were improved; except Young's modulus. Also the sodium hydroxide improves the interactions between Alfa fiber and matrix and promoted better dispersion of the fiber in the PVC matrix. The thermal characterization of PVC filled with 15%w/w of Alfa revealed a decrease in the onset temperatures of decomposition for the treated composites.

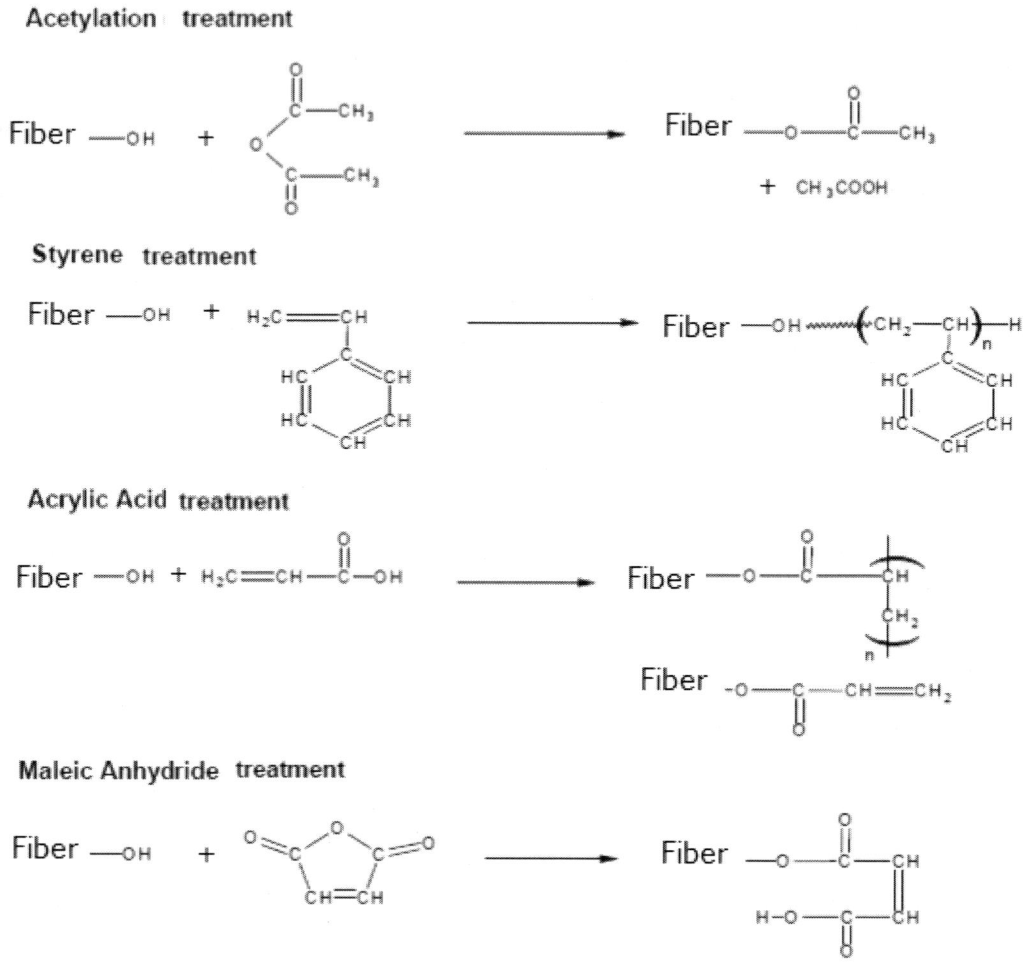

Figure 4. Scheme of the chemical treatments applied on Alfa fiber: maleic anhydride, styrene, acrylic acid and acetic anhydride.

Other method was discussed as silane treatment. Silanes are coupling agents used to improve the adhesion between fiber and matrix. In silane treatment, one end group attached to silicon reacts with hydrophilic fiber and the other with matrix. Thus, it improves the degree of cross-linking in the interface region and hence leads to perfect bonding between matrix and fiber. The hydrocarbon chains in silane groups can influence the wettability of the fibers and thus chemical affinity toward the non polar polymer matrices [56].

With regards to the silane treatment technique and conditions, Ghallabi et al., [57] prepared polymer composites of a polyester resin (UP) matrix with randomly dispersed Alfa fibers; their dielectric spectra were measured in the frequency range 0.1 Hz–0.1 MHz and temperature interval from ambient to 180 °C. Cellulose fibers were incorporated into the matrix, as such or after surface modification involving three silane coupling agents, namely c-methacryloxypropyltrimethoxy (MPS), hexadecyltriméthoxy-silanes (HDS), c-mercaptoproyl trimethoxy-silane (MRPS). The presence of celluloses fibers in the composite give rise to other relaxation associated to the MWS relaxation. The treatments of the fibers with reactive silane coupling agent likely to react with the matrix leads to a decrease of the magnitude of water relaxation and increase the activation energy of the α relaxation. The second relaxation observed around 180 °C, known as the Maxwell-Wagner-Sillars (MWS) relaxation is shown to be sensitive to the treatment with reactive silane coupling agent especially with MRPS.

Maleic Anhydride treatment was also the most widely used and efficient way for coupling fibers with polymer matrix. Treatment of natural fiber with anhydride falls in the category of esterification of the hydroxyl groups present on fiber into an ester group.

Bessadok et al., [58] investigated the grafting of maleic anhydride (MA) onto the surfaces of Alfa fiber. Composite materials were prepared using unsaturated polyester resins reinforced by Alfa fibers. The fibers were previously modified by chemical treatments with MA. The preliminary S or MA treatment of the fibers allowed a decrease of the water permeability of the composite. A slight increase of the stiffness of the composite was observed due to the presence of the fibers (treated or untreated).

Other chemical treatments were also carried out on surface of Alfa fiber. Cyanoethylation of long Alfa fibers was studied by Ben Abderrahmane and Ben Cheikh [59], and these chemically modified fibers were used to manufacture Alfa-polyester composites. It was found that there are large differences in modulus values between the modified and the unmodified Alfa fiber; the modified.

Alfa fibers are markedly less stiff than the unmodified ones. It appears that the fragility of the used fibers has been compensated by the greater interfacial bond strength between the matrix resin and the fiber. The hydrophilic nature of the Alfa induces the poor wettability and adhesion characteristics with unsaturated polyester resin, and the presence of moisture at the Alfa-resin interface promotes the formation of voids at the interface. Thus, the presence of moisture and voids at the interface weakens the bonds and produces a composite of lower stiffness and strength. The presence of moisture in the system may result from diffusion of atmospheric moisture through the matrix on the subsequent aging. On the other hand, owing to cyanoethylation, the moisture regain capacity of the Alfa fiber is much reduced; moreover, the compatibility with unsaturated polyester resin is improved and a strong interfacial bond with the matrix resin is produced. It was observed that the scanning electron micrographs of the fracture surface of the samples reveal the improved bonding at the interface between the cyanoethylated fiber and the polyester resin.

As it can be perceived, chemical treatments increase the interface adhesion between the fiber and matrix, and decrease the water absorption of fibers. Therefore, chemical treatments can be considered in modifying the properties of Alfa fibers. Some compounds are known to promote adhesion by chemically coupling the adhesive to the material. Fiber modification methods discussed previously have different efficacy in causing adhesion between the matrix and the fiber. Yet most chemical treatments have achieved various levels of success in improving fiber strength, fiber fitness and fiber-matrix adhesion in natural fiber reinforced composites.

Other smooth chemical modification techniques have to be explored in future to improve or at least maintain the mechanical characteristics of the fibers.

3.2. Effect of Compatibilizers

Other technique to improve surface compatibilization is based on using surfactant type of structure like maleated polymers. Maleated coupling agents are widely used to strengthen composites containing fillers and fiber reinforcements. The difference with other chemical treatments is that maleic anhydride is not only used to modify fiber surface but also the matrix to achieve better interfacial bonding and mechanical properties in composites [60-62]. Compatibilizers are mostly functionalized polymers, which are incorporated to lower the interfacial tension and to improve the interfacial interactions between the polymer and the fiber.

There are two ways by which the maleic anhydride compatibilizer chemistry can be applied during composite fabrications: They are discussed below

(1) Natural/fibers are pretreated with maleated polymer, and such treated fibers are then reinforced with the desired polymer matrix to obtain composites.
(2) During extrusion processing chopped fiber, polymer matrix, and maleic anhydride can be added with peroxide initiator in one-step processing to get the comptabilized composite product for further compression molding/injection molding.

Several compatibilizers for improving adhesion in natural fiber/polymer composites have been described in the literature, but in the case of the composites based on Alfa fiber, only one author used the compatibilizer to improve the properties of the interfaces of these composites.

Hammiche et al., [63] studied the suitability of using PVC modified with maleic anhydride (PVC-g-MA) as new compatibilizer for poly (vinyl chloride)/Alfa composites. The main objective of this research was to synthesize a new compatibilizer (PVC-g-MA), which was grafted from the maleic anhydride on the PVC chains in order to solve the problem of the incompatibility between the hydrophobic polymeric matrix (PVC) and hydrophilic fiber (Alfa). Composites samples were prepared with different Alfa fiber loading (10, 20, and 30 wt %) and incorporating PVC-g-MA (1, 3, and 5 wt %). Based on the result of this study, mechanical properties such as tensile strength and hardness of treated composites are found to be much higher than those of the untreated ones. This result was evident from SEM study. The whole of the data collected concerning the thermal properties show clearly that the temperature of degradation of the polymer matrix increased about 16 °C in comparison to the noncoupled composite, indicating that PVC-g-MA improved the thermal stability of the polymer. The results obtained make it possible to conclude that the introduction of a compatibilizer seems crucial. However, for the sight of the results, it should be noted that the study showed that the best adhesion with the interface was obtained with PVC-g-MA; this work confirms that the choice of the compatibilizer must be made according to the matrix and to the application aimed for composite material.

It must be noted that the inherent properties of maleated coupling agents as described below significantly influence their effectiveness as a compatibilizer. For example, the chemical structure and molecular weight affect the miscibility and entanglement with the base resin and the degree of grafting determines the level of functionality present in the compatibilizer. Thus choice of a suitable compatibilizer is an important issue in developing polymer composites. The degree of grafting effect was investigated by Hammiche et al., [64], this work studied the poly (vinyl chloride) (PVC) chemically modified with maleic anhydride (MA) through reactions in solution, using benzoyl peroxide as an initiator. Quantities of the grafted MA were determined by the titration of carboxylic acid groups derived from the anhydride functionsionalty. Increases in reaction time led to higher levels of grafted MA. The effects of three different PVCs grafted with maleic anhydride (PVC-g-MAs) on the morphological, mechanical and thermal properties of PVC/Alfa (fiber) composites were examined. Enhancements of the properties of the PVC-g-MA treated composites were found to be strongly dependent on the amount of MA grafting on PVC. The effect of the concentration of three distinct coupling agents was simulated with three PVC-g-MA samples, which mutually differed in the MA content bound to PVC chains. From the results of tensile strength variation, the best coupling agent for this system was PVC-g-MA_3. This coupling agent promoted an increase of 61% in tensile strength, compared with that of the untreated composite. The coupling agent PVC-g-MA_3 had more anhydride groups in its structure than the other coupling agents tested and therefore was more efficient in improving the interactions between the matrix and the Alfa fiber. It was observed that the weight loss of PVC during degradation was slightly higher than those of treated composites. This result indicated that the thermal stability of the PVC-g-MA- treated composites was higher than that of unfilled PVC. The best results were achieved with the sample PVC-g-MA_3, which contained a higher quantity of anhydride groups.

4. Nanocomposites Based on Cellulose Nanowhiskers Extracted from Alfa Fiber

The use of renewable materials for industrial applications is becoming impellent due to the increasing demand of alternatives to scarce and unrenewable petroleum supplies. In this regard, nanocrystalline cellulose (NCC) derived from cellulose, the most abundant biopolymer, is one of the most promising materials [65].

In essence, the principle reason to utilize cellulose nanofibers in composite materials is because one can potentially exploit the high stiffness of the cellulose crystal for reinforcement. This can be done by breaking down the hierarchical structure of the plant into individualized nanofibers of high crystallinity, therefore reducing the amount of amorphous material present. Several processes have been used to extract highly purified nanofibers from cellulosic materials. All these methods lead to different types of nanofibrillar materials, depending on the cellulose raw material and its pretreatment, and more importantly, depending on the disintegration process itself [66].

There are basically two families of nanosized cellulosic particles. The first one consists of cellulose nanocrystals and the second one is microfibrillated cellulose (MFC) [67, 68]. The nanocellulose, also known as nanocrystalline cellulose or cellulose crystallites are produced from wood (nanofribrillated cellulose-NFC), from bacteria (bacterial nanocellulose-BNC), and from high crystalline cellulose sources using acids or cellulases (biological enzymes) as catalysts for hydrolysis (nanocrystalline cellulose-NCC). This nanocellulose received increasing attention due to their extraordinary mechanical properties such as high Young's modulus and tensile strength.

The terminology used to describe cellulose nanoparticles extracted from grass fiber by hydrolysis process is named cellulose nanowhiskers [68].

An important field of cellulose application is in composite materials as reinforcement of engineering polymer systems. Alfa fiber, given its high cellulose content, around 45%, application as filler in thermoplastic matrix based composites can be considered [51, 10, 63, 64, 69].

The research conducted by Maghchiche et al., [33] had the objective to determine the effects of different low concentration (< 1%) polyacrylamide, Alfa fiber and its mixtures with the polymer at water retention in arid and semi-arid soil. The cellulose was extracted from Alfa plant and was used to prepare polymer blend for soil treatment.

The results showed that polymer blend in soil could improve better soil physical proprieties decreased evaporation and increase water retention in arid soils compared with application of any other blend at the same concentration. The use of polyacrylamide-cellulose blend appears to promise for reduction the labor cost of irrigation at arid and semi-arid soils, and offers safe and environmentally friendly inexpensive materials. The importance of polyacrylamide-cellulose blends is to alleviate poor physical properties and retain water in these arid regions to sustain plant growth.

It is well known that cellulose is a linear polysaccharide macromolecule aggregated in the form of thin microfibrils with typical cross sections around 3.5-10 nm consisting of elementary crystallites that still aggregate in the microfibrils which may be viewed as a string of cellulose crystals linked along the microfibril axis by disordered amorphous domains. Partial release of these microfibrils can be obtained by submitting aqueous dispersions of lignocellulosic fibers to high shear forces in a conventional homogenizer in which the suspension is passed through a thin slit. The ensuing material is usually called microfibrillated cellulose (MFC). Cellulose nanocrystals (CNC) or whiskers can be prepared by dissolving the amorphous or less ordered regions of the microfibrils by acid degradation.

The extraction of crystalline cellulosic regions, in the form of nanowhiskers, is a simple process based on acid hydrolysis. The raw fibers are first milled and then submitted to alkali and bleaching treatments with $NaClO_2$. These steps allow elimination of lignin and hemicelluloses, while leaving cellulose moieties intact if optimal conditions are respected. The bleached fibers are then ready to be hydrolyzed (acid hydrolysis treatment). However, it has been reported that cellulose whiskers prepared by sulphuric acid hydrolysis are more stable than those prepared using hydrochloric acid [70].

In the work conducted by Ben Mabrouk et al., [21], another approach toward the valorization of Alfa fibers, consisting of the extraction of nanosized cellulosic particles is presented. The reinforcing potential of the ensuing nanoparticles was investigated by casting a mixture of acrylic latex and aqueous dispersion of cellulose nanoparticles. The transmission electron microscopy (TEM) results reveal a needle-like structure with a tendency to partial agglomeration by alignment of the whiskers along their main axis as the water is going off. The average length of Alfa whiskers was estimated to be around 200 ± 20 nm.

To investigate the reinforcement efficiency of the ensuing nanoparticles, nanocomposite films were prepared by casting mixture of an aqueous suspension of CNC and commercial latex of poly (styrene-co-butyl acrylate). Films were obtained by casting/evaporation. The thermo-mechanical properties of the ensuing nanocomposite films have been investigated using DMA. A significant increase of the stiffness of the acrylic polymer beyond the glass transition temperature was shown.

Contrary to traditional composites, polymer nanocomposites generally involve low content of well-dispersed nanofillers. The main reason is that it is not necessary to fill the polymer with high quantities of filler to achieve high mechanical properties. In other words, the nanoparticles are generally considered as excellent opportunities for the development of high-performance multifunctional composites.

The recent resurgence of interest in nanocomposites is related to several reasons. The first one is the significant industrial impact related to the possibility to design and create new materials and structures with unprecedented flexibility and physical properties. Secondly, nanoscale fillers are almost free of defects and their application in the composite area opens a window of opportunity to overcome the limitations of traditional micrometer scale. Finally, due to the high specific surface area, nanocomposites present a large volume of interfacial matrix material (interphase) with properties different from those of the bulk polymer. A uniform dispersion of nanoparticles leads to a very large matrix-filler interfacial area, changing the molecular mobility, relaxation behavior and ensuing thermal and mechanical properties [71, 72].

Cellulose nanofibers are environmentally benign and have unique features compared to macro scales. Nowadays, many efforts have been made in developing potential composite products with improved properties. However, the manufacturing of nanocomposites with natural fibers has been a challenging task due to the difficulty encountered in order to aggregate the nanofibers. Indeed, one drawback related to the use of cellulose whiskers for polymer nanocomposites is their inherent difficulty to disperse in non-polar medium, because of their polar surface [73]. In other words, the incorporation of cellulose nanocrystals as a reinforcement material has so far been mainly limited to aqueous or polar environments.

To get a more accurate idea about the dispersion of nanocellulose, it is worth noting the work carried out by Hammiche et al., [74]. The aim of this work was to isolate cellulose nanowhiskers from the Alfa fiber, using chemical treatment. They compared properties of nanocomposites based on cellulose nanowhiskers (CNW) which were dispersed with two different methods. They tried the dispersion of CNW in clay (Cloisite 30B) (1) and in polyethylene glycol (PEG) (2). The morphological, mechanical, thermal and dynamical properties were carried out.

The size distribution of CNW was measured using a Zeta-sizer method. From observing the results shown in Figure 5, the size range of samples was 30-170 nm. More than 20% of CNW had a diameter lower than 30 nm, 15% of CNW was particles of diameter size distribution from 70 to 120 nm, and however, 60% of CNW was obtained with a diameter between 40-60 nm.

It has been found that the dispersion method and the organization of the reinforcements play a key role in the nanocomposites properties. The results showed higher organization with the Cloisite 30B/CNW system. This system also showed higher enhancements. The dispersion in clay was advantageous for improved Young's modulus of nanocomposites (Figure 6). However, the presence of PEG leads to the decrease in this property due to the effect plasticization of PEG. CNW addition restricted the polymer deformation and offered good stress transfer. Tensile modulus and elasticity of nanocomposites was increased based on mechanical results. Homogenous dispersion of CNW in PVC matrix also influences the properties of nanocomposites.

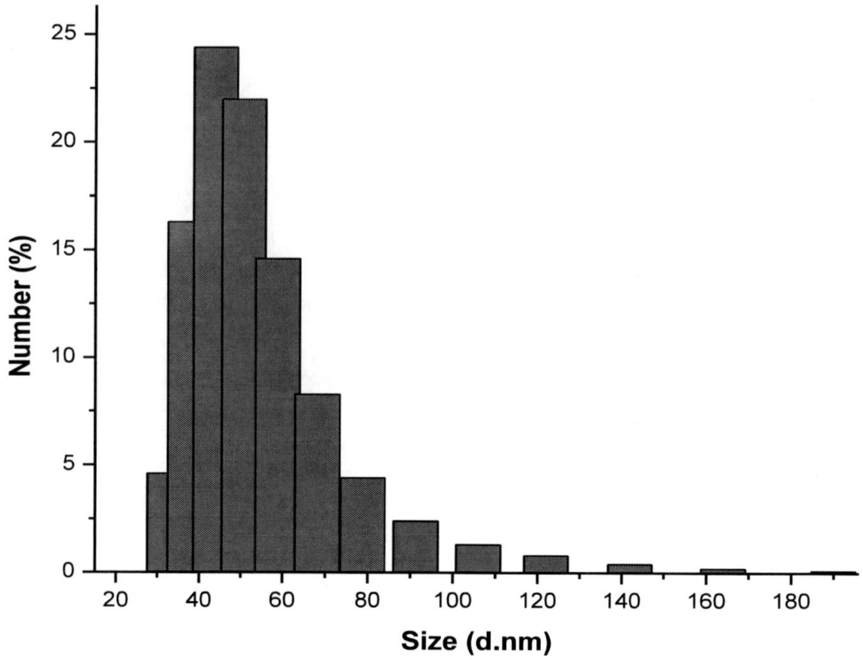

Figure 5. Diameter distribution of CNW extracted from Alfa fiber.

In the end, the obtained CNW may have potential applications in alternative power sources as a reinforcement phase.

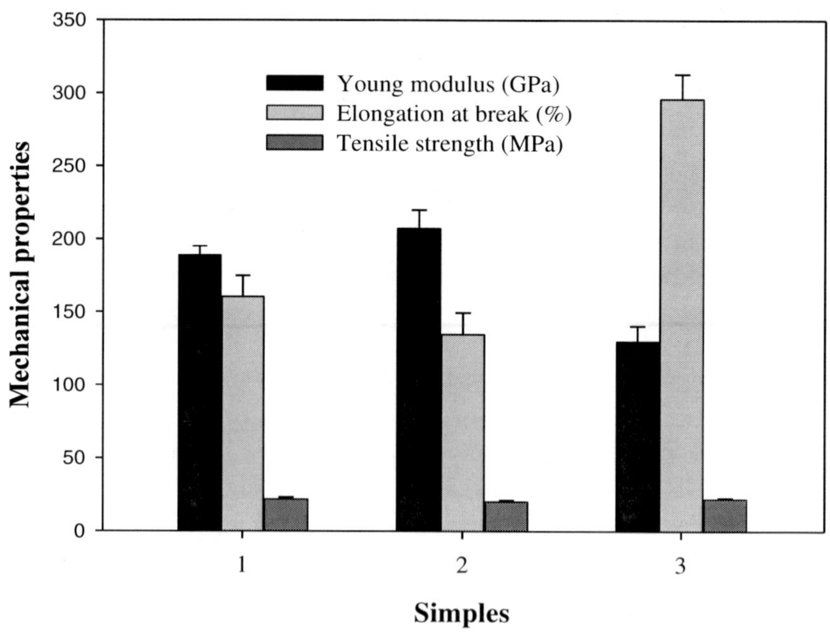

Figure 6. Mechanical properties of PVC and nanocomposites.

The properties of nanocomposites depend on the chemistry of polymer matrices, nature of the nanofiller and also the processing technique. Nanotechnology shows up today as an inherently interdisciplinary field encompassing chemistry, physics, biology and engineering. This area has made significant scientific advances in recent years. It has been reported that nanoreinforcements can increase the resistance without sacrificing the elongation of the material. In some cases, the nanoreinforcements have shown increasing resistance and modulus at the same time. The reason for the improvements is reported that nano-sized reinforcements have a surface area much larger than the conventional microcomposites and defects in the array are reduced at the nano-scales [75].

CONCLUSION

Composites reinforced with natural fibers developed significantly over the past years because of their significant processing advantages, biodegradability, low cost, low relative density, high specific strength and renewable nature. In order to use the lignocellilosic fibers in various applications in structural components, selection of suitable fibers with consistent strength values with given dimensions (diameter and length) should be applied. Hence it is necessary to revisit the characterization of all lignocellulosic fibers and analyze the corresponding extensive data generated so far in respect of mechanical properties.

The present chapter reports the chemical composition, morphological, mechanical properties and thermal behavior of Alfa fiber and its use as reinforcements and nanoreinforcement in polymer matrices. Many studies are examined, reviewed and highlighted in this chapter regarding the importance of the interface, the influence of various types of surface modifications, different types of matrices used for the composites, as well as fabrication methods, and the performance of composites based on Alfa fiber. Based on the literature discussed in this chapter, it seems that surface modification of Alfa fibers is absolutely necessary to improve their thermal stability, dispersion in the polymer matrix, and compatibility.

The present contribution summarizes also some selected works in the field of cellulose nanofillers extracted from Alfa fiber. The influence of the dispersion of cellulose on the properties of nanocomposites was reported.

Finally, the literature is very poor concerning the characteristics of Alfa grass and its composites. Some of the issues that need to be considered for future studies on the degradability of Alfa fiber composites. Biodegradation is a relatively new and very interesting recovery option for Alfa fiber reinforced composites. The behavior of decomposition has to be analyzed in order to obtain reliable data on the degree of biodegradation and the total time for decomposition.

REFERENCES

[1] Faruk, O., Bledzki, A.K., Fink, H.P., Sain, M., (2012). Biocomposites with natural fibers: 2000-2010. *Prog. Polym. Sci.* 37, 1552.

[2] Summerscales, J., Dissanayake, N.P.J., Virk, AS., Hall, W., (2010). A review of bast fibres and their composites. Part 1- Fibres as reinforcements, *Compos. part A-Appl. S.* 41, 1329.

[3] Loh, Y.R., Sujan, D., Rahman, M.E. Das, C.A., (2013). Sugarcane bagasse-The future composite material: A literature review. *Resour. Conserv. Recy.* 75, 14.

[4] Akil, H.M., Omar, M.F., Mazuki, A.A.M., Safiee, S., Ishak, Z.A.M., Abu Bakar, A., (2011). Kenaf fiber reinforced composites: A review. *Mater. Design.* 32, 4107.

[5] Blezki, A.K., Sperber, V.E., Farouk, O., (2002). Natural and wood fibre reinforcements in polymers, *report 152. RAPRA Technology*, 138.ISBNL:1-85957635962.

[6] Rahman, Md.R., Huque, M., Islam, Md.N., Hasan, M., (2009). Mechanical properties of polypropylene composites reinforced with chemically treated abaca. *Compos. part A-Appl. S.* 40, 511.

[7] Vilaseca, F., Valadez-Gonzalez, A., Herrera-Franco, P.J., Pèlach, M.À., López, J.P., Mutjé, P., (2010). Biocomposites from abaca strands and polypropylene. Part I: Evaluation of the tensile properties. *Bioresource. Technol.*101, 387.

[8] Zainudin, E.S., Sapuan, S.M., Abdan, K., Mohamad, M.T.M., (2009). Thermal degradation of banana pseudo-stem filled unplasticized polyvinylchloride (UPVC) composites. *Mater. Design.* 30, 557.

[9] Jawaid, M., Abdul Khalil, H.P.S., Abu Bakar, A., (2010). Mechanical performance of oil palm empty fruit bunches/jute fibres reinforced epoxy hybrid composites. *Mater. Sci. Eng.* 527, 7944.

[10] Hammiche, D., Boukerrou, A., Djidjelli, H., Corre, Y.M., Grohens, Y., Pillin, I., (2013). Hydrothermal ageing of alfa fiber reinforced polyvinylchloride composites. *Constr. Build. Mater.* 47, 293.

[11] George, J., Sreekala, M.S., Thomas, S., (2001). A review on interface modification and characterization of natural fiber reinforced plastic composites. *Polym. Eng. Sci.* 41(9), 1471.

[12] Kokta, B.V., Kamdem, D.P., Beshay, A.D., Daneault, C., (1986). Use of grafted aspen fibers in thermoplastic composites: IV. Effect of extreme conditions on mechanical properties of polyethylene composites. *Polym. Composite.* 7, 337.

[13] Bledzki, A.K., Gassan, J., Theis, S., (1998). Wood-filler thermoplastic composites. *Mech. Compos. Mater.* 34, 563.

[14] Felix, M.J., Gatenholm, P., (1991). The nature of adhesion in composites of modified cellulose fibers and polypropylene. *J. Appl. Polym. Sci.* 42, 609.

[15] Zafeiropoulos, E.N., Baille, C.A., Hodgkinson, J.M., (2002). Engineering and characterisation of the interface in flax fibre/polypropylene composite materials. Part II. The effect of surface treatments on the interface. *Compos. part A-Appl. S.* 33, 1185-90.

[16] Viksne, A., Bledzki, A.K., Rence, L., Berzina, R., (2006).Water uptake and mechanical characteristics of wood fiber-polypropylene composites. *Mech. Compos. Mater.* 42(1), 73.

[17] Portail de l'agriculture marocaine, (2002). L'alfa: Importance écologique et socio-économique, *Terre et Vie* 61-62, 1-3.

[18] Pouget, M., (1980). Les relations sol-végétation dans les steppes sud algérien, *Ph.D. Dissertation*, Paris.

[19] Reguant, J. Rinaudo, M., (1999). Etude bibliographique sur les matériaux issus de la biomasse végétale, *CERMAV-CNRS*, France.

[20] Hammiche, D., Boukerrou, A., Djidjelli, H., Kervolen, A., Grohens, Y., (2014). Etude des Propriétés Physico-chimiques, Thermiques et Mécaniques des Fibres d'Alfa Grasses. *Revue des Composites et des Matériaux avancés.* Accepted.

[21] Ben Mabrouk, A., Kaddami, H., Boufi, S., Erchiqui, F., Dufresne, A., (2012). Cellulosic nanoparticles from alfa fibers (Stipa tenacissima): extraction procedures and reinforcement potential in polymer nanocomposites. *Cellulose* 19, 843.

[22] Bessadok, A., Marais, S., Gouanve, F., Colasse, L., Zimmerlin, I., Roudesli, S., Metayer, M., (2007). Effect of chemical treatments of alfa (Stipa tenacissima) fibres on water-sorption properties. *Compos. Sci. Technol.* 67, 685.

[23] Parham, R.A. Kaustinen, H.M., (1974). Paper making materials. An atlas of electron micrograph, The Institute of Paper Chemistry, *Appleton Wisconsin*, p. 23.

[24] Hattalli, S., Benaboura, A., F., Ham-Pichavant, Nourmamode, A., (2002). Castellan, A. Adding value to Alfa grass (Stipa tenacissima L.) soda lignin as phenolic resins, *Polym. Degrad. Stabil.* 75, 259.

[25] Bendahou, A., Habibi,Y., Kaddami, H., Dufresne, A., (2009). Physico-chemical characterization of palm from phoenix dactylifera-L, preparation of cellulose whiskers and natural rubber-based nanocomposites. *Journal of Biobased Materials and Bioenergy* 3, 1.

[26] Bendahou, A. Kaddami, H. Dufresne, A., (2010). Investigation on the effect of cellulosic nanoparticles morphology on the properties of natural rubber based nanocomposites. *Eur. Polym. J.* 46, 609.

[27] Khiari, R., Mhenni, M.F., Belgacem, M.N. Mauret, E., (2010). Chemical composition and pulping of date palm rachis and Posidonia oceanic - A comparison with other wood and non-wood fibre sources. *Bioresource Technol.* 101, 775.

[28] El ghazi, M., Haloui, B., Janin, G., Ory, J.M., Clerment, A., (1997). Propriétés papetières du mélange alfa-eucalyptus. *Revue du papier et carton*, 5, 65.

[29] Dittenber, D.B., GangaRao, H.V.S., (2012). Critical review of recent publications on use of natural composites in infrastructure. *Compos. part A-Appl. S.* 43, 1419.

[30] Suardana, N.P.G., Ku, M.S., Lim, J.K., (2011). Effects of diammonium phosphate on the flammability and mechanical properties of bio-composites. *Mater. Design* 32, 1990.

[31] Sen, T., Jagannatha Reddy, H., (2011). Application of sisal, bamboo, coir and jute natural composites in structural upgradation. *International Journal of Innovation and Technology Management* 2, 186.

[32] Methacanon, P., Weerawatsophon, U., Sumransin, N., Prahsarn, C., Bergado, D.T., (2010). Properties and potential application of the selected natural fibers as limited life geotextiles. *Carbohyd. Polym.* 82, 1090.

[33] Maghchiche, A., Haouam, A., Immirzi, B., (2012). Effect of plant fiber-polyacrylamide blend on retention and evaporation water at arid and semi-arid soils of Algeria. *Journal of Chemestry and Chemical Engineering,* 6, 7-17.

[34] Ammar, I., Ben Cheikh, R., Campos, A.R., Cunha, A.M., (2006). Injection molded composites of short Alfa fibers and biodegradable blends. Polym. Composite. 27, 341.

[35] Geethamma, V.G., Mathew, K.T., Lakshminarayana, R.., Thomas, S., (1998). Composite of short coir fibers and natural rubber: effect of chemical modification, loading and orientation of fiber. *Polymer* 39, 6-7, 1483.

[36] Paul, A., Thomas, S., Pavithran, C., (1997). Electrical properties of natural fiber reinforced low density polyethylene composites: a comparison with carbon black and glass-fiber filled low density poly-ethylene composites. *J. Appl. Polym. Sci.* 63, 247.

[37] Rao, K.M.M., Rao, K.M., (2007). Extraction and tensile properties of natural fibers: Vakka, date and bamboo. *Compos. Struct.* 77, 288.

[38] Romildo, D.T.F. Karen, S. George, L.E. Ghavami, K., (2000). Durability of alkali-sensitive sisal and coconut fibers in cement mortar composites. *Cement Concrete Compos.* 22(2), 127.

[39] Hepworth, D.G., Vincent, J.F.V., Jeronimidis, G., Bruce, D.M., (2000).The penetration of epoxy resin into plant fibre cell walls increases the stiffness of plant fibre composites. *Compos. part A-Appl. S.* 31, 599.

[40] Alawar, A. Hamed, A.M. Al-Kaabi, K., (2009). Characterization of treated date palm tree fiber as composite reinforcement. *Compos Part B Eng.*, 40(7), 601.

[41] Rosa, M.F. Chiou, B., Medeiros, E.S., Wood, D.F. Williams, T.G., Mattoso, L.H.C., (2009). Effect of fiber treatments on tensile and thermal properties of starch/ethylene vinyl alcohol copolymers/coir biocomposites. *Bioresource Technol.* 100, 5196.

[42] Fu, S-Y., Lauke, B., (1996). Effects of fiber length and fiber orientation distributions on the tensile strength of short-fiber-reinforced polymers. *Compos. Sci. Technol.* 56, 1179.

[43] Tungjitpornkull, S., Sombatsompop, N., (2009). Processing technique and fiber orientation angle affecting the mechanical properties of E-glass fiber reinforced wood/PVC composites. *J. Mater. Process. Technol.* 209, 3079.

[44] Duval, A., Bourmaud, A., Augier, L., Baley, C., (2011). Influence of the sampling area of the stem on the mechanical properties of hemp fibers. *Mater. Lett.* 65, 797.

[45] Rouison, D., Sain, M., Couturier, M., (2006). Resin transfer molding of hemp fiber composites: optimization of the process and mechanical properties of the materials. *Compos. Sci. Technol.* 66, 895.

[46] Beg, M.D.H., Pickering, K.L., (2008). Mechanical performance of Kraft fibre reinforced polypropylene composites: influence of fibre length, fibre beating and hygrothermal ageing. *Compos. part A-Appl. S.* 39, 1748.

[47] Joseph, P.V., Joseph, K., Thomas, S., (1999). Effect of processing variables on the mechanical properties of sisal–fiber-reinforced polypropylene composites. *Compos. Sci. Technol.* 59, 1625.

[48] Rokbi, M., Osmani, H., Imad, A., Benseddiq, N., (2011). Effect of chemical treatment on flexure properties of natural fiber-reinforced polyester composite. *Proceeding Engineering* 10, 2092.

[49] Venkateshwaran, N., Elayaperumal, A., Sathiya, G.K., (2012). Prediction of tensile properties of hybrid-natural fiber composites. *Compos. Part B Eng.* 43, 793.

[50] Yang, H., Luo, R, (2011). A novel bronze-impregnated carbon strip containing Al_2O_3 particles for subway current collectors. *Wear* 270, 675.

[51] Boukerrou, A., Hamour, N., Djidjelli, H., Hammiche, D., (2012). Effect of the Different Sizes of the Alfa on the Physical, Morphological and Mechanical Properties of PVC/Alfa Composites. *Macromol. Symp.* 321-322, 191.

[52] Belhassen, R., Boufia, S., Vilaseca, F., Lopez, J.P., Mendez, J.A., Francob, E. Pelach, M.A., Mutje, P., (2009). Biocomposites based on Alfa fibers and starch-based biopolymer. *Polym. Advan. Technol.* 20, 1068.

[53] George, j., Sreekala, M.S., Thomas, S., (2001). A review on interface modification and characterization of natural fiber reinforced plastic composites. *Polym. Eng. Sci.* 41(9) 1471.

[54] Xue, L., Lope, G., Satyanarayan, P., (2007). Chemical treatments of natural fiber for use in natural fiber-reinforced composites: A Review. *Journal Polymer Environment* 15, 25.

[55] Boukerrou, A. Rahmoune, M.A. Kaci, M., (2010). Influence of alkaline treatment of Alfa fiber on the mechanical, viscoelastic, thermal and morphological properties of PVC/Alfa composites. *Academic Journal of Manufactoring Engineering* 8, 1.

[56] Kalia, S., Kaith, B., Kaur, I., (2009). Pretreatment of natural fibers and their application as reinforcing material in polymer composites: a review. *Polym. Eng. Sci.* 49(7), 1253.

[57] Ghallabi, Z., Rekik, H., Boufi, S., Arous, M., Kallel, A., (2010) Effect of the interface treatment on the dielectric behavior of composite materials of unsaturated polyester reinforced by Alfa fiber. *J. Non-Cryst. Solids* 356, 684.

[58] Bessadok, A., Roudesli, S., Marais, S., Follain, N., Lebrun, L., (2009). Alfa fibres for unsaturated polyester composites reinforcement: Effects of chemical treatments on mechanical and permeation properties. *Compos. part A-Appl. S.* 40,184.

[59] Ben Abderrahmane, M., Ben Cheikh, R., (2008). Study of cyanoethylated Alfa fiber reinforced composites by dynamic mechanical analysis. *Surface Engineering and Applied Electrochemistry* 44, 6, 484.

[60] Gassan, J., Bledzki, A.K., (1997). The influence of fiber-surface treatment on the mechanical properties of jute-polypropylene composites, *Compos. part A-Appl. S.* 28, 1001.

[61] Van den Oever, M., Peijs, T., (1998). Continuous-glass-fibre-reinforced polypropylene composites II. Influence of maleic-anhydride modified polypropylene on fatigue behaviour. *Compos. part A-Appl. S.* 29(3), 227.

[62] Joseph, P.V., Joseph, K., Thomas, S., Pillai, C.K.S., Prasad, V.S., Groeninckx, G., Sarkissova, M., (2003). The thermal and crystallisation studies of short sisal fibre reinforced polypropylene composites. *Compos. part A-Appl. S.*, 34(3), 253.

[63] Hammiche, D., Boukerrou, A., Djidjelli, H., Beztout, M., Krim, S., (2012). Synthesis of a new compatibilisant agent PVC-g-MA and its use in the PVC/Alfa composites. *J. Appl. Polym. Sci.* 124, 4352.

[64] Hammiche, D., Boukerrou, A., Djidjelli, H., Djerrada, A., (2013). Effects of some PVC-grafted maleic anhydrides (PVC-g-MAs) on the morphology, and the mechanical and thermal properties of (Alfa Fiber)-reinforced PVC composites. *J. Vinyl Addit. Techn.* 19, 225.

[65] Brinchi, L., Cotana, F., Fortunati, E., Kenny, J.M., (2013) Production of nanocrystalline cellulose from lignocellulosic biomass: Technology and applications. *Carbohyd.Polym.* 94(1), 154.

[66] Abdul Khalil, H.P.S., Bhat, A.H., Ireana Yusra, A.F., (2012). Green composites from sustainable cellulose nanofibrils: A review. *Carbohyd. Polym.* 87, 963.

[67] Dufresne, A. In Encyclopedia of Nanoscience and Nanotechnology, 2nd ed.; Nalwa, H.S., Ed.; American Scientific Publisher: Valencia, CA, USA; in press.

[68] Lu, J., Askeland, P., Drzal L.T., (2008). Surface modification of microfibrillated cellulose for epoxy composite applications, *Polymer* 49, 1285.

[69] Hammiche, D., Boukerrou, A., Djidjelli, H., Beztout, M., (2011). Effet d'un nouvel agent de couplage PVC-g-Ma sur les propriétés physiques et diélectriques des composites PVC/alfa. *Annales de Chimie et Science des Matériaux* 36/3, 125.

[70] Rodriguez, D., Thielemans, N.L.G., Dufresne, A., (2006). Sisal cellulose whiskers reinforced polyvinyl acetate nanocomposites. *Cellulose* 13, 261.

[71] Azeredo. M.C., (2009). Nanocomposites for food packaging applications. *Food Res. Int.*42, 1240.

[72] Schadler, L.S., Brinson, L., C., Sawyer, W.G., (2007). Polylmer nanocomposites: A small part of the story. *Nanocomposite Materials* 59, 53.

[73] Oksman, K., Mathew, A.P., Bondeson, D., Kvien, I., (2006). Manufacturing process of cellulose whiskers/polylactic acid nanocomposites. *Compos. Sci. Technol.* 66, 2776.

[74] Hammiche, D., Boukerrou, A., Grohens, Y., Bendahou, A., Seantier, B., *(2013).* Characterization of cellulose nanowhiskers extracted from alfa fiber and the effect of their dispersion methods on nanocomposites properties. Cel. Chem. Technol. Under *review.*

[75] Souza, S.F., Leão, A.L., Cai, J.H., Wu, C., Sain, M. Cherian, B.M., (2010). Nanocellulose from Curava fibers and their nanocomposites, *Mol. Cryst. Liqu. Cryst.* 522, 42/[342]-52/[352].

In: Polymer-Matrix Composites
Editor: Rakesh Kumar
ISBN: 978-1-61122-005-6

Chapter 6

INJECTED POLYMER-MATRIX NANOCOMPOSITES: MORPHOLOGY-PERFORMANCE RELATIONSHIP

Valeria Pettarin*
Polymer Science & Engineering, Institute of Materials Science and Technology, University of Mar del Plata CONICET, Mar del Plata, Argentina

ABSTRACT

Polymer nanocomposites are new materials that show great potential in a variety of applications. By incorporating nanoscale particles of a filler material, the performance and properties of the bulk material can be drastically altered. Indeed, quite exceptional improvements can be achieved with small amounts of filler. To make nanocomposites economically viable, their production must use common processing equipment, and additional processing steps should be avoided. Therefore, many thermoplastic-composites are processed by an injection molding process. It is generally accepted that the performance of polymer nanocomposites is intimately related to the degree of dispersion of the nanofiller: good dispersion and exfoliation may lead to the expected improvement in properties. This dispersion depends on factors such as the affinity and compatibility of the filler with the matrix, the matrix viscosity and the applied thermo-mechanical level, which depends on the processing conditions. The understanding of the fracture, micro-deformation and mechanics of failure of nanocomposites is crucial for engineers. In this chapter the relationship between processing and performance is reviewed for different injected thermoplastic pieces. The effects of the molding process and of the nanoclay filler on the properties of the polymer composites are reported. Also, the effect of the occurrence of in-homogeneities, such as weld lines or flow lines in microstructure and therefore in performance are summarized.

Keywords: Polymer, Nanoclay, Carbon nanotubes, Injection molding, Morphology

* Corresponding Author address: University of Mar del Plata CONICET, Mar del Plata, Argentina; Email: pettarin@fi.mdp.edu.ar

1. INTRODUCTION

Polymer nanocomposites are relative new materials developed during the last 20 years that show great potential in a variety of applications. By incorporating nanoscale particles of a filler material, the performance and properties of the bulk material can be drastically altered. Indeed, quite exceptional improvements can be achieved with small amounts of filler.

Thermoplastic nanocomposites containing different fillers have been widely studied. The three main methods used for inorganic nanofiller incorporation in a matrix are in situ polymerization, solution intercalation, and melt compounding. In situ polymerization has been claimed as the most efficient method for producing a nanocomposite with high filler dispersion and enhanced properties. Solution intercalation usually requires dissolving the polymer resin or prepolymer in an organic solvent. Because of the complexities of the in situ and solution intercalation procedures and/or the undesirable environmental effects of the solvents used, these methods are often regarded as inappropriate especially for high volume manufacture. To make nanocomposites economically viable, their production must use common processing equipment, and additional processing steps should be avoided. For industrial applications, the most convenient and preferred method is melt compounding, because it is a simple, low cost, and environmentally friendly (i.e., no solvents needed) technique and, moreover, compatible with existing processing infrastructures, offering thus industrial and economic advantages [1-6]. Thus, the challenging area is the production of good polymer nanocomposites by compounding the nanofiller with polymer matrix by melt mixing.

Nanoclays and nanotubes are typical fillers. Only well-dispersed and well-exfoliated (in the case of clays) nanoparticles can lead to the expected improvement of properties [7-9]. Therefore raw material producers, converters, and end-users have to tackle both compounding and processing issues. Surface modification of nanofillers with organic surfactant and adaptation of compounding conditions (high shear, high residence time, special screw profile design in case of melt compounding for example) may help to get rid of most of compounding issues.

Many thermoplastic-composites are processed by an injection molding process. The injection molding process involves the injection of a polymer melt flow into a mould impression where the melt cools and solidifies to form a plastics product. Polymeric injection moldings are currently used in many application areas. The increasing use of polymers is greatly associated to cost reduction resulting from lower weight, elimination of secondary operations, and use of less energy. Nonetheless, the influence of processing parameters is critical in the performance of these products.

Features like mechanical properties and morphology of injection molded composites, strongly depend on thermomechanical processing variables [10]. Whilst not restricted to polymer nanocomposites, the quality (i.e., the properties) of the processed article of a polymer is determined not only by its morphology but also its primary structures; accordingly, the quality design of a processed article can – and should – be achieved by controlling its morphology. For any processed article, the morphology is determined by the raw material's characteristics, the processing methods used, and the processing conditions, which naturally determine also the article's quality. Therefore a fundamental and important aim in the polymer composites is control of the morphology [11-16]. In the case of injection molded semi-crystalline polymers, crystal properties are highly influenced by processing conditions [17]. It is noteworthy that by compounding and processing (thermally, mechanically) of different polymers and reinforcements can be explored diversity of morphologies. Change of the structure on molecular or macroscopic level may improve dramatically mechanical properties, e.g. modulus, stiffness and impact strength [18-22] as well as toughness and flexural modulus. Processing can be regarded as a polymer structuring tool for enhanced and tailored mechanical performance.

The understanding of the fracture, microdeformation and mechanics of nanocomposites is crucial for engineers. Over many decades, research has dealt with the question of how mechanical properties are related to morphological features [23-27] and processing conditions [28-36], as well as how molding parameters influence those features [37-41]. Much work has been done characterizing injection molded thermoplastic nanocomposite specimens. Aim of work should be directed to obtain reinforced structure through enrichment by nano particles and by technique as well. In this chapter the relationship between processing and performance is reviewed for different injected thermoplastic pieces. The effects of the molding process and of the nanoclay filler on the properties of the polymer composites are reported. Also, the effect of the occurrence of in-homogeneities, such as weld lines or flow lines in microstructure and therefore in performance are summarized.

2. Injection Molding

2.1. Injection Molding Techniques

The injection molding technique is a high throughput process adequate to manufacture thermoplastic components of complex geometry with tight dimensional tolerances. Injection molding of polymeric materials is an intricate dynamic and transient process, involving convoluted melting-flow-pressure-solidification phases and a complex material behavior strongly affecting the quality and properties of the final molded component.

An important advantage of injection molding is that processing of complex geometries becomes available in one automated production step.

In injection molding, the thermomechanical environment imposed to the polymer melt is controlled by the definition of the operative processing variables (e.g. plasticating temperatures, cycle time, injection and holding pressures) and/or system geometry (e.g. plasticating screw, injection gate location, water lines layout). These thermomechanical conditions control the microstructure and morphology of the final molded component [40,19], which determines their dimensions (shrinkage), dimensional stability (distortion and warpage) and properties (e.g., mechanical behavior, permeability, appearance) [42, 43].

The presence of weld lines in injection moldings is a major design concern. Weld lines are formed when two separate melt fronts rejoin during injection molding. Weld lines often occur because of multiple gating, obstacles, variable part thickness, and jetting and therefore are almost impossible to avoid in injection molding. It has been observed that weld lines weaken the strength of injection molded parts and so designers often need to use larger safety factors in the design analysis to compensate for that weakness. Reduced strength of a weld line results largely from four factors, including: incomplete molecular entanglement or diffusion, formation of V-notches at the weld surface, presence of contamination or micro voids at the weld interface, and unfavorable molecular or filler orientation at the weld [44, 45]. Mielewski et al. who studied the formation of weld lines in unfilled PP showed that the depth of the V-notch generated at the molding surface is related to the injection temperature, mold temperature, and injection speed [46]. Wu and Liang determined that the melt temperature was the most influential factor [47]. They also stated that the cross-sectional dimensions of microstructures have a higher effect on the mechanical characteristics than the process parameters do. Whereas the complete elimination of weld lines is not always successful, their adverse effects on part appearance and mechanical performance can be minimized.

While weld lines can be deleterious in homopolymer moldings, the problem can be amplified in two-phase systems, such as reinforced thermoplastics.

The establishment definition of the adequate processing conditions, to mould a high quality plastic component, is a complex task because of the significant number of processing variables, to the high level of interactions between these variables and the numerous molding features and end-use properties to improve. Therefore, various strategies using different methodologies to optimize the injection molding process have been reported in the literature [48-54].

Additionally, hot runner molds are adequate to mass production. These molds lead to the reduction of production costs by saving material of the nonrequired gating system and by shortening the cycle time. Furthermore, lower injection pressure is required, the process temperature is controllable with greater precision, more uniform filling is achievable in multicavity molds, and thus, the final mechanical properties of the moldings are improved [55].

Another improvement is the development of special non-conventional injection molding techniques based on the melt manipulation during the solidification phase. Such processes include Shear Controlled Injection Molding – SCORIM [28, 37], Push-Pull [56-58] and Vibration Assisted Injection Molding [59]. The main difference of these non-conventional melt manipulation injection molding processes in comparison to conventional technique (CIM) is the way the melt is treated during the solidification stage. In CIM, during one cycle, the polymer melt is injected, packed and cooled in the mould cavity. After melt injection there is limited possibility to manipulate the morphology. The non-conventional melt manipulation injection molding gives possibility to control the highly oriented shear zone in the molded components that allows enhancing the mechanical properties of polymeric material systems. The equipment conformed to non-conventional injection techniques may also be based on the CIM machine. In the case of SCORIM, the main difference is in the mould, where the typical fixed mould plate has been replaced by the special unit combined with two hydraulic pistons. This SCORIM head device introduces additional operative variables (such as the stroke time, the stroke pressure, the temperature of the manifold and 3 independent operative modes) enlarging the possibilities of morphology manipulation. The melt, provided through one hot-runner channels, fills out the cavity from one side through one gate. Once injected melt material fills the cavity the SCORIM stages initiate. While during one movement of the pistons, the melt directed in one direction is getting solidified, next movement initiates the creation of next layer directed parallel with opposed flow direction. As a consequence, the SCORIM moldings develop a typical microstructure, resulting from the high shear stress fields applied at the solidifying interfaces and treated by actuated external pistons.

2.2. Structure Developed In Injection Molding

In injection molding the polymer melt is subject to a complex thermomechanical environment characterized by high cooling rates and important stress fields. This environment varies along the flow path and through the thickness of the molding. As a consequence, a molded polymer shows an intrinsic heterogeneous microstructure, featuring a gradual and hierarchical variation of the morphology that evolves throughout the spatial domain of the part. For a semi-crystalline polymer a typical layered arrangement is developed, normally referred to as the skin-core microstructure, which can be observed by polarized light microscopy (PLM) (Figure 1). The number of identified layers depends on the resolution of the experimental technique used and on the level of discrimination considered. A simple analysis assumes a three-layer model (two external skins and an interior core) [60-64], but other layers may be discerned (e.g. two outer skins, two sub-skin regions, two intermediate shear zones and an inner core). The intrinsic molecular nature of the polymer together with the magnitude of these layers and their morphological state determine the mechanical behavior of the injection molded engineering components.

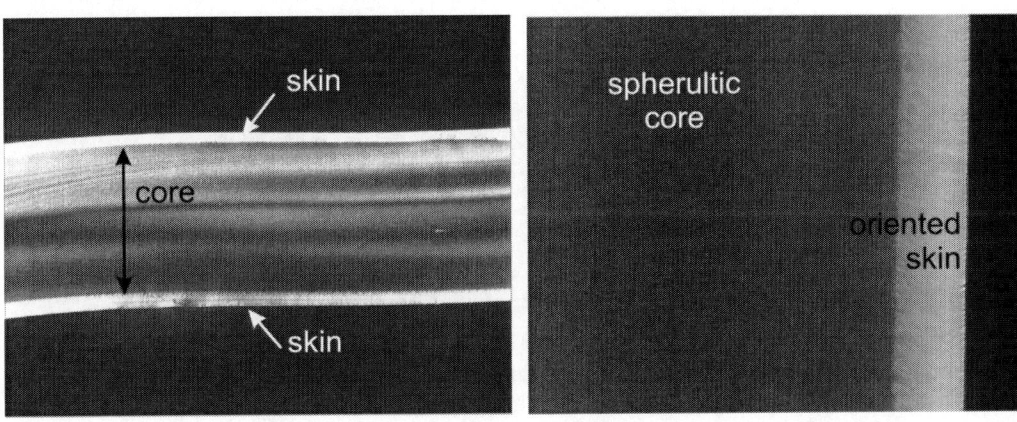

Figure 1. Skin-core morphology of injected PP pieces as revealed by polarized optical microscopy.

In SCORIM a layered structure is formed instead of a skin-core-skin morphology. The SCORIM samples featured a thick shear zone which contains many oriented layers in an onion-like microstructure (Figure 2). This microstructure is comprised of external highly oriented multilayers and a spherulitic core. The thickness of these layers, as well as their number, depends on the selected process variables setting (e.g., stroke number and time). Also shear may induce the formation of a shish kebab structure. It is well known that a superstructure of polymer crystals can be created by crystallization during flow [65]. Specific is also fact that in crystalline structure of polyolefins, subjected to stress (like during injection molding process) appears shish kebab structure, firstly described by Penning [66]. In his work he presented a model of developed structure, where extended chain crystals are surrounded by folded chain crystals called shish and kebab respectively (Figure 3). Firstly is formed long fibrelike core (shish) during a crystallization process and under stress. Then plate like lamellar crystals (kebabs) overgrow this shish structure epitaxially. Injection molding as technique for obtaining such structures and commonly used technique for wide range of applications with domestic and engineering purpose has been used.

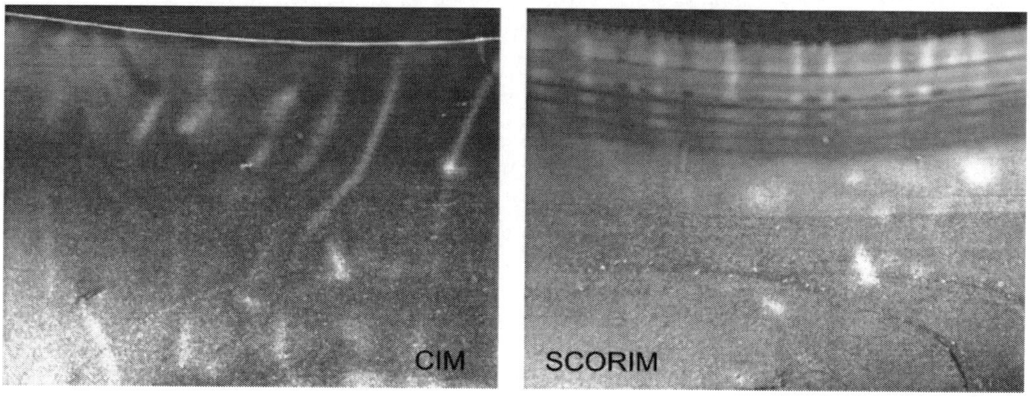

Figure 2. Comparison between morphology of conventional injection molding (CIM) and shear controlled injection molding (SCORIM) PP samples as revealed by polarized optical microscopy.

Figure 3. Models for polyethylene shish-kebabs with different distributions of interlamellar fibrils (Reprinted from Ref. [66] Pening, A.J., (1980). Journal of Crystal Growth 48, 574).

Viana et al. [67] identified the significant processing variables affecting the development of the morphological parameters. They found out that the morphology (level of molecular orientation and degree of crystallinity) is mostly governed by the melt temperature and the degree of crystallinity of the core is essentially dominated by the mould temperature. Demiray et al. [41] studied the relationship between processing conditions and skin–core morphology of PP. They concluded that the melt temperature is the most important variable affecting the skin layer thickness and spherulite size distribution.

The thickness of the skin layer has been evidenced as a highly significant microstructural parameter [64, 68, 40, 69-72]. Let’s think about the mechanism of formation of the skin. The skin of the plastic in contact with the cool mold freezes rapidly, while the central core remains molten. When additional material is injected, it flows into this central core, displacing the material already there, which then forms a new flow front. The flow of this displaced material is a combination of forward flow and outward flow. The outward flow contacts the wall, freezes, and forms the next section of skin while the forward flow forms the new molten core. When more material enters the mold, it flows along a channel lined with these frozen walls of plastic. This flow pattern is often called fountain flow or bubble flow because the flow front is like a bubble being inflated with hot plastic from the center. The frozen layer is formed by the flow front inflating, and so is subject to only a low shear stress and, therefore, has a very low level of molecular orientation. Once it is frozen it cannot be orientated any further, so the frozen layer in the finished part has a low level of orientation. Now, consider what happens upstream. Hot plastic is continuously flowing, bringing new hot material along and generating significant frictional heat. At the same time, heat is being lost through the frozen layer to the cold mold surface. Initially, the frozen layer is very thin, so heat is lost very rapidly. This results in more plastic freezing and the frozen layer getting thicker, cutting down the heat flow. After a time, the frozen layer will reach a thickness such that the heat lost by conduction is equal to the heat input from plastic flow and frictional heating, i.e., an equilibrium condition is reached. It is interesting to do same calculations on the time taken to reach this state of equilibrium. The actual rate of heat flow is very large in comparison with the small heat content of the plastic in the frozen layer. The result is that equilibrium is reached very quickly, often in a time measured in a few tenths of a second. As the total filling time is measured in seconds, the frozen layer reaches an equilibrium state early in the filling cycle [73]. It is useful to think about how the thickness of this frozen layer will vary. If the injection rate were slowed, less heat would be generated by friction along the flow path, with less heat input from the flow. The heat loss would be at the same rate, and with less heat input the frozen layer would grow in thickness. If the injection rate were raised, the frozen layer would be thinner. Similarly, higher melt and mold temperatures would reduce the thickness of the frozen layer. Results supporting these assumptions were reported by Viana [19], who has analyzed the development of the skin layer in injection molding for different molding geometries and processing conditions. He related the skin ratio to the local bulk temperature and maximum shear stress at the end of the filling phase and proposed the mechanisms of formation of the skin layer, which he connected to the time to reach the crystallization temperature and the material relaxation time.

3. POLYMER BASED NANOCOMPOSITES

In nature nanostructures occur commonly. Many examples of biological micro and nanostructure systems like bones, sinews, cells, organs or tissues can be found. Equipped by nature, systems are manifold role models designed by engineers and architects. Macromolecules and biological molecules build nanostructures and nanocomposites in biological systems [74-76]. Based on these discoveries many composites (including laminate, dispersed and fiber morphology) are created.

The incorporation of rigid fillers to a polymer often leads to a stiffness increase with a concomitant decrease on the energy absorption capabilities [77]. The incorporation of nanosized particles (with dimensions 100 nm and a high surface area as compared with conventional microparticles) into polymers is expected to lead to simultaneous improvements on stiffness and toughness. Polymer based nanocomposites received major attention in recent years, because of their extremely interesting properties [78]. The modification of the macromolecular interactions (interfacial effects) by the nanoparticles affects the polymer behavior. The addition of small amounts (1–3%) of nanosized particles to a polymer nanocomposite is enough to increase considerably its mechanical behavior (stiffness, strength and scratch resistance), barrier properties, thermal stability, chemical resistance, electrical properties, aesthetic/optical properties (higher gloss and clarity) and fire resistance/ flame retardancy. A much better understanding of the structure properties relationships and deformation mechanism of polymer nanocomposite is still necessary. In this work polymer nanocomposites based on clays and carbon nanotubes will be explained as they are the most reported, even though nanocomposites based on other fillers have been reported and well developed, as for example composites with silica nanoparticles [79], carbon nanofibers, carbon black, nanocalcium carbonate, polyhedral oligomeric silsesquioxane (POSS), and others that continue to attract a great deal of attention.

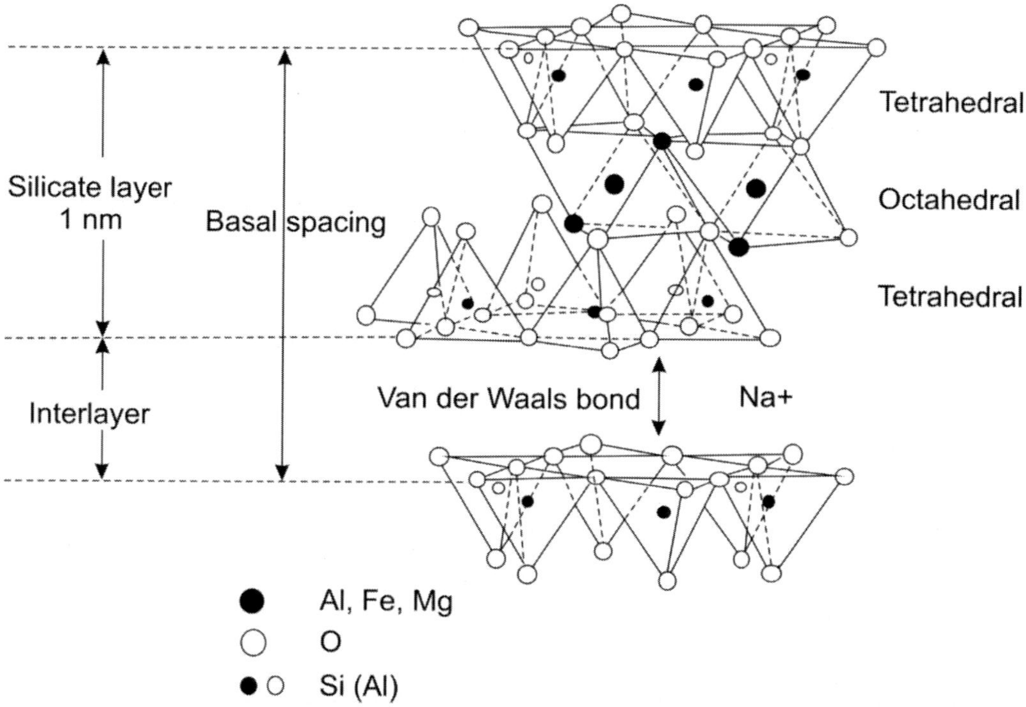

Figure 4. Schematic illustration of atoms arrangements in a typical MMT layer.

Nanoclay based polymer nanocomposite are a class of emergent multiphase organic/inorganic hybrid polymeric systems [81, 1, 82] pioneered by researchers at Toyota [83-85], which may exhibit significantly improved mechanical, flammability, and permeability properties relative to the base polymer matrix at very low clay loading. Although first demonstrated for nylon, polymer clay nanocomposites have since been prepared for a range of thermoplastic composites [86-90]. Clays themselves are layered silicate minerals with charged surfaces neutralized by interlayer counter-ions. The most used clays are layered aluminosilicates and among them, montmorillonite (MMT). MMT, a layered silicate, is a mica-type silicate mineral, composed of extremely small crystalline particles. Its structure consists of two fused silica tetrahedral sheets sandwiching an edge-shared octahedral sheet aluminum hydroxide (Figure 4). Stacking of the layers by secondary interactions leads to interlayers or galleries between the layers. The galleries are normally occupied by inorganic cations, balancing the charge of the oxide layers. MMT particles have only its thickness in the nanometer range. To improve dispersion and polymer compatibility the nanoclays are surface modified (organo modified nanoclays) [81]. Intercalated or exfoliated nanocomposites can be achieved. In intercalated polymer nanocomposites (Figure 5a), the insertion of polymer chains into the clay layered structure occurs in a crystallographically regular fashion, regardless of the clay-to-polymer ratio. Intercalated polymer nanocomposites normally feature interlayers composed of a few molecular layers of polymer. The properties of such composites typically resemble those of ceramic materials. In an exfoliated nanocomposite (Figure 5b), the individual clay layers are separated in a continuous polymer matrix by average distances that depend on clay loading. Usually, the clay content of an exfoliated nanocomposite is much lower than that of intercalated polymer nanocomposites. To achieve true nanocomposites the clay should be intercalated or exfoliated during processing, but in most cases it is partially exfoliated [82]. Clay structure in polymer nanocomposites can be characterized as a combination of exfoliated platelets and intercalated tactoids. The anisometric shape and approximately 1 nm width of the clay platelets dramatically increase the amount of interfacial contact between the clay and the polymer matrix. Thus the clay surface can mediate changes in matrix polymer conformation, crystal structure, and crystal morphology through interfacial mechanisms that are absent in classical polymer composite materials. In spite of increments on stiffness and strength, ductility is in general reduced in these composites, with tendency for embrittlement [1, 91-93]. The impact toughness of MMT based polymer nanocomposite was found to decrease [1, 92, 94-95]. However, if polymer microstructure can be changed combining processing with nanoclay, improvements in ductility and toughness can also be found [96, 97].

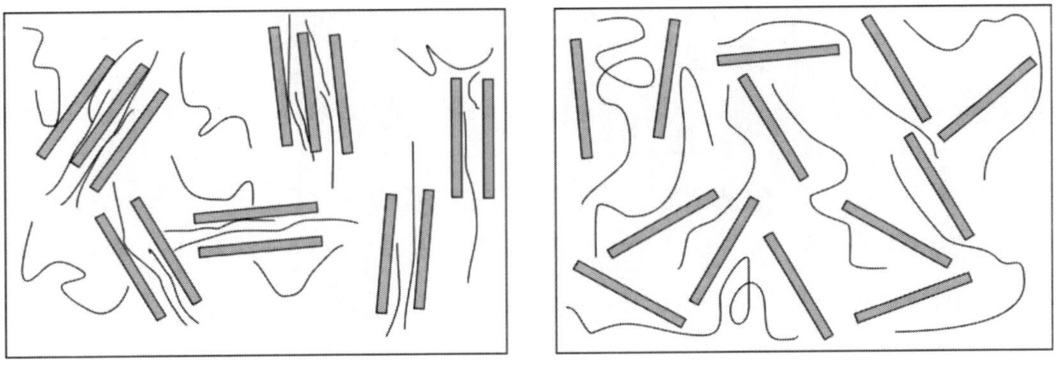

Figure 5. Schemes for (a) intercalated and (b) exfoliated nanocomposites.

Carbon nanotubes (CNT) based polymer nanocomposite are very promising materials [98], still under development. Due to their extraordinary mechanical properties like an extremely high Young's modulus (≈1.2 TPa), stiffness and flexibility [99, 100], CNTs are ideal structural reinforcements for polymer nanocomposites. They have diameters of 1–50 nm and length of few micrometers, featuring aspect ratios>1000 [98] (Figure 6).The high aspect ratio and nanoscopic dimensions of carbon nanotubes give CNT-based nanocomposites specific properties differing from those achieved in classical composites [78, 101]. Carbon nanotubes can have two forms: single or multiwalled (SWCNT and MWCNT) respectively. A good dispersion of CNT in the polymeric matrix is also essential to obtain an optimized performance, mainly to improve strength and toughness. The surface of CNT is modified usually by chemical, electrochemical or plasma treatments [102]. A controlled degree of alignment of the CNT during processing will ensure maximized mechanical behavior. Concomitant increases on the stiffness, yield strength, fracture toughness and wear resistance of high density polyethylene (HDPE) with increasing percentage of MWCNT content were reported [103]. The interface layer of crystallized polymer enveloping the fiber is an important micro structural parameter determining the mechanical properties of semicrystalline polymer nanocomposite. The yield stress of MWCNT filled PE was found to increase slightly with the amount CNT, but both the stress and the strain at break decreased [102]. The impact toughness of poly (methyl-methacrylate) (PMMA) has been improved considerably by the addition of CNT that was attributed to a weak interfacial adhesion [104]. The modulus, yield stress and toughness of MWCNTs reinforced PP exhibited a maximum at low CNT amounts (1 wt-%) [105, 106]. Significant increases in the quasistatic tensile properties of starch were found from the incorporation of very small quantities of MWCNTs [88]. Stiffness increased almost 70% and ultimate tensile strength ~35% with only 0.055 wt% of MWCNTs, keeping deformations higher than 80% without break. As a consequence, tensile toughness also increased (~50%). Further understanding of the relationships between the structure and properties of CNT reinforced polymer nanocomposites is still needed.

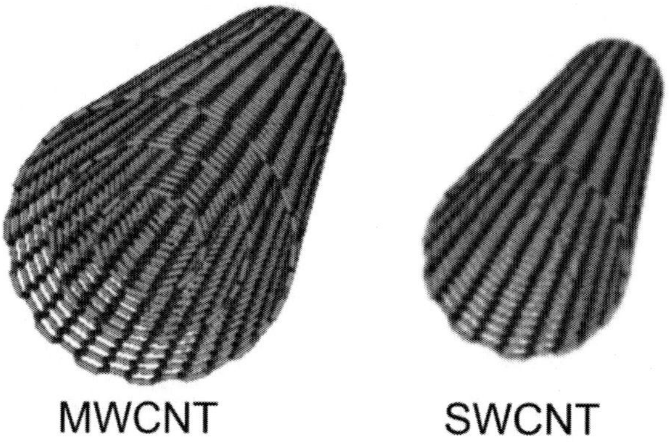

Figure 6. Scheme of carbon nanotubes: multi-walled (MWCNT) and single walled (SWCNT).

4. Injected Nanocomposites

As said in previous sections, the mechanical properties of molded polymers are extremely dependent upon the processing method and conditions used to produce them. The processing thermomechanical conditions imposed to the melt governs the morphology development that affects the mechanical response of the molded product. For example, a thicker skin layer results in a high stiffness, strength and enhanced impact response [70, 71]. A high degree of crystallinity results in a higher stiffness, but it is generally detrimental for the capability of the material to absorb energy in very short time intervals [107]. A high level of molecular orientation is also beneficial in terms of impact strength, but it reduces the deformation capabilities of the moldings [70, 71]. Trough this section a deeper insight into these relationships will be done.

4.1. Dispersion of Nanofiller in Injected Moldings

Only homogeneously dispersed and sufficiently exfoliated (in the case of clays) nanoparticles can lead to the expected improvement of properties in nanocomposites. Exfoliation plays a key role in mechanical properties. In fact, a good dispersion of nanofillers in the polymer tends to improve the tensile strength and strain at break [108], although on occasion there may be significant reductions in tensile ductility and impact strength compared to the virgin polymer [109-111]. Polyolefins (such as polypropylene (PP) and polyethylene (PE)) are the most widely used commodity plastics and are thus an interesting option for nanocomposites. Generally, inorganic nanofillers have insufficient adhesion or interaction with hydrophobic polyolefins for good dispersion. For uniform nanofiller dispersion, elimination of agglomerates and creation of proper adhesion between matrix and nanofiller pose major challenges. Surface modification of filler particles and choice of adhesion promoter or coupling agent are the key factors in prohibiting and breaking down filler agglomerates, in assisting proper filler dispersion, and improving interactions at the filler–matrix interface [1-6, 112, 113]. Following this idea, novel masterbatches (MB) have been developed. MB producers claim that they include all compatibilizers needed to promote complete nanoclay dispersion, easy processability and compatibility with standard equipment, including extruders, mixers, and injection molders. The final injection—or extrusion—molded part should be easily obtained by mixing/diluting the MB in the appropriate polymeric matrix. This seem to be the direct way to obtain a polymer nanocomposite from commercial materials, and so some authors have reported the preparation of injected pieces by using MB [96, 97, 114, 115].

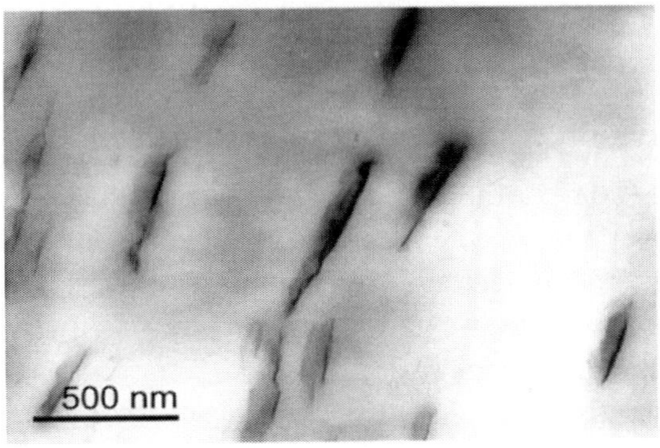

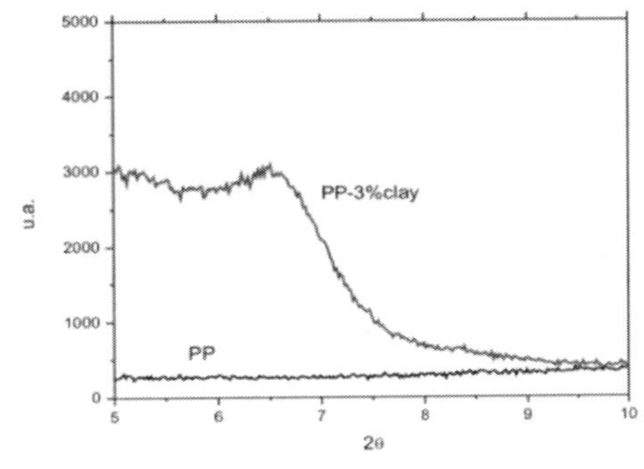

Figure 7. TEM micrograph and XRD patterns of injected PP-3% nanoclay pieces. Intercalation is evident.

For example, injected PP/nanoclay moldings produced from commercial PP and masterbatch of 50/50 PP/nanoclay were studied [96]. These moldings were injected in two points, i.e. they have a weld line. A uniform dispersion of clay in PP nanocomposites was observed by transmition electron microscopy (TEM), with apparently no indication of aggregation of clay (big tactoids) (Figure 7). The XRD patterns indicated that the nanoclay platelets were not exfoliated, they were only intercalated. These results of nanoclay dispersion and exfoliation are comparable to what is generally reported in the literature for PP/clay nanocomposites obtained using a MB. Rajesh et al. [116] claimed that exfoliation does not occur during injection molding leading to intercalated nanocomposites with a dispersion degree insufficient to achieve a percolation effect, in which clay platelets are dispersed as tactoids. Rodríguez-Llamazares et al. [117] also reported clays dispersed in the PP matrix in the form of small aggregates. Similar results were reported for poly(ethylene terephthalate) (PET)-nanoclay composites, in which both XRD patterns and TEM images reveal that some clay tactoids remain in all nano- and multiscale composites, and this dispersion improved the stiffness properties of PET/glass fibres/nanoclay composites, as revealed by the increase of Young's modulus values [118]. Another interesting feature is that clay platelets were oriented in the flow direction inside the impression of the mold [97], this direction being predicted by mold filling simulation using commercial software. Similar results of clay orientation were reported by Chow et al. [119] for polyamide 6/polypropylene-clay nanocomposites. There is a global tendency to believe that all composites' properties must be enhanced if the particle size is very small. Conversely, it has been found that intercalated layered silicates were more effective than exfoliated layered silicates in improving fracture toughness of nanocomposites [120, 121]. Small particle size has a positive effect on many of the functional properties of polymer composites. However this is not completely true for toughness. There are many toughening mechanisms in composites which cannot be effective with nanoparticles. For example nanoparticles are too small to cause significant crack bridging or crack deflection. On the other hand the very large surface area of nanoparticles does provide the possibility of large energy absorption if they delaminate. However, even here there is an optimum particle size for toughening because the stress necessary to cause delamination is inversely proportional to the square root of the particle size [122]. According to Cotterel et al. [122] statements, from a mechanical point of view there are two main potential sources of toughness in semicrystalline intercalated polymer nanocomposites: delamination or splitting of particles and matrix deformation where the major energy absorbing mechanism is the formation of multiple craze-like bands [122]. In fact, intercalated PP/nanoclay composites exhibited an improvement in fracture toughness [97]. Samples in which nanoclay platelets were oriented approximated at 45° with relation to the crack line, were more effective in initiating multiple crazes and particle delamination. A failure mechanism which involves nanoclay delamination and multiple matrix crazing was proposed to explain the toughening of PP in the directions where the macromolecules and nanoparticles orientation with respect to loading was adequate [96]. Authors concluded that the orientation induced by processing controlled the toughness distribution in the injection moldings.

Weld lines are not only aesthetically unattractive, but it is also in the weld line region where the properties are different from the bulk [123, 124]. There exists a large amount of documentation of studies on the mechanical characterization of molded items at the weld line and on the weld line morphology, especially in fiber reinforced polymers [125]. High aspect ratio particles were shown to orientate in the plane of the weld line and parallel to the weld line in micro [126] and nanocomposites [126, 96]. It is been reported that in PP/nanoclay composites under tensile conditions, the amount of absorbed energy was lower at the weld line than away from it and in the flow direction [96]. The impact toughness tended to increase from a minimum associated to the sample in which there was a frontal hit of the flow fronts, to the reference sample in a location well away from the weld line in the flow direction. Thus, in the sample in which the angle of the weld line plane with respect to flux direction was 90°, the polymer chains and the nanoclay platelets were oriented in a plane transversal to the applied force, thus contributing to the weakness of the weld line. In the samples with angles at 70° and 55°, the weld line was not exactly perpendicular to the imposed tensile loading, which attenuated the depreciative effect of the weld line. The samples without weld line showed the best performance because of the load being applied in the flow direction (and, expectantly, along the highest molecular orientation). In general the incorporation of MB was found to be beneficial for the tensile impact toughness with weld lines [96]. On the contrary, under biaxial conditions the weld line zone of PP moldings exhibited better impact toughness than nearby the injection point showing a more ductile behavior at the weld line (similar findings were reported by Bucknall [128]. Biaxial tests do not impose a preferential direction of failure of the specimen and isotropic or orthotropic materials are expected to display a symmetric deformation pattern with respect to the impact site. The failures are originated at the weakest point in the sample and propagate from there due to the high radial and circumferential tensile stresses [129]. In samples impacted near the injection point of unfilled PP moldings, bending along the clamping ring was seen and the plate failed by single splitting along the melt flow direction. Conversely, in samples fractured at the weld line zone some cracks were seen running radially from the central point of impact, while others follow a circular path around the same point (circular cracks occur where the deflection caused by the striker produces a large bending moment), and there was evidence of plastic deformation shown by the stress whitening (Figure 8). It was therefore concluded that the molecular orientation induced by melt flow prevailed, and the weld line contribution was not the determinant of PP molding toughness. Regarding PP/nanoclay moldings, it was found that samples exhibited similar toughness near the injection point and at the weld lines and there was not only a main crack path along the melt flow direction but also a main crack path along the weld line.

This result is a consequence of the lower flow interpenetration and chain entanglement at the weld line, because of the viscosity increment resulting from the nanoclay addition, and unfavorable orientation of those particles in the weld line plane. However, there was not a depreciation of the impact toughness due to weld line formation. In biaxial tests, the difference in impact toughness with nanoclay content was negligible away from the weld line. In uniaxial tests and away from the weld line, impact toughness increased with clay content, but the toughening effect of clay depends on the test direction. It was seen that toughening effect of clay on PP is not isotropic, and it depends on nanoclay orientation.

When non-conventional injection molding techniques were used, it was reported that in SCORIM the galleries of nanoparticles can be easily broken thanks to high shear rates in the flow orientation during strokes [130], so the technique can be efficiently used to obtain high performance nanocomposites without pre-treatment. As a matter of fact, Costantino et al. [114] reported that higher exfoliation/intercalation of the nanoclay was seen in SCORIM PP/nanoclay samples than in the same CIM samples (evidenced by TEM) may be due to the high shear stresses exerted during SCORIM processing.

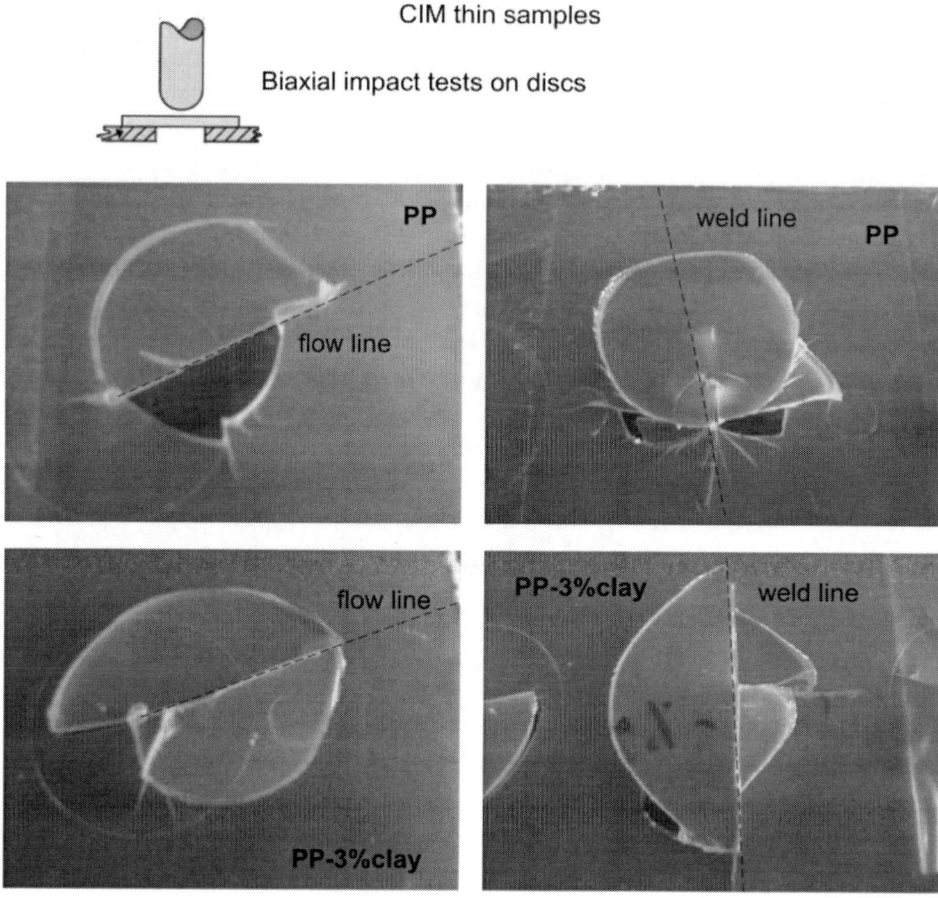

Figure 8. Typical fracture surfaces of PP nanoclay injected samples under biaxial impact tests.

Regarding CNT based nanocomposites, it is been found that the flow fields of injection molding induce ordered aggregates close to the sample walls with a different crystal packing that native SWCNT ropes in poly(butylene terephthalate)-SWCNT nanocomposites [90]. These authors found that in some specific positions of the WAXS pattern of the nanocomposite, in addition to the structural features of the PBT matrix, there appeared new reflections that can neither be indexed into the α form nor can be indexed into the β form of PBT [131]. The extra reflections, as described elsewhere [90], were indexed according to a rectangular centered 2D crystal lattice, which is not the expected hexagonal one for native SWCNT. Considering that this unexpected packing of SWCNT appeared in the regions close to the walls of the sample, they were attributed to the shear stress which reaches a maximum in these regions and tends to zero in the centre of the sample [132]. This hypothesis was supported by Raman spectroscopy experiments. In summary, a combination of X-ray and Raman spectroscopy, both working on microfocus fashion, was used to show that the shear flow tends to self-assemble the nanotubes into a novel lattice, different from that of native SWCNT, the maximum concentration of ordered aggregates being 1 mm to the wall edge of the injection molded bar.

4.2. Skin-Core Structure in Nanocomposites

As previously said, the skin-core structure of the moldings is generally revealed by polarized optical microscopy (PLM). It is been found that near the cold mold surface of PP-nanoclay composites, a frozen layer is formed adjacent to a shish-kebab or row structure developed, probably, in the region of highest shear rate. The shish-kebab structure is common in PP and other polyolefins, as previously described, and is a consequence of the fast crystallization of the polymer chains under high stresses near the skin region, as described by Fujiyama et al. [133]. In the vicinity of this sheared layer, β-spherulites can be found as a result of the crystallization under shear. In the core, α-spherulites are formed under quiescent crystallization conditions. A decrease of the skin thickness with the increase in clay content and the distance from the injection point was reported [96]. The latter is in accord with Fujiyama's work that stated that the crystalline structure and thickness of the skin change along the dimensions of the molded plaque [134-136]. The surprising finding is that the skin thickness decreases with clay content. It seems that the nanoclays reduce PP macromolecules relaxation time and therefore skin thickness. Also shear dissipation is higher further reducing the skin thickness. As a consequence of these differences, a toughening effect was observed for PP/nanoclay injected composites [97]. It was observed that skin and core underwent different deformations and failure mechanisms. The mid-thickness region, i.e. the core, underwent semibrittle fracture with little contribution of plastic deformation. The external skin layers still connected the two halves of the specimen after the core fractured, and further increase in displacement was caused by elongation, necking and gradual fracture of the skin layers.

This behavior was not seen for the same injected polymers without nanoclays (Figure 9). Similar delamination between skin and core in injection molded nanocomposites has been reported by other researchers [137-140]. The observed behavior was explained by proposing a failure model [97]: during fracture, the stress field in the core is intensified by the thermal residual stresses generated during injection [141, 142]. Therefore cracks nucleate first in the core at sub-micron size particles and subsequently crazes appear perpendicularly to the direction of loading. Due to the morphology differences between skin and core (mostly crystallite structure and molecular orientation), they may delaminate. Some of the crazes in the core develop as voids that grow in the load direction, producing the fracture of the core. This fracture develops as a sheet-like structure. Meanwhile the very oriented skin is still able to sustain load and to deform plastically.

The nanoclay toughening of polymers were attributed to the local and global conformation of the polymers within the host galleries of nanoparticles being significantly different from that observed in the bulk. Assuming that the nanoclays are intercalated, this is due to the confinement of the polymer chain. An additional dissipative mechanism appears as a result of the mobility of the nanoparticles [143]. Since the nanofiller is interacting at the polymer chain level, the time scales for motion of both filler and polymer are comparable. As a result, during the deformation process the filler can create temporary polymer chain crosslinks, thereby creating a local region of enhanced strength and consequently retarding the growth of microdefects.

Hernandez-Luna et al. (2002) reported that during the manufacturing of PP nanocomposites the low viscosity of the compatibilizer (maleic PP) led to a mixed structure where the bulk of the nanocomposite has an exfoliated structure and this is gradually lost at the edges [144]. However, this effect does not affect considerably the final properties of the material. This skin-core effect seems to lead to anisotropic behavior of the materials.

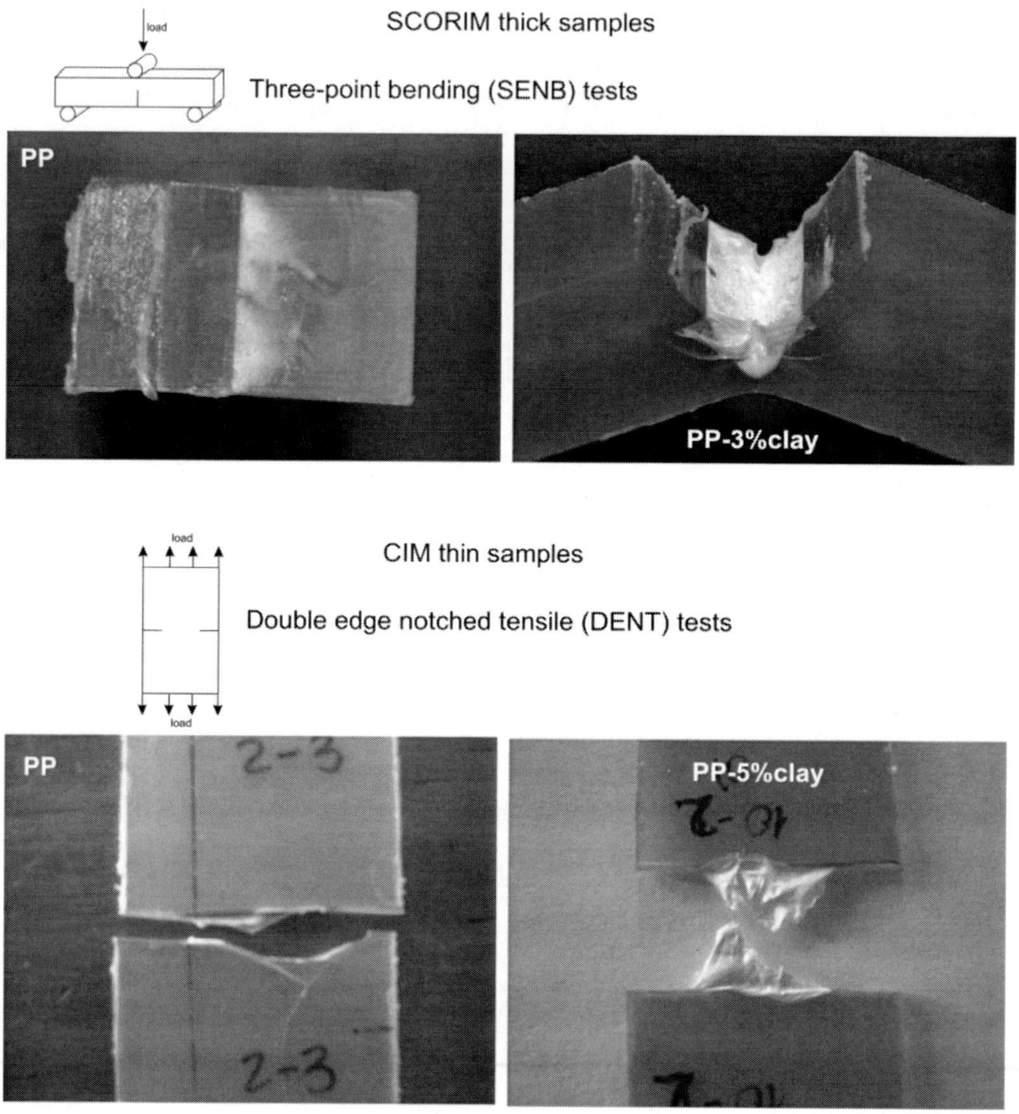

Figure 9. Skin effects found during deformation of PP-nanoclay samples injected both by conventional and shear controlled injection molding.

Several authors have studied the effect of SCORIM on morphology and mechanical properties of PP and PC/clay nanocomposites [13, 145, 114, 146]. General difference between CIM and N-CIM structures is the shrinkage (Figure 10). Cross section area of whole specimen obtained by CIM contains more shrinkage than relative of SCORIM. Addition of nanofillers diversified the morphology of injected pieces, especially developing the shear zone.

The manipulation of the processing parameters changes the lamination of the layers, depending upon the type of reinforced polymer. High setting variables contain much more layers than low settings.

The skin thickness for the same nanocomposite is more than 2.5 times thicker for the lowest settings. Addition of nanoparticles to PP did not change extremely the dimensions of this zone, but increased the number of layers. There was confirmation in literature of existence shish-kebab structure in the skin zone for SCORIM technique, where not just skin, but much more layers are obtained [145].

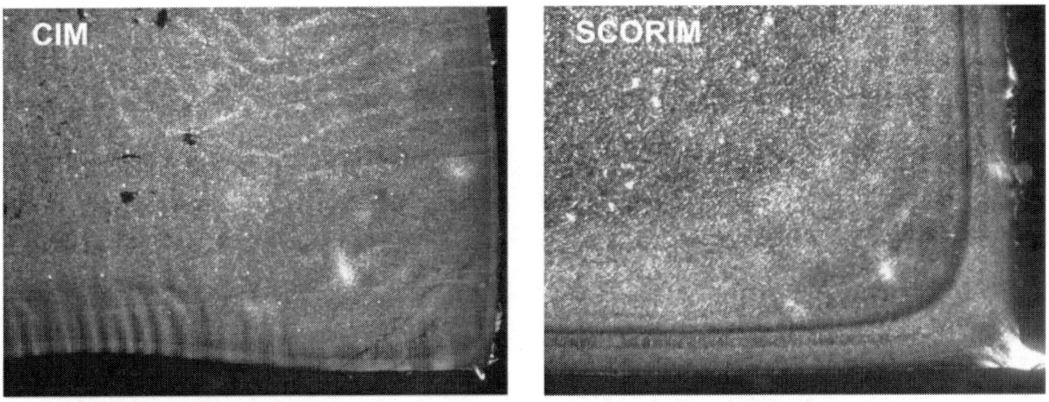

Figure 10. Polarized optical microscopy of PP injected samples in which differences in shrinkage are evident.

Regarding thick pieces injected by SCORIM, the effect of the SCORIM processing variables was reported to be much higher on the PP nanocomposite than on the neat PP [114]. For the former system, the skin layer thickness increased with stroke time and number, as would be expected as a result of a high applied shear level. Also, the skin layer thickness decreased with the melt temperature due to the high allowed relaxation time and low expected level of shear at high temperatures (lower melt viscosity). In every case of nanocomposites produced by non-conventional injection techniques, mechanical properties increased considerably by the combined effect of nanoclay and SCORIM processing technique (see some examples in Figure11).

PP/nanoclay thick pieces produced by SCORIM exhibited quasi-stable behavior in fracture tests [114]. Moreover, a large deformation of the skin was observed. This large deformation of the skin layer can be related to the high level of molecular and nanoclay orientation which favors the sliding of the macromolecules (fostered by the oriented nanoclays), thus improving the deformation capability. In fact, it has been pointed out that the development of necking is probably a combined effect of skin–core ratio, skin orientation and possible morphology rearrangement in the skin under different loading directions influenced also by heat evolution and dissipation processes. Since SCORIM induces a thicker multi-layered skin in the PP/nanoclay moldings, the skin effect is favored. This phenomenon leads to differences in the load–line displacement curves promoting a more stable crack propagation followed by a necking phase of highly different extension (skin effect) [114, 147].

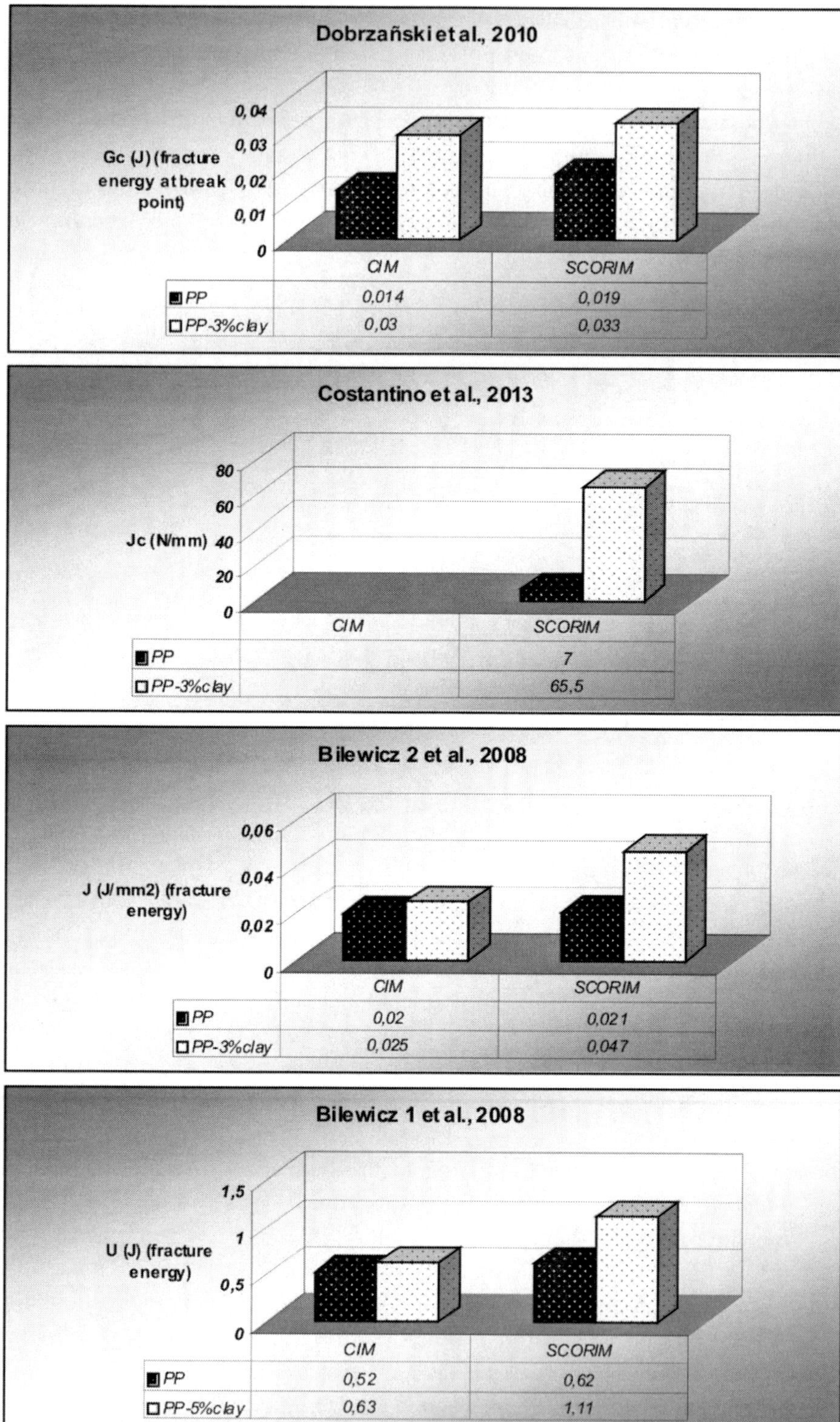

Figure 11. Comparison between fracture performance of CIM and SCORIM PP/clay nanocomposites.

4.3. Morphology of Matrix Polymer

As said, features like mechanical properties and morphology of injection molded composites, strongly depend on thermomechanical processing variables [10]. Processing modifications designed to enhance the orientation of polymers, such as low injection temperature or dynamic packing operation, can generate modest improvements in tensile modulus, even without the addition of conventional or nanoscopic filler. However, the filler can induce morphological changes in the crystalline structure of the polymer itself [148]. Despite that properties target can be achieved by manipulation of ratio and type of components, the processing is also fundamental and responsible for the mechanical response of the moldings. It is noteworthy that by compounding and processing (thermally, mechanically) of different polymers and reinforcements can be explored diversity of morphologies. These morphological changes will in turn have implications for mechanical properties such as modulus, strength, and toughness. The most fundamental and important aim in the polymer composites is control of the morphology [11-16]. This control significantly influences properties of materials. Change of the structure on molecular or macroscopic level may improve dramatically mechanical properties, e.g. modulus, stiffness and impact strength [18-22] as well toughness and flexural modulus. In this scenario, the nanoscopic scale of the filler becomes relevant because smaller fillers, with their larger surface area-to-volume ratio, reside in close proximity to a greater percentage of the bulk polymer than conventional fillers. There are a number of ways in which nanoscopic fillers combined with processing parameters may influence the morphology of polymers.

4.3.1. Crystalline Unit Cell

The mechanical properties of the pristine polymers and polymer–clay nanocomposites are mainly dependent on crystal properties, as the packing parameters will determine the inter-chain interactions in the different type of crystals. The crystalline unit cell may change by the addition of nanofillers.

This effect has been observed in, for example, nylon 6 nanocomposites where addition of clay induces a change from α to γ crystalline form [149]. Polyamide-6 is a semicrystalline polymer consisting of amorphous and three crystalline phases: α, γ and β, the latter considered as an intermediate state between the former two [150] [Xenopoulos and Clark, 1995]. The occurrence of the α or γ structure depends mainly on the packing of the alkyl chains during crystallization and on the length of the hydrogen bonds. The α structure is monoclinic, with molecules in a fully extended zig-zag conformation and grouped into planar, H-bonded sheets. The γ structure is monoclinic or pseudo-hexagonal, with the amide group tilted with respect to the chain axis and lower density packing compared to the α structure. The H-bond lengths are higher, not collinear and disposed on extended sheets [151, 152]. As a consequence, α and γ structures present different properties, such as a higher Young's Modulus in the α structure derived from the more extended organization of the molecules and a higher ductility in the γ due to the longer distance between adjacent amidic groups [153]. It is been reported that the crystallization of the γ structure is favored by clay mineral platelets [154-158] while maintaining the percent of crystallinity. Due to the fact that the distance between adjacent amidic groups is longer in the γ phase, the ductility of this structure is supposed to be higher than in the α. On the other hand, the fracture energy has to be higher in the α structure, that causing the enhancement of the toughness. The most extended unit cell of the α structure improves properties like Young's modulus with respect to the γ structure. Yebra-Rodríguez et al. [159] found that injected polyamide-6 (PA6) presented both α and γ structures, whereas in injected PA6-clay only the γ structure appeared, as determined by XRD and DSC. Although the crystal properties of polymers are influenced by processing conditions [160], the main factor affecting the occurrence of the γ structure in these injected PA6-clay pieces was the presence of montmorillonite, considering that all samples were manufactured under the same conditions. According to the literature [161, 162], the occurrence of α or γ structure depends on (1) the packing density of the alkyl chains, and (2) the length of hydrogen bonds. In the α structure the H-bond was shorter (and consequently more stable), and the packing density higher than in the surfaces of montmorillonite platelets that interacted with the polyamide macromolecules allowing efficient packing of the chains, but at larger distances compared to those required for the H-bonds in the α form. The montmorillonite impeded the shifting of H-bonds associated with the transformation from γ to thermodynamically more stable α structures. Moreover, the literature reports that the epitaxial nucleation of the polymer on the clay surface (i.e. montmorillonite) induces the formation of the less stable γ structure. The explanation for the epitaxial mechanism is the good correspondence between the disposition of the amide groups in the polyamide-6 and the OH groups in the basal plane of the montmorillonite along its [100] direction. In fact, epitaxy triggers the formation of H-bonds between amide and OH ions and further stabilizes the polymer-clay nanocomposites structure [163,164]. Taking into account the processing conditions, a high rate of crystallization in the nanocomposite samples could also have played an important role in the preservation of the γ structure [165].

Three crystalline lattices for isotactic PP are known and there is evidence for preferential formation of the γ-phase in some PP nanocomposites [166, 167]. Regarding injected nanocomposites, the effect on clay as α-phase nucleating agent in PP has been reported [96, 97]. In XRD patterns of PP moldings a small amount of β-form (peaks at $2\theta = 16.2$ and 21.2°, associated to (300) and (301) planes) was found, besides the main α-form crystallites (peaks at $2\,\theta = 14.2$, 17, 18.8 y 25.5°, corresponding to (110), (040), (130), and (131) planes, respectively [28]. In the nanocomposite moldings, the intensities of peaks at 16.2° and 21.2° (β polymorph) were attenuated, while the intensity peaks at 17° and 25.5° (α polymorph) were enhanced. It seems therefore that clay acted as α nucleation site, reducing the amount of β-form PP [96]. Similar results of β-suppressor effect of filler were reported by Varga et al for PP-talc composites [168]. The addition of 2% masterbatch of PP-nanoclay caused an increase in the elastic modulus and hardness as assessed by depth sensing indentation [96], due to the α-nucleating effect of the nanoclay and the differences in crystalline structure induced by the presence of the masterbatch. Further incorporation of MB did not affect the stiffness, probably due to the compensation effect between clay and compatibilizer (PP-MAN) present in MB. MB content larger than 2% induced only a slight decrease in hardness, indicating a slight reduction in yield stress. The nanoclay stacks provide microvoid nucleating sites at their surfaces and therefore release the constraints for shear yielding [169].

Marked differences in crystalline structure at the core and the surface of injected PP/nanoclay composites were observed, being more intense in the case of SCORIM thick pieces than in CIM pieces [114]. Three PP polymorphs, α, β and γ, were detected in the core while only α and γ polymorphs were detected in the skin. A γ–phase nucleating effect was seen in the skin of nanocomposites, the doublet (111)-(041) being smaller in the skin than in the core, indicating a higher orientation of PP molecules in the skin layer. In the X-ray spectra of the skin of PP/nanoclay the intensities of peaks at 16.2° and 21.2° (β polymorph) were attenuated, while the intensity peaks at 17° and 25.5° (α polymorph) and 20° (γ polymorph) were enhanced. Given that the specific diffraction peaks of α and γ phases correspond to (130)α and (117)γ, respectively, the content of crystalline γ-phase (*G*) was evaluated using the Turner-Jones equation:

$$G = \frac{I\gamma_{(117)}}{I\gamma_{(1117)} + I\alpha_{(130)}} \qquad (1)$$

where in a generic sense *Ii*(*abc*) is the intensity of the (*abc*) plane for each phase *i* [170]. It was seen that the amount of γ increased with clay content in an asymptotic way to a maximum value, this finding in agreement with other authors who claimed that nanoclay [171, 172] and SCORIM promote the formation of γ-phase PP crystallites [173]. They proposed that the nanoclay changes the equilibrium state of the polymer (conformation) and it provides favorable sites for possible epitaxial growth of γ-phase because the lattice mismatch is <10%. In summary, in these composites the nanoclay acted as α and γ nucleation site, reducing the amount of β-form PP. In flexural tests, the mechanical response is governed by the most exterior fiber, i.e. the skin of injected samples. As this skin in very changed for nanocomposites with respect to matrix PP, the flexural modulus increased approximately 50% with the addition of 3% nanoclay [114]. The addition of clay improved the fracture performance of PP SCORIM samples thanks to the differences induced by nanoclay in the microstructure described above: it refines the crystal structure and promotes epitaxial growth of the γ phase. This γ phase induces a large-scale plastic deformation with consequent tearing of matrix ligaments leading to fibrillation, responsible for higher toughness in relation to neat PP [171].

4.3.2. Crystalline Morphology

The crystalline morphology of the polymer matrix may be affected by nanofillers. For example, a transformation from spherulitic to oriented structure may occur.

When crystallized at high shear rates, PP forms oriented structures that could be modified by the presence of dispersed clay [174, 96, 114]. A way to measure the orientation of PP by XRD is to observe the (040)/(110) ratio, which indicates the orientation of the crystalline-phase of PP in the (040) direction [175]. From the wide angle X-ray diffraction (WAXD) pattern [176], the PP α-phase orientation indices can be determined as

$$A_{110} = \frac{I_{\alpha_1}}{I_{\alpha_1} + I_{\alpha_4}} \qquad (2)$$

$$A_{040} = \frac{I_{\alpha_2}}{I_{\alpha_1} + I_{\alpha_2} + I_{\alpha_3}} \qquad (3)$$

where I is the peak height after background subtraction, α_1 corresponds to the (110) reflection, α_2 to the (040) reflection, α_3 to (130) reflection and α_4 to the (111) and (041) reflections. The A_{110} index is based on the extinction of (hk1) reflections (in which case, A_{110} tends to 1) relatively to the (110) reflections. It thus gives an indication of an orientation roughly parallel to the c-axis of the crystallites corresponding to the flow direction in the injected samples [177]. The A_{040} index is based on the extinction of the (hk0) reflections relatively to the (0k0) reflections and gives an indication of a preferential orientation perpendicular to the b-axis, *i.e.* pretty much parallel to the surface sample [177]. Orientation indices based on the reflection intensities of PP were reported in literature as a function of clay content, and showed that the presence of nanoclay caused the orientation of α-PP crystals in the (040) direction which means preferential orientation perpendicular to the b-axis of the crystallites, i.e. parallel to the surface sample.

Fibrous crystalline morphology has also been reported in PP nanocomposites [178].

In SCORIM thick samples, a high increase of the (040)/(110) ratio in XRD patterns was reported [114]. This indicates that in SCORIM moldings the nanoclay induced the orientation of the crystalline-phase of PP in the (040) direction – i.e. increase of degree of epitaxiality [175] – and imparting a low level of crystalline phase orientation in the flow direction. This was also revealed by the high values of the epitiaxial indices. The analysis of results indicated that the most relevant morphology feature in these pieces was the skin orientation and crystallinity in the flow direction: higher skin orientation levels and crystallinity induced a lower toughness (see Figure 12).

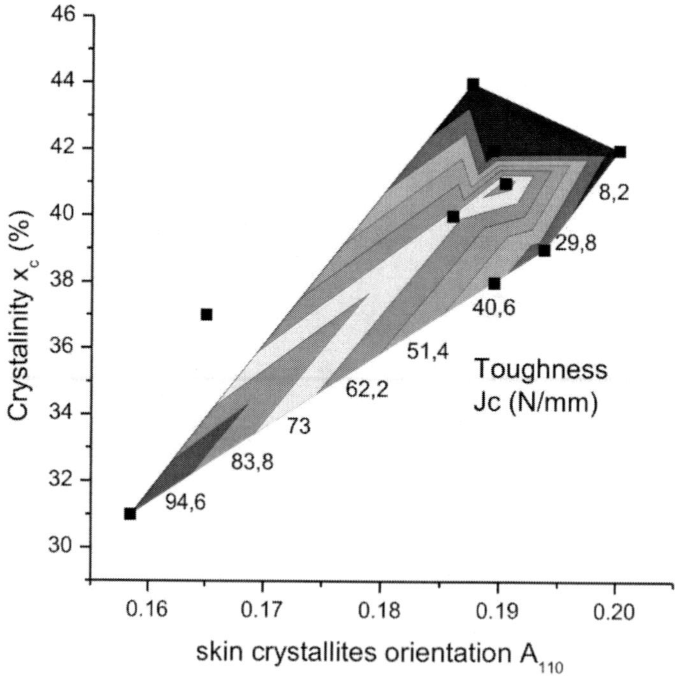

Figure 12. Correlation between fracture energy values (Jc) and morphology parameters. A_{110} indicates skin orientation (Reprinted from Ref. [114] Costantino, A., Pettarin, V., Viana, J.C., Pontes, A., Pouzada, A., Frontini, P., (2013). Polym. Int. 62, 1589).

Vasanthan in 2007 demonstrated that crystallinity has a direct correlation with the terminal modulus and extension-at-break in polyamide-6 fibres [179]. The orientation of both the crystalline and the amorphous components of the polyamide-6, as well as that of the clay mineral platelets, are keys to enhance the mechanical properties of nanocomposites. According to Muratoglu et al., fracture mechanisms are highly dependent on the orientation of the clay mineral platelets in the polymeric matrix [180]. The change of orientation affects the mechanical properties such that under uniaxial tension the Young's Modulus and flow stress drop with increasing thickness, i.e. with decreasing orientation. Yebra-Rodríguez et al. [159] explained observations in XRD and 2D-XRD patterns [164] of polyamide 6-clay injected samples taking into account that clay mineral platelets acted as nucleation agents. Their results indicated that the polyamide crystals in the PA6-clay injected samples were more oriented than in the PA6. Authors argued that the sharpened XRD peaks resulted from a larger quantity of preferentially oriented crystals enhancing the diffraction intensity. Also the 2D-XRD pattern for injected PA6 samples exhibited only one Debye–Scherrer ring resulting from a homogeneous intensity produced by randomly orientated polyamide-6 crystals. On the contrary the PA6-clay samples exhibited two Debye–Scherrer incomplete rings (arcs). The disposition of these arcs indicated that polyamide-6 crystals had a preferential orientation defined by the flow direction. The uniaxial orientation of the polyamide-6 suggested a fibre texture of the crystals aligned with the molding direction. Presumably during the molding process the montmorillonite platelets aligned parallel to the (h00) planes of the polyamide-6 crystals, thus inducing the occurrence of the γ structure. The skin-core morphology resulted from the different shear flow across the thickness of the mould during the injection molding process. The shear stress was higher in the skin, and thus the molecular chains were preferentially oriented in the flow direction. Consequently the thicker samples with better developed cores tended to be scarcely oriented, irrespectively on the presence of clay mineral platelets. Nonetheless it cannot be ruled out that an increase in cooling rate could have enlarged the effect of flow velocity during the manufacture process [160].

Regarding CNT based nanocomposites, in order to transfer the outstanding properties of CNT to the composite material from nano- to microscale, one essential step involves assembling and processing of the nanocomposite. By combining microfocus X-ray diffraction and Raman spectroscopy techniques, García-Gutiérrez et al. have studied the effect of flow fields on the structure of SWCNT and matrix for an injection molded poly(butylene terephthalate) nanocomposite [90]. They found that a modest amount of SWCNT is enough to template the crystalline morphology of the nanocomposite during processing. The intensity of all the three reflections (100, 010 and 001) in the Wide Angle X-ray Scattering (WAXS) pattern in the PBT/SWCNT nanocomposite presented a clear azimuthal dependence, such dependence for the 100 reflection in PBT being different when compared to that in the nanocomposite. This fact in addition to those differences found between the isotropic small-angle X-ray scattering (SAXS) pattern of PBT and the oriented one of the nanocomposite indicated that SWCNT firstly induce orientation of the polymer matrix and secondly template the morphology of crystallization during flow. Considering that pure PBT revealed isotropic SAXS patterns then the orientation observed in the nanocomposite can be thought to be induced by the presence of SWCNT in the sample. Authors found that the SAXS pattern exhibited both equatorial streaks along with the meridian maxima. The equatorial intensity can be attributed to bundles of parallel chain segments preferentially oriented along the flow direction following the orientation of SWCNT in this direction (shish structures), while the meridian maximum observed is consistent with the presence of layered lamellar assemblies (kebab) growing epitaxially from the shish structures in a direction perpendicular to their axis [181, 182]. This situation was represented by a model of a hybrid shish-kebab structure which is in agreement with reports in similar systems [183, 184], where the SWCNT act mainly as the shish and the epitaxial grown polymer crystals perpendicular to the SWCNT are the kebabs (Figure 13). These authors also found that in the nanocomposite subjected to an elongational flow field (i.e. injected), a fraction of the SWCNTs tends to self-assemble into a rectangular lattice which is not the expected hexagonal one for native SWCNTs. Considering that this unexpected packing of SWCNT appeared in the regions close to the walls of the they attribute it to the shear stress which reaches a maximum in these regions and tends to zero in the center of the sample [184].

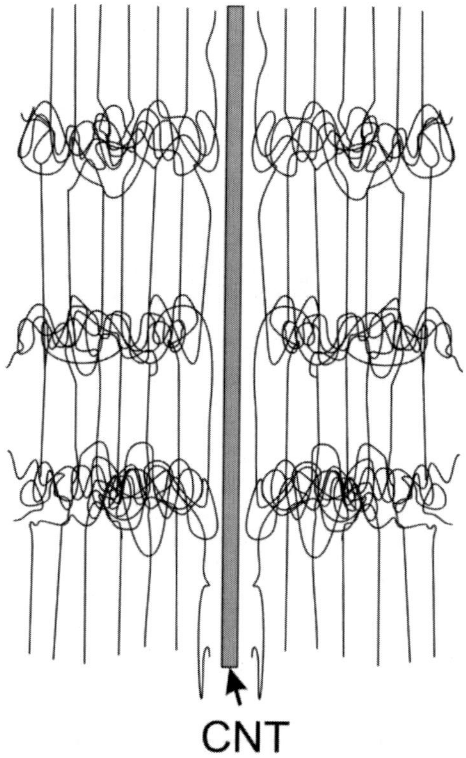

Figure 13. Schematic model of a hybrid SWCNT-PBT shish-kebab (Reprinted from Ref. [184] García-Gutiérrez, M.C., Nogales, A., Rueda, D.R., Domingo, C., García-Ramos, J.V., Broza, G., Roslaniec, Z., Schulte, K., Davies, R.J., Ezquerra, T.A. (2006). Polymer 47, 341).

The previous explained orientation of crystalline phases influences mechanical performance of injection moldings. To obtain oriented materials is a technological goal because of the significant improvements of their mechanical properties such as Young's modulus. Fujiyama and Wakino studied different injected moldings of PP and stated that the higher the crystalline orientation, the higher the flexural modulus, flexural strength, and mold shrinkage [185].

4.3.3. Degree of Crystallization

The degree of crystallization may change by the addition of nanoreinforcements. As quantified by differential scanning calorimetry, it has been reported both that the crystallinity of PP nanocomposites decreases modestly with increased clay content [186] and that it increased modestly with clay content [96]. The addition of nanoclay to PET matrices increases the degree of crystallinity [118], the clay platelets possibly playing the role of nucleating agent [187, 188], as revealed by DSC. Another reason is that MMT nano-structure itself can help the PET molecules stack on each other to grow into crystallites [188]. This result was confirmed by FTIR [116]. Also PA6-clay injected samples present higher crystallinity than PA6 samples, irrespectively of the processing conditions (i.e. cooling temperature after injection molding) [162]. Higher degree of crystallinity implies higher stiffness and modulus, but lower toughness.

Also, differences in crystallinity between skin and core have been reported for PP-nanoclay composites as assessed by depth sensing indentation [96]. The indentation data through the thickness showed that the core region is stiffer and harder than the skin region due to the higher crystallinity of the core in comparison with the skin. Moreover, Bilewicz et al. reported that specimens molded with low melt temperature and presented a low degree of crystallinity evidenced superior fracture toughness [189].

PP/nanoclay pieces obtained by SCORIM showed that the amount of overall crystallinity of PP moldings increased with the addition of nanoclay, so the nanoclay seems to act as a nucleating agent [114]. Analysis of variance (ANOVA) was used to verify the effect of the processing variables in morphology development. It emerges that both high shear rate and high temperature induce a high degree of crystallinity in the PP nanocomposite.

4.3.4. Size of Crystalline Morphology

The characteristic size of matrix crystalline morphology may also be affected by nanofillers. Svoboda et al. found that for nanocomposites with spherulitic structure, the average crystallite size decreases significantly with increasing clay content, possibly because of a role of clay in nucleation [186]. Time-resolved light scattering and optical microscopy at the early stages of crystallization have also identified differences in the number of point nuclei in PP and PP nanocomposites [190]. Hernandez Luna et al. reported that the introduction of clay into (PP + maleic PP) resulted in smaller spherulites of injected samples indicating nucleation effects [144]. However, higher concentrations, around 5%, showed a distribution of spherulite sizes. While the initial presence of the clay would serve as a nucleating agent, additional clay would serve as barriers to spherulite growth. Huang et al. reported that the number of crystalline spot sites in the PA6/clay nanocomposite part is much larger than that of the spherulites in the PA6 part, owing to the nucleation and confinement effects imposed by the clay [191]. This suggests that the clay is an effective nucleating agent for PA6.

CONCLUSION

Interest in nanocomposites is motivated by the challenge to provide light-weight, low-cost materials for industrial applications, with high mechanical performances. In the case of polymer nanocomposite containing nanoclays and carbon nanotubes, the performance of a polymer is largely improved while its functionalities – such as gas-barrier properties and flame retardancy – can be endowed and improved by the addition of very small amounts of fillers. Accordingly, in recent years the research and development of polymer nanocomposites has accelerated. Recently, thermoplastic nanocomposites have attracted much attention due to advances in the methods of their manufacture.

By adding small amounts of nanofiller into a polymer so as to prepare a nanocomposite, both its melt characteristics (processabilities) and solid characteristics (product properties) are greatly altered.

In this chapter we have described the morphology developments governing the solid characteristics of polymer nanocomposites, and details have been provided of the morphologies that most affect the products' properties. The morphology of composites, and the intercalation, exfoliation and dispersion of the MMT and dispersion of CNT, are important factors in general polymer/filler composites. Yet, the crystalline structures and the orientation and crystallization behavior of the base polymer may also affect the quality of the nanocomposites.

Clearly, today investigations with polymer nanocomposites will continue to be actively pursued, and as the importance of research and development in this area becomes increasingly recognized major advances are to be expected. During the past two decades, much effort has been expended in attempts to understand the relationship between the preparation methodologies, the microstructures and the mechanical properties of various nanocomposites. During coming years investigations should be directed towards the development of new moldings with optimized performances.

REFERENCES

[1] Alexandre, M., Dubois, P., (2001). Polymer-layered silicate nanocomposites: Preparation, properties and uses of a new class of materials. *Mater. Sci. Eng.* 28, 1.

[2] Ray, S.S., Okamoto, M., (2003). Polymer/layered silicate nanocomposites: a review from preparation to processing. *Prog. Polym. Sci.* 28, 1539.

[3] Karian, H.G., (2003). Ed., Handbook of Polypropylene and Polypropylene Composites, 2nd ed., Marcel Dekker, New York.

[4] Lau, A.K.-T., Hussain, F., K. Lafdi, (2010). Eds. Nano- and Biocomposites, CRC Press, Taylor & Francis Group, Boca Raton.

[5] Baniasadi, H., Ramazani, A.S.A., Javan Nikkhah, S., (2010). Investigation of in situ prepared polypropylene/clay nanocomposites properties and comparing to melt blending method. *Materials & Design* 31, 76.

[6] Gupta R.K., Kennel, O., Kim, K-J., (2010). Eds. Polymer Nanocomposites Handbook, CRC Press, Boca Raton.

[7] Ding, C., Jia, D., He, H., Guo, B., Hong, H., (2005). How organo-montmorillonite truly affects the structure and properties of polypropylene. *Polym. Test.* 24, 94.

[8] Zhang, Y.-Q., Lee, J.-H., Rhee, J.M., Rhee K.Y., (2004). Polypropylene–clay nanocomposites prepared by in situ grafting-intercalating in melt. *Compos. Sci. Technol.* 64, 1383.

[9] Chen, L., Wong, S.-C., Liu, T., Lu, X., He, C., (2004). Deformation mechanisms of nanoclay-reinforced maleic anhydride-modified polypropylene. *J. Polym. Sci. Polym. Phys.* 42, 2759.

[10] Viana, J.C., Cunha, A.M., Billon, N., (1997). The tensile behaviour of an injection-moulded propylene-ethylene. *Polymer International* 43, 159.

[11] Utracki, L.A. (Ed), (2002). Polymer Blends Handbook, Kluwer Academic Publishers.

[12] Cunha, M., Fakirov, S., (2000). Structure development during polymer processing, Kluwer Academic Publishers.

[13] Bilewicz, M., Viana, J.C., Cunha, A.M., Dobrzanski, L.A., (2006). Morphology diversity and mechanical response of injection moulded polymer nanocomposites and polymer-polymer composites. *Journal of Achievements in Materials and Manufacturing Engineering* 15, 159.

[14] Wróbel, G., Rdzawski, Z., Muzia, G., Pawlak, S., (2009). The application of transient thermography for the thermal characterisation of carbon fibre/epoxy composites. *Journal of Achievements in Materials and Manufacturing Engineering* 36/1, 49.

[15] Zibowicz, B., Drak, M., Dobrzanski, L.A., (2009). Corrosion resistance of the composite materials: nanocrystalline powder – polymer type in acid environment. *Journal of Achievements in Materials and Manufacturing Engineering* 36/2, 126.

[16] Dobrzanski, L.A., Tomiczek, A., Tomiczek, B., Nlawska-Waniewska, A., Iesenchuk, O., (2009). Polymer matrix composite materials reinforced by $Tb_{0.3}Dy_{0.7}Fe_{1.9}$ magnetostrictive particles. *Journal of Achievements in Materials and Manufacturing Engineering* 37/1, 16.

[17] Van der Beek, M.H.E., Peters, G.W.M., Meijer, H.E.H., (2005). A dilatometer to measure the influence of cooling rate and melt shearing on specific volume. *Int. Polym. Proc.* 20, 111.

[18] Powell, C.E., Beall, G.W., (2006). Physical properties of polymer/clay nanocomposites. *Solid State and Materials Science* 10, 73.

[19] Viana, J.C., (2004). Development of the skin layer in injection moulding: Phenomenological model. *Polymer* 45, 993.

[20] Rojek, M., Stabik, J., (2009). The influence of X-rays on strength properties of polyester vascular system prosthesis, *Journal of Achievements in Materials and Manufacturing Engineering* 35/1, 47.

[21] Dobrzanski, L.A., Grajcar, A., Borek, W., (2008). Microstructure evolution and phase composition of high-manganese austenitic steels. *Journal of Achievements in Materials and Manufacturing Engineering* 31/2, 218.

[22] Wróbel, G., Kaczmarczyk, J., Stabik, J., Rojek, M., (2009). Numerical models of polymeric composite to simulate fatigue and ageing processes. *Journal of Achievements in Materials and Manufacturing Engineering* 34/1, 31.

[23] Aurrekoetxea, J., Sarrionandia, M.A., Urrutibeascoa, I., Maspoch, M., (2003). Effects of injection moulding induced morphology on the fracture behaviour of virgin and recycled polypropylene. *Polymer* 44, 6959.

[24] Godinho, J.S., Cunha, A.M., Crawford, R.J., (2000). Prediction of mechanical properties of polyethylene mouldings based on laminate theory and thermomechanical indices. *Plast. Rubber. Compos.* 29, 329.

[25] Housmans, J-W., Gahleitner, M., Peters, G.W., Meijer, H.E., (2009). Structure–property relations in molded, nucleated isotactic polypropylene. *Polymer* 50, 2304.

[26] Viana, J.C., Billon, N., Cunha, A.M., (2004). The thermomechanical environment and the mechanical properties of injection moldings. *Polym. Eng. Sci.* 44, 1522.

[27] Viana, J.C., Cunha, A.M., Billon, N., (2000). In: ANTEC 2000 plastics: annual technical conference, Orlando, USA, May 7–11, 2000. Society of Plastics Engineers, 617.

[28] Kalay, G., Bevis, M.J., (1997). Processing and physical property relationships in injection-molded isotactic polypropylene. 1. Mechanical properties. *J. Appl. Polym. Sci. Part B* 35, 241.

[29] Cermak, R., Obadal, M., Ponizil, P., Polaskova, M., Stoklasa, K., Heckova, J., (2006). Injection-moulded α- and β-polypropylenes: II. Tensile properties vs. processing parameters. *Eur. Polym. J.* 42, 2185.

[30] Davé, P., Chundury, D., (1997.) Effect of injection molding conditions on the properties of filled and reinforced polypropylene. *J. Inject. Mold. Technol.* 1, 181.

[31] Guerrica-Echevarría, G., Eguiazábal, J.I., Nazábal, J., (1998). Influence of molding conditions and talc content on the properties of polypropylene composites. *Eur. Polym. J.* 34, 1213.

[32] Ozcelik, B., Ozbay, A., Demirbas, E., (2010). Influence of injection parameters and mold materials on mechanical properties of ABS in plastic injection molding. *Int. Commun. Heat Mass Transfer* 37, 1359.

[33] Sahin, S., Yayla, P., (2005). Effects of processing parameters on the mechanical properties of polypropylene random copolymer. *Polym. Test.* 24, 1012.

[34] Shie J-R., (2008). Optimization of injection-molding process for mechanical properties of polypropylene components via a generalized regression neural network. *Polym. Adv. Technol.* 19, 73.

[35] Yang, Y-K., (2006). Optimization of Injection-Molding Process of Short Glass Fiber and Polytetrafluoroethylene Reinforced Polycarbonate Composites via Design of Experiments Method: A Case Study. *Mater. Manuf. Process* 21, 915.

[36] Zhou, Y., Mallick, P., (2005). Effects of melt temperature and hold pressure on the tensile and fatigue properties of an injection molded talc-filled polypropylene. *Polym. Eng. Sci.* 45, 755.

[37] Kalay, G., Bevis, M., (1997). Processing and physical property relationships in injection-molded isotactic polypropylene. 2. Morphology and crystallinity. *J. Appl. Polym. Sci. Part B* 35, 265.

[38] Cermak, R., Obadal, M., Ponizil, P., Polaskova, M., Stoklasa, K., Lengalova, A., (2005). Injection-moulded α- and β-polypropylenes: I. Structure vs. processing parameters. *Eur. Polym. J.* 41, 1838.

[39] Pantani, R., Coccorullo, I., Speranza, V., Titomanlio, G., (2005). Modeling of morphology evolution in the injection molding process of thermoplastic polymers. *Prog. Polym. Sci.* 30, 1185.

[40] Viana, J.C., Cunha, A.M., Billon, N., (2002). The thermomechanical environment and the microstructure of an injection moulded polypropylene copolymer. *Polymer* 43, 4185.

[41] Demiray, M., Isayev, A., (1996). Effect of proccesing conditions on crystallinity and microstructure of injection moldings of polypropylene of varios molecular weights. SPE ANTEC Tech Papers 42, 1576.

[42] Chang, T.C., Faison III, E., (2001). Shrinkage behaviour and optimization of injection molded parts studied by the taguchi method. *Polym. Eng. Sci.* 41, 703.

[43] Viana J.C., Billon, N., Cunha, A.M., (2004). The thermomechanical environment and the mechanical properties of injection moldings. *Polym. Eng. Sci.* 44, 1522.

[44] Chang, T.C., Faison III, E., (1999). Optimization of weld line quality in injection molding using an experimental design approach. *J. Inject. Mold Technol.* 3, 61.

[45] Selden, R., (1997). Effect of processing on weld line strength in five thermoplastics. *Polym. Eng. Sci.* 37, 205.

[46] Mielewski, D.F., Bauer, D.R., Schmitz, P.J., Van Oene, H., (1998). Weld line morphology of injection molded polypropylene. *Polym. Eng. Sci.* 38, 2020.

[47] Wu, C-H., Liang, W-J., (2005). Effects of geometry and injection-molding parameters on weld-line strength. *Polym. Eng. Sci.* 45, 1021.

[48] Kim, S.J., Lee, K., Kim, Y.I., (1996). Proceedings of SPIE – The 4th Int. Conf. on Computer-Aided Design and Computer Graphics. 2644, 173.

[49] Lotti, C., Bretas, R.E.S., (2003). The influence of morphological features on mechanical properties of injection molded PPS and sits prediction using neural networks. *Proceedings of PPS, European Conference*, Athens, 184-P.

[50] Castro, C., Bhagavatula, N., Cabrera-Rios, M., Lilly, B., Castro, J.M., (2003). Identifying the best compromises between multiple performance measures in injection molding (IM) using data envelopment analysis. *Proceedings SPE Technical Papers ANTEC'03, Nashville.*

[51] Turng, L.S., Peic, M., (2002). Computer aided process and design optimization for injection moulding. *Journal of Engineering Manufacture. Proceedings of the Institution of Mechanical Engineers Part B*. 216, 1523.

[52] Alam, K., Kamal, M.R., (2003). A Genetic Optimization of Shrinkage by Runner Balancing. *Proceedings SPE Technical Papers ANTEC'03, Nashville.*

[53] Cunha-Gaspar, A., Viana, J.C., (2005). Using Multi-Objective Evolutionary Algorithms to Optimize Mechanical Properties of Injection Moulded Parts. *Int. Polym. Proc.* XX, 3, 274.

[54] Fernandes, C., Pontes, A.J., Viana, J.C., Gaspar-Cunha, A., (2010). Using multiobjective evolutionary algorithms in the optimization of operating conditions of polymer injection molding. *Polym. Eng. Sci.* 50, 1667.

[55] Demirer, A., Soydan, Y., Kapti, A.O., (2007). An experimental investigation of the effects of hot runner system on injection moulding process in comparison with conventional runner system. *Mater. Des.* 28, 1467.

[56] Bevis, M.J., (1999). Structure development and associated physical property enhancement by design in molded and extruded plastics. Proceed. PPS-16, Xangai.

[57] Kech, A., Ludwig, H.C., Eyerer, P., (2000). Improvement of material properties of unreinforced isotatic polypropylene using push-pull processing. Proceed. 3rd ESAFORM Conf. Mat. Froming, Stutgard, Germany, v. 35-38.

[58] Becker, H., Fischer, G., Mliller, U., (1993). Push-pull injection molding of industrial products. *Kunststoffe German Plastics* 83(3), 3.

[59] Youbing, L., Xuequin, G., Shaoyan, Y., Yi, Y., Jie, Z., Wenti, K., Kaizhi, S., (2005). The relationship between mechanical properties and morphology of vibration-injection-molded polyethylene. *Polymer International* 54, 240.

[60] Kantz, M.R., Newman, H.D., Stigale, F.H., (1972). The skin-core morphology and structure–property relationships in injection-molded polypropylene. *J. Appl. Polym. Sci.* 16, 1249.

[61] Schultz, J.M., (1984). Microstructural aspects of failure in semicrystalline polymers. *Polym. Eng. Sci.* 24, 770.

[62] Trotignon, J-P., Verdu. (1987). Skin-core structure–fatigue behavior relationships for injection-molded parts of polypropylene. I. Influence of molecular weight and injection conditions on the morphology. *J. Appl. Polym. Sci.* 34, 1.

[63] Fujiyama, M., Wakino, T., (1992). Molecular Orientation in Injection-molded Polypropylene Copolymers with Ethylene. *Intern. Polym. Proc.* VII(1), 97.

[64] Phillips, R., Herbert, G., News, J., Wolkowicz. J., (1994). High modulus polypropylene: Effect of polymer and processing variables on morphology and properties. *Polym. Eng. Sci.* 34, 1731.

[65] Keller, A., Kolnaar, H.W.H., (1997). Flow-induced orientation and structure formation. In: Cahn RW, Haasen P and Kramer EJ (eds.) *Materials Science and Technology: A comprehensive treatment,* Volume 18: *Processing of Polymers,* Wiley VCH, New York, 189.

[66] Pening, A.J., (1980). Polymer crystallization. *Journal of Crystal Growth* 48, 574.

[67] Viana, J.C., Cunha, A.M., Billon, N., (2002). In: Annual conference of the polymer processing society (PPS-18), Guimaraes, Portugal, June 16–20, 2002, *Polymer Processing Society.*

[68] Fujiyama, M., (1995). In: Karger-Koscis J, Editor. Polypropylene structure, blends and composites: structure and morphology, Vol. 1. London: Chapman and Hall, 167.

[69] Cunha, A.M., Godinho, J.S., Viana, J.C., (2000). Structure development during polymer processing. In: Cunha, A.M., Fakirov, S., Editors. Nato Science Series, Series E: Applied Science, Vol. 370. Dordrecht: Kluver Academic; 255.

[70] Viana, J.C., Cunha, A.M., Billon, N., (1999). Prediction of the tensile impact behavior of injection molded samples from quasi-static data. *Polym. Eng. Sci.* 39, 1463.

[71] Cunha, A.M., Pouzada, A.S., (1995). In: Williams JG, Pavan A, Editors. Impact and dynamic fracture of polymers and composites, ESIS19. London: Mech. Eng. Public, 315.

[72] Viana, J.C., Cunha, A.M., Billon, N., (2001). The effect of the skin thickness and spherulite size on the mechanical properties of injection mouldings. *J. Mat. Sci.* 36, 4411.

[73] Shoemaker, J., (2006). Moldflow Design Guide - A Resource for Plastics Engineers, Hanser.

[74] Sousa, R.A., Reis, R.L., Cunha, A.M., Bevis, M.J., (2003). Processing and properties of bone-analogue biodegradable and bioinert polymeric composites. *Compos. Sci. Technol.* 63, 389.

[75] Niemeyer, C.M., (2001). Nanoparticles, proteins, and nucleic acids: Biotechnology meets materials science. *Angewandte Chemie-International Edition In English* 40/22, 4128.

[76] Drechsler, U., Erdogan, B., Rotello, V.M., (2004). Nanoparticles: Scaffolds for molecular recognition. *Chemistry - A European Journal* 10/22, 5570.

[77] Viana, J.C., (2006). Polymeric materials for impact and energy dissipation. *Plast. Rubber. Compos.* 35, 260.

[78] Vaia, R.A., Wagner, H.D., (2004). Framework for nanocomposites. *Materials Today* November, 32.

[79] Chau, J.L.H., Hsu, S.L-C., Chen, Y-M., Yang, C-C., Hsu, P.C.F., (2010). A simple route towards polycarbonate–silica nanocomposite, *Advanced Powder Technology* 21, 341.

[80] Kato, M., Usuki, A., (2000). Polymer-Clay Nanocomposites, John Wiley and Sons, New York, 98.

[81] LeBaron, P., Wang, Z., Pinnavaia, T. J., (1999). Polymer-layered silicate nanocomposites: an overview. *Appl. Clay Sci.* 15, 11.

[82] Zeng, O.H., Yu, A.B., Lu, G.Q., Paul, D.R., (2005). Clay-based polymer nanocomposites: research and commercial development. *J. Nanosci. Nanotechnol.* 5, 1574.

[83] Kojima, Y., Usuki, A., Kawasumi, M., Okada, A., Kukushima, Y., Kurauchi, T., Kamigaito, O., (1993). Mechanical properties of nylon-6 clay hybrid. *J. Mater. Res.* 8, 1185.

[84] Usuki, A., Kawasumi, M., Kojima, Y., Okada, A., Kurauchi, T., Kamigaito, O., (1993). Swelling behavior of montmorillonite cation exchanged for ω-amin acids by ε-caprolactam. *J. Mater. Sci.* 8, 1174.

[85] Usuki, A., Kojima, Y., Kawasumi, M., Okada, A., Fukushima, Y., Kurauchi, T., Kamigaito, O., (1993). Synthesis of nylon 6-clay hybrid. *J. Mater. Res.* 8, 1179.

[86] Solomon, M.J., Somwangthanaroj, A., (2004). Intercalated Polypropylene Nanocomposites, Dekker Encyclopedia of Nanoscience and Nanotechnology, Marcel Dekker Inc., 1483.

[87] Pettarin, V., Frontini, P.M., Pita, V.J.R.R., Dias, M.L., Valenzuela, Diaz, F., (2008). Polyethylene/(organo-montmorillonite) composites modified with ethylene/methacrylic acid copolymer: morphology and mechanical properties. *Composites Part A: Applied Science and Manufacturing* 39, 1822.

[88] Famá, L., Pettarin, V., Goyanes, S.N., Bernal, C.R., (2011). Starch/multi-walled carbon nanotubes composites with improved mechanical properties. *Carbohydrate Polymers* 83, 226.

[89] Costantino, A., Pettarin, V., Viana, J., Pontes, A., Pouzada, A., Frontini, P., (2012). Microstructure of PP/clay nanocomposites produced by shear induced injection moulding. *Procedia Materials Science* 1, 34.

[90] García-Gutiérrez, M.C., Nogales, A., Rueda, D.R., Domingo, C., García-Ramos, J.V., Broza, G., Roslaniec, Z., Schulte, K., Ezquerra, T.A., (2007). X-ray microdiffraction and micro-Raman study on an injection molding SWCNT-polymer nanocomposite. *Compos. Sci. Technol.* 67, 798.

[91] Weon, J.I., Sue, H.J., (2005). Effects of clay orientation and aspect ratio on mechanical behavior of nylon-6 nanocomposite. *Polymer* 46, 6325.

[92] Gianeli, W., Ferrara, G., Camino, G., Pellegatti,G., Rosenthal, J., Trombini,R.C., (2005). Effect of matrix features on polypropylene layered silicate nanocomposites. *Polymer* 46, 7037.

[93] Vlasveld, D.P., Vaidya, S.G., Bersee, H.E.N., Picken, S.J., (2005). A comparison of the temperature dependence of the modulus, yield stress and ductility of nanocomposites based on high and low MW PA6 and PA66. *Polymer* 46, 3452.

[94] Hong, C.H., Lee, Y.B., Bae, J.W., Jho, J.Y., Nam, B.U., Hwang, T.W., (2005). Preparation and mechanical properties of polypropylene/clay nanocomposites for automotive parts application. *J. Appl. Polym. Sci.* 98, 427.
[95] Yoon, P. J., Hunter, D.L., Paul, D.R., (2003). Polycarbonate nanocomposites. Part 1. Effect of organoclay structure on morphology and properties. *Polymer* 44, 5323.
[96] Pettarin, V., Viau, G., Fasce, L., Viana, J.C., Pontes, A.J., Frontini, P.M., Pouzada, A.S., (2013). Uni- and biaxial impact behavior of double-gated nanoclay-reinforced polypropylene injection moldings. *Polym. Eng. Sci.* 53, 724.
[97] Pettarin,[2] V., Brun, F., Viana, J.C., Pouzada, A.S., Frontini, P.M., (2013). Toughness distribution in complex PP/nanoclay injected mouldings. *Compos. Sci. Technol.* 74, 28.
[98] Szleifer, I., Yerushalmi-Rozen, R., (2005). Polymers and carbon nanotubes—dimensionality, interactions and nanotechnology. *Polymer* 46, 7803.
[99] Terrones, M., (2003). Science and Technology of the Twenty-First Century: Synthesis, Properties, and Applications of Carbon Nanotubes. *Ann. Rev. Mater. Res.* 33, 419.
[100] Lu, J.P., (1997). Elastic properties of carbon nanotubes and nanoropes. *Phys. Rev. Lett.* 79, 1297.
[101] Li, Z., Dharap, P., Nagarajaiah, S., (2004). Carbon nanotube film sensors. *Adv. Mater.* 16, 640.
[102] McNally, T., (2005). Polyethylene multiwalled carbon nanotube composites. *Polymer* 46, 8222.
[103] Tang, W., Santare, M.H., Advani, S.G., (2003). Melt processing and mechanical property characterization of multi-walled carbon nanotube/high density polyethylene (MWNT/HDPE) composite films. *Carbon* 41, 2779.
[104] Cooper, C.A., Ravich, D., Lips, D., Mayer, J., Wagner, H.D., (2002). Distribution and alignment of carbon nanotubes and nanofibrils in a polymer matrix. *Compos. Sci. Technol.* 62, 1105.
[105] Dondero, W.E., Gorga, R.E., (2006). Morphological and mechanical properties of carbon nanotube/polymer composites via melt compounding. *J. Polym. Sci. B* 44, 864.
[106] Manchado, M.A.L., Valentini, L., Biagiotti, J., Kenny, J.M., (2005). Thermal and mechanical properties of single-walled carbon nanotubes–polypropylene composites prepared by melt processing. *Carbon* 43, 1499.
[107] van der Wal, A., Mulder, J.J., Gaymans, R.J., (1998). Fracture of polypropylene: The effect of crystallinity. *Polymer* 39, 5477.
[108] Aït Hocine, N., Méderic, P., Aubry, T., (2008). Mechanical properties of polyamide-12 layered silicate nanocomposites and their relations with structure. *Polym. Test.* 27 (3), 330.
[109] Tjong, S.C., Meng, Y.Z., (2003). Tuning the photophysical properties of an ionic conjugated polymer through interactions with conventional polyelectrolytes. *J. Polym. Sci. Part B: Polym. Phys.* 41, 1676.
[110] Lee, J.H., Jung, D., Hong, C.E., Rhee, K.Y., Advani, S.G., (2005). Properties of polyethylene-layered silicate nanocomposites prepared by melt intercalation with a PP-g-MA compatibilizer. *Compos. Sci. Technol.* 65, 1996.
[111] Tjong, S.C., Meng, Y.Z., Xu, Y., (2002). Structure and properties of polyamide-6/vermiculite nanocomposites prepared by direct melt compounding. *J. Polym. Sci. Part B: Polym. Phys.* 40 (24), 2860.

[112] Zhu, L., Xanthos, M., (2004). Effects of process conditions and mixing protocols on structure of extruded polypropylene nanocomposites. *J. Appl. Polym. Sci.* 93, 1891.

[113] Bureau, M.N., Perrin-Sarazin, F., Ton-That, M.-T., (2004). Polyolefin nanocomposites: Essential work of fracture analysis. *Polym. Eng. Sci.* 44, 1142.

[114] Costantino, A., Pettarin, V., Viana, J.C., Pontes, A., Pouzada, A., Frontini, P., (2013). Morphology−performance relationship of polypropylene−nanoclay composites processed by shear controlled injection moulding. *Polym. Int.* 62, 1589.

[115] Nevalainen, K., Auvinen, S., Orell, O., Eteläaho, P., Suihkonen, R., Vuorinen, J., Järvelä, P., (2013). Characterization of melt-compounded and masterbatch-diluted polypropylene composites filled with several fillers. *Polym. Compos.* 34, 1.

[116] Rajesh, J.J., Soulestin, J., Lacrampe, M.F., Krawczak. P., (2012). Effect of injection molding parameters on nanofillers dispersion in masterbatch based PP–clay nanocomposites. *Express Polym. Lett.* 6, 237.

[117] Rodríguez-Llamazares, S., Rivas, B.L., Pérez, M., Perrin-Sarazin, F., Maldonado, A., Venegas, C., (2011). The effect of clay type and of clay-masterbatch product in the preparation of polypropylene/clay nanocomposites. *J. Appl. Polym. Sci.* 122, 2013.

[118] Barbosa, C.N., Chabert, F., Nassiet, V., Viana, J.C., Pereira, P., (2011). Effect of clay amounts on morphology and mechanical performances in multiscale pet composites. *AIP Conf. Proc.* 1353, 779.

[119] Chow, W.S., Mohd., Ishak, Z.A., Ishiaku, U.S., Karger-Kocsis, J., Apostolov, A.A., (2004). The effect of organoclay on the mechanical properties and morphology of injection-molded polyamide 6/polypropylene nanocomposites. *J. Appl. Polym. Sci.* 91, 175.

[120] Sun, L., Gibson, R.F., Gordaninejad, F., Suhr, J., (2009). Energy absorption capability of nanocomposites: A review. *Compos. Sci. Technol.* 69, 2392.

[121] Bousmina, M., (2006). Study of Intercalation and Exfoliation Processes in Polymer Nanocomposites. *Macromolecules* 39, 4259.

[122] Cotterell, B., Chia, J.Y.H., Hbaieb, K., (2007). Fracture mechanisms and fracture toughness in semicrystalline polymer nanocomposites. *Eng. Fract. Mech.* 74, 1054.

[123] Morelli, C.L., de Sousa, J.A., Pouzada A.S., (2007). Assessment of weld line performance of PP/Talc moldings produced in hot runner injection molds. *J. Vinyl Addit. Tech.* 13, 159.

[124] Ersoy, O.G., Nugay, N., (2004). A new approach to increase weld line strength of incompatible polymer blend composites: selective filler addition. *Polymer* 45, 1243.

[125] Martinez, Gamba, M., Antonio, Sergio, P., Patricia, Marıa, F., (2009). Impact properties and microhardness of double-gated glass-reinforced polypropylene injection moldings. *Polym. Eng. Sci.* 49, 1688.

[126] Vaxman, A., Narkis, M., Siegmann, A., Kenig S., (1991). Weld-line characteristics in short fiber reinforced thermoplastics. *Polym. Composite* 12, 161.

[127] Pozsgay, A., Papp, L., Fráter, T., Pukánszky, B., (2002). Polypropylene /montmorillonite nanocomposites prepared by the delamination of the filler, in Adsorption and Nanostructure, I. Dékány, Editor. Springer Berlin / Heidelberg. 120.

[128] Bucknall, C.B., Impact testing of polypropylene mouldings. (1986). *Pure & Appl. Chem.* 58, 985.

[129] Alcock, B., Cabrera, N.O., Barkoula, N.M., Peijs, T., (2006). Low velocity impact performance of recyclable all-polypropylene composites. *Compos. Sci. Technol.* 66, 1724.

[130] Bilewicz, M., Dobrzański, L.A., Viana, J.C., (2009). Morphological aspect of multilaminar PP composite. *Journal of Achievements in Materials and Manufacturing Engineering*, 37, 598.

[131] Nitzsche, S.A., Wang, Y.K., Hsu, S.L., (1992). Application of the molecular simulation technique for clarification of the □↔□ phase transformation in poly(butylenes terephthalate). *Macromolecules* 25, 2397.

[132] Kumaraswamy, G., Verma, R.K., Kornfield, J.A., (1999). Novel flow apparatus for investigating shear-enhanced crystallization and structure development in semicrystalline polymers. *Rev. Sci. Instrum.* 70, 2097.

[133] Fujiyama, M., Wakino, T., Kawasaki, Y., (1988). Structure of skin layer in injection-molded polypropylene. *J. Appl. Polym. Sci.* 35, 29.

[134] Fujiyama, M., (1995). Structures and Properties of Injection Moldings of β-Crystal Nucleator-Added Polypropylenes. *Int. Polym. Proc.* 10, 172.

[135] Fujiyama, M., (1995). Structure and Properties of Injection Moldings of β-Crystal Nucleator-Added Polypropylene. *Int. Polym. Proc.* 10, 251.

[136] Fujiyama, M., (1999). ''Processing-induced Morphology'' in Polypropylene: An AZ Reference, J. Karger- Kocsis, Ed., Kluwer Academic, Dordrecht, The Netherlands, 668.

[137] Saminathan, K., Selvakumar, P., Bhatnagar, N., (2008). Fracture studies of polypropylene/nanoclay composite. Part II: Failure mechanism under fracture loads. *Polym. Test.* 27, 453.

[138] Uribe-Arocha, P., Mehler, C., Puskas, J.E., Altstädt, V., (2003). Effect of sample thickness on the mechanical properties of injection-molded polyamide-6 and polyamide-6 clay nanocomposites. *Polymer* 44, 2441.

[139] Kwon, H.J., Jar, P.Y.B., (2007). New energy partitioning approach to the measurement of plane-strain fracture toughness of high-density polyethylene based on the concept of essential work of fracture. *Eng. Fract. Mech.* 74, 2471.

[140] Choi, B.H., Chudnovsky, A., Zhou, Z., (2009). Observation of failure characteristics of notched injection molded PP/TPO/MMT nanocomposites. In: ICCE-17, Honololu, Hawaii 149.

[141] Isayev, A.I., Crouthamel, D.L., (1984). Residual stress development in the injection molding of polymers. *Polym–Plastics Technol. Eng.* 22, 177.

[142] Zoetelief, W.F., Douven, L.F.A., Housz, A.J.I., (1996). Residual thermal stresses in injection molded products. *Polym. Eng. Sci.* 36, 1886.

[143] Gersappe, D., (2002). Molecular mechanisms of failure in polymer nanocomposites. *Phys. Rev. Lett.* 89, 058301.

[144] Hernandez-Luna, A., D. Souza, N.A., Ranade, A., Drewniak, M., (2002). Skin-core effects in polypropylene nanocomposites; Society of Plastics Engineers; ANTEC 2002 conference proceedings, 1470.

[145] Bilewicz, M., Viana, J.C., Dobrzañski, L.A., (2008). Development of microstructure affected by in-mould manipulation in polymer composites and nanocomposites, *Archives of Materials Science and Engineering* 30, 71.

[146] Dobrzañski, L.A., Bilewicz, M., Viana, J.C., (2010). Mechanical approach of PP/MMT polymer nanocomposite. *Archives of Materials Science and Engineering* 43/2, 94.

[147] Heino, M., Hietaoja, P., Seppälä, J., Harmia, T., Friedrich, K., (1997). Studies on fracture behavior of tough PA6/PP blends. *J. Appl. Polym. Sci.* 66, 2209.

[148] Fujiyama, M., (2010). Morphology development in polyolefin nanocomposites, In: Optimization of Polymer Nanocomposite Properties. Edited by Vikas Mittal WILEY-VCH Verlag GmbH & Co. KGaA, Weinheim .

[149] Lincoln, D.M., Vaia, R.A., Wang, Z.-G., Hsiao, B.S., Krishnamoorti, R., (2001). Temperature dependence of polymer crystalline morphology in nylon 6/montmorillonite nanocomposites. *Polymer* 42, 9975.

[150] Xenopoulos, A., Clark, E.S., 1995. Physical Structures. In: Kochan, M.I. (Ed.), Nylon Plastics Handbook. Hanser Publishers, Munich, pp. 108–137.

[151] Vogelsong, D.C., (1963). Crystal structure studies on the polymorphic forms of nylon 6 and 8 and other even nylons. *J. Polym. Sci., Part A: Polym. Chem.* 1, 1055.

[152] Arimoto, H., Ishibashi, M., Hirai, M., (1965). Crystal structure of the γ-form of nylon 6. *J. Polym. Sci., A* 3, 317.

[153] Ito, M., Mizuochi, K., Kanamoto, T., (1998). Effects of crystalline forms on the deformation behaviour of nylon-6. *Polymer* 39, 4593.

[154] Kojima, Y., Usuki, A., Kawasumi, M., Okada, A., Kurauchi, T., Kamigaito, O., (1993). One-pot synthesis of nylon 6-clay hybrid. *J. Polym. Sci. Polym. Chem.* 31, 1755.

[155] Giannelis, E.P., (1996). Polymer layered silicate nanocomposites. *Adv. Mater.* 8, 29.

[156] Liu, L., Qi, Z., Zhu, X., (1999). Studies on nylon 6/clay nanocomposites by melt intercalation process. *J. Appl. Polym. Sci.* 71, 1133.

[157] Fornes, T.D., Yoon, P.J., Hunter, D.L., Keskkula, H., Paul, D.R., (2002). Effect of organoclay structure on nylon 6 nanocomposite morphology and properties. *Polymer* 43, 5915.

[158] Wang, K., Wang, C., Li, J., Su, J., Zhang, Q., Du, R., Fu, Q., (2007). Effects of clay on phase morphology and mechanical properties in polyamide 6/EPDM-g-MA/organoclay ternary nanocomposites. *Polymer* 48, 2144.

[159] Yebra-Rodríguez, A., Alvarez-Lloret, P., Cardell, C., Rodríguez-Navarro, A.B., (2009). Crystalline properties of injection molded polyamide-6 and polyamide-6/montmorillonite nanocomposites. *Appl. Clay Sci.* 43, 91.

[160] Van der Beek, M.H.E., Peters, G.W.M., Meijer, H.E.H., (2006). Influence of shear flow on the specific volume and the crystalline morphology of isotactic polypropylene. *Macromolecules* 39, 1805.

[161] Fornes, T.D., Paul, D.R., (2003). Crystallization behavior of nylon 6 nanocomposites. *Polymer* 44, 3945.

[162] Garofalo, E., Russo, G.M., DiMaio, L., Incarnato, L., (2007). Study on the effect of uniaxial elongational flow on polyamide based nanocomposites. *Macromol. Symp.* 247, 110.

[163] Yebra-Rodríguez, A., Alvarez-Lloret, P., Rodríguez-Navarro, A.B., Martín-Ramos, J.D., Cardell, C., (2009). Thermo-XRD and differential scanning calorimetry to trace epitaxial crystallization in PA6/montmorillonite nanocomposites. *Mater. Lett.* 63, 1159.

[164] Yebra-Rodriguez, A., Alvarez-Lloret, P., Yebra, Y., Cardell, C., Rodriguez-Navarro, A.B., (2011). Influence of processing conditions on the optical and crystallographic properties of injection molded polyamide-6 and polyamide-6/montmorillonite nanocomposites, *Applied Clay Science* 51, 414.

[165] Kamal, M.R., Borse, N.K., García-Rejón, A., (2002). The effect of pressure and clay on the crystallization behavior and kinetics of polyamide-6 in nanocomposites. *Polym. Eng. Sci.* 42, 1883.

[166] Maiti, P., Nam, P.H., Okamoto, M., Kotaka, T., Hasegawa, N., Usuki, A., (2002) The effect of crystallization on the structure and morphology of polypropylene/ clay nanocomposites. *Polym. Eng. Sci.* 42, 1864.

[167] Nam, P.H., Maiti, P., Okamoto, M., Kotaka, T., Hasegawa, N., Usuki, A., (2001). A hierarchical structure and properties of intercalated polypropylene/clay nanocomposites. *Polymer* 42, 9633.

[168] Varga, J., (2002). β-modification of isotactic polypropylene: preparation, structure, processing, properties, and application. *J. Macromol. Sci. B*, 41, 1121.

[169] Jiang, L., Zhang, J., Wolcott, M.P., (2007). Comparison of polylactide/nano-sized calcium carbonate and polylactide/montmorillonite composites: Reinforcing effects and toughening mechanisms. *Polymer* 48, 7632.

[170] Turner-Jones, A., (1971). Development of the γ-crystal form in random copolymers of propylene and their analysis by dsc and x-ray method. *Polymer* 12, 487.

[171] Yuan, Q., Chen, J., Yang, Y., Misra, R.D.K., (2008). Nanoparticle effects on spherulitic structure and phase formation in polypropylene crystallized at moderately elevated pressures: The influence on fracture resistance. *Mat. Sci. Eng. A-Struct.* 480, 181.

[172] Yuan, Q., Chen, J., Yang, Y., Misra, R.D.K., (2010). Nanoparticle interface driven microstructural evolution and crystalline phases of polypropylene: The effect of nanoclay content on structure and physical properties. *Mat. Sci. Eng. A-Struct.* 527, 6002.

[173] Kalay, G., Zhong, Z., Allan, P., Bevis, M.J., (1996). The occurrence of the γ-phase in injection moulded polypropylene in relation to the processing conditions. *Polymer* 37, 2077.

[174] Liedauer, S., Eder, G., Janeschitz-Kriegl, H., (1995). On the limitations of shear induced crystallization in polypropylene melts. *Int. Polym. Process.* X, 243.

[175] Kim, B., Lee, S.H., Lee, D., Ha, B., Park, J., Char, K., (2004). Crystallization kinetics maleated polypropylene/clay hybrids. *Ind. Eng. Chem. Res.* 43, 6082.

[176] Trotignon, J.P., Lebrun, J.L., Verdu, J., (1982). Crystalline polymorphism and orientation in injection-moulded polypropylene. *Plast. Rub. Proc. Appl.* 2, 247.

[177] Zhu, P.-W., Edward, G., (2003). Studies of Injection-Moulded Isotactic Poly(propylene) by Synchrotron WAXD/SAXS: Effects of Nucleating Agent on Morphological Distribution. *Macromol. Mater. Eng.* 288, 301.

[178] Hambir, S., Bulakh, N., Kodgire, P., Kalgaonkar, R., Jog, J.P., (2001). PP/clay nanocomposites: A study of crystallization and dynamic mechanical behavior. *J. Polym. Sci. B Polym. Phys.* 39, 446.

[179] Vasanthan, N., (2007). Effect of the microstructure on the dye diffusion and mechanical properties of polyamide-6 fibers. *J. Polym. Sci., Part B: Polym. Phys.* 45, 349.

[180] Muratoglu, O.K., Argon, A.S., Cohen, R.E., (1995). Crystalline morphology of polyamide-6 near planar surfaces. *Polymer* 36, 2143.

[181] Samon, J.M., Schultz, J.M., Hsiao, B.S., (1999). Structure development during the melt spinning of polyethylene and poly(vinylidene fluoride) fibers by in situ synchrotron small- and wide-angle X-ray scattering techniques. *Macromolecules* 32, 8121.

[182] Somani, R.H., Yang, L., Hsiao, B.S., (2002). Shear-Induced Precursor structures in isotactic polypropylene melt by in-situ rheo-SAXS and rheo-WAXD studies. *Macromolecules* 35, 9096.

[183] Li, C.Y., Li, L., Cai, W., Kodjie, S.L., Tenneti, K.K., (2005). Nanohybrid shishkebabs: periodically functionalized carbon nanotubes. *Adv. Matter.* 17, 1198.

[184] García-Gutiérrez, M.C., Nogales, A., Rueda, D.R., Domingo, C., García-Ramos, J.V., Broza, G., Roslaniec, Z., Schulte, K., Davies, R.J., Ezquerra, T.A. (2006). Templating of crystallization and shear-induced self-assembly of single-wall carbon nanotubes in a polymer-nanocomposite. *Polymer* 47, 341.

[185] Fujiyama, M., Wakino, T., (1991). Structures and properties of injection moldings of crystallization nucleator-added polypropylenes.1. structure-property relationships. *J. Appl. Polym. Sci.* 42, 2739.

[186] Svoboda, P., Zeng, C., Wang, H., Lee, J., Tomasko, D.L., (2002). Morphology and mechanical properties of polypropylene/organoclay nanocomposites. *J. Appl. Polym. Sci.* 85, 1562.

[187] Sanchez-Solis, A., Romero-Ibarra, I., Estrada, M.R., Calderas, F., Manero, O., (2004). Mechanical and rheological studies on polyethylene terephthalate-montmorillonite nanocomposites. *Polym. Eng. Sci.* 44, 1094.

[188] Wang, Y., Gao, J., Ma, Y., Agarwal, U., (2006). Study on mechanical properties, thermal stability and crystallization behavior of PET/MMT nanocomposites. *Composites Part B: Engineering* 37, 399.

[189] Bilewicz, M., Viana, J.C., Dobrzański, L.A., (2008). Polymer composite strengthening by developed injection moulding technique, *Archives of Materials Science and Engineering*, 30, 69.

[190] Somwangthanaroj, A., Lee, E.C., Solomon, M.J., (2003). Early stage quiescent and flow-induced crystallization of intercalated polypropylene nanocomposites by time-resolved light scattering. *Macromolecules* 36, 2333.

[191] Huang, H-X., Wang, B., Zhou, W-W., (2012). Polymorphism in polyamide 6 and polyamide 6/clay nanocomposites molded via water-assisted injection molding. *Composites: Part B* 43, 972.

In: Polymer-Matrix Composites
Editor: Rakesh Kumar

ISBN: 978-1-61122-005-6

Chapter 7

EVALUATION OF ELASTIC MODULUS OF UNIDIRECTIONAL FIBRE REINFORCED POLYMER COMPOSITE REBAR BY ALTERNATIVE METHOD

Moyeenuddin Ahmad Sawpan*
Composite Materials Research, Pultron Composites Limited, Gisborne, New Zealand

ABSTRACT

Glass fibre reinforced polymer (GFRP) composite rebars as strengthening material of concrete elements have become a promising alternative to steel. This is because GFRP rebar has non corrosive, non conductive and non magnetic properties. The required service life of concrete reinforcement is usually in the order of 50 to 100 years. So the durability analysis of GFRP rebar is an important research topic in recent years. Preferably, GFRP rebar that are intended for long term use should be examined in real time and in-service environment. However, the time involved is not viable and therefore, accelerated ageing (or weathering) techniques are required. The most common approach to accelerate ageing uses elevated temperature tests combined with reasonable in-service environment. This research presents elastic modulus of alkaline concrete environment (pH □ 13) aged pultruded GFRP rebar as evaluated by three different methods, namely, quasi-static tensile, quasi-static flexural and dynamic mechanical thermal tests. It was found that elastic modulus of the GFRP rebar samples did not change significantly due to exposure in alkaline environment at 60 oC for 1, 2, 3, 4, 6, 14 and 24 months when compared with that of control sample. Elastic modulus was found to be in the range of 52.5 to 57 GPa irrespective to testing methods and ageing time. In addition, it was estimated from the long time projected results that quasi-static tensile, quasi-static flexural and dynamic mechanical moduli will be retained by about 93%, 95% and 85%, respectively, after 100 years in alkaline environment at 60 oC. Microscopic analysis indicated that quasi-static tensile and flexural failure was mainly due to matrix cracking and shear failure of fibre/matrix interface.

* Corresponding author: Composite Materials Research, Pultron Composites Limited, Gisborne, New Zealand Email: moyeen@pultron.com; moyeenbd@gmail.com.

Keywords: Elatic modulus, Polymer composite rebar, Dynamic mechanical thermal analysis, Alkaline ageing

1. Introduction

Fibre reinforced polymer (FRP) composites offer a high strength, lightweight alternative to traditional methods of structural upgrade. Typically, FRP composites would consist of reinforcing fibre (e.g. glass, carbon and aramid fibres) which is the hardest, strongest and stiffest component, embedded in a continuous matrix (e.g. polyester, vinyl ester, epoxy and urethane acrylate resins). The main functions of the matrix are to transmit externally applied loads to the reinforcement and to protect the latter from external mechanical and environmental damages [1]. Pultrusion is a continuous process for the manufacture of FRP composites of continuous length and constant cross-section such as rods, tubes and beams. In a typical process, continuous strands roving, mat, or surface veil is impregnated in a resin bath, and then pulled through a hot steel die. The steel die consolidates the saturated reinforcement, sets the shape of the stock, and controls the fibre/resin ratio. Reinforcements (i.e. roving and mat) can be highly oriented in the axial direction since the automated pultrusion process provides good uniformity across profile thickness and can lead to the highest level of materials efficiency [2].

Glass fibre reinforced polymer (GFRP) composite rebars as strengthening material of concrete elements have become a promising alternative to steel. This is because GFRP has non corrosive, non conductive and non magnetic properties [3]. In addition, GFRP's low cost-to-performance advantage is driving its worldwide use and acceptance. However, GFRP rebars are susceptible to changes of both strength and stiffness in the presence of environments prior to, during, and after construction. As mentioned earlier, the polymer matrix of GFRP rebar provides a chemical barrier for the glass fibres from the corrosive effects of moisture and alkalis in concrete. In moist environment, water molecules react with the ester group (-O-CO-) of polymer matrix (e.g. polyester resin) which cause irreversible chemical change by hydrolytic reaction and diminishes toughness and fracture strain. Vinyl ester resin is less vulnerable to hydrolytic reaction because of less ester group present in the polymer molecule compared to polyester resin. Epoxy resin is also resistant to chemical attack; however, price of this resin is reasonably higher than the vinyl ester resin. So vinyl ester resin is being commonly used for manufacturing of GFRP rebar. Both vinyl ester resin and epoxy resin are permitted for fibre reinforced rebar products for permanent applications in concrete [4]. In the case of glass fibre, the chemical reaction involves a breakdown of the silica (-Si-O-Si-) network by hydroxide ions (OH^-) of concrete pore solution and eventually glass fibres lose weight and strength when they are in contact with strong alkalis. The rate of deterioration of the fibres depends on the alkalinity of the matrix (i.e. the availability of OH^- ions). Boron free E-CR glass ("Electrical" grade "Corrosion Resistant" glass) is more resistant to strong alkalis than E-glass as reported in the literature [5].

Strength and stiffness of GFRP rebar may increase, decrease, or remain the same, depending on the particular material and exposure conditions [6]. So it is important to know whether GFRP rebar used in a structural application will continue to perform satisfactorily over the lifetime of the part or structure. Due to the commercial potential for GFRP composites in construction sector much research has gone into characterizing and predicting material properties during short and long term exposure to various environments. For instance, Chu et al. [7] reported degradation of tensile strength of pultruded E-glass/vinyl ester resin composites in deionised water and alkaline environments at different temperature. Tensile strength retention of GFRP rebars of different resins (e.g. polyester and vinyl ester resin) and reinforcements (e.g. electric resistance glass, electric and corrosion resistance glass, and alkali resistance glass) exposed in high pH pore water solution and in concrete are reported by Benmokrane et al. [5]. They found that, under a 30% stress level and 140 days ageing, tensile strength of alkali resistance glass/vinyl ester rebar was not changed noticeably. On the other hand, tensile strength of E-glass/vinyl ester decreased by 16-17%. In another work, Dejke [8] also investigated durability of alkali resistance glass and E-glass reinforced vinyl ester and polyester rebars. It was found that tensile strength of all types of rebars dropped by about 50% after 1 year exposure in concrete at 60 °C. It was also observed that alkaline solution is more aggressive environment than concrete at 100% relative humidity for GFRP rebars. Apicella et al. [9] showed that glass fibre/polyester composites immersed for 15 days in water experienced a reduction in elastic modulus with increasing test temperature due to greater plasticization effects, and an increase in the elongation at break up to 60 °C followed by a decrease at 90 °C, which was correlated with an increase in glass transition temperature (T_g) at the higher temperatures. Later in another report, Apicella et al. [10] showed that the short-term exposure of glass fibre/polyester composites to water at 20 °C resulted in a decrease of the elastic modulus due to plasticization, whereas exposure to water at 90 °C led to an increase in the elastic modulus and T_g due to embrittlement of the matrix. The latter increase was explained by the effect of the degradation of resin due to dissolution of low-molecular-weight ingredients overcoming the effect of water plasticization. Pantuso et al. [11] conducted a study on polyester resin based GFRP bars of three different diameters 12, 16 and 20 mm. The bars were exposed to water cyclically, one full day of immersion and the other day air dried at a temperature of 23.2 °C. After 2 months, they observed that there was small reduction in tensile strength and elastic modulus in the order of 1 to 7% and 1 to 10%, respectively. Tensile properties evaluated by Liao et al. [12] on E-glass/vinyl-ester composite specimens aged in room temperature water showed a substantial decrease in tensile modulus (23%) and tensile strength (29%) after 120 h of immersion. Abbasi et al. [13] used E-glass reinforced polyester, vinyl ester and urethane modified vinyl ester rebars for durability analysis in concrete environment for up to 240 days at different temperatures. They reported that both tensile strength and modulus of GFRP rebar deteriorated with temperature and time of exposure to alkaline environments. Chen et al. [14] also reported a decrease in tensile strength of E-glass/vinyl ester rebars due to exposure in various alkaline environments at 40 and 60 °C for up to 120 days. In another report, Masmoudi et al. [15] exposed E-glass/vinyl ester rebars in deionised water and alkaline solution under a stress level of 20 to 29% for 104 days with temperature varying from 45 to 63 °C. They found that tensile strength was dropped by 15% and 17% due to exposure in deionised water and alkaline solution, respectively. In addition, they also reported that tensile modulus of elasticity did not change significantly after ageing. Sen et al. [16] exposed E-glass/ vinyl ester bars to simulated concrete pore solution with a pH ranging from 13.35 to 13.5 for periods of 1, 3, 6 and 9 months under unstressed and stressed conditions. They observed that specimens under unstressed conditions lost 63% of their strength while the specimen stressed to 10% of their ultimate short-term tensile strength lost 70% of their strength. The specimens exposed to 180 days, then more than 10% of ultimate strength failed. Ellyin and Maser [17, 18] have characterized the hygrothermal damage induced in filament wound E-glass/epoxy composites conditioned in distilled water at two temperatures, namely room temperature (approximately 20 ºC) and an elevated 50 ºC condition. The samples were tested under various bi-axial load ratios, as well as pure hoop loading. It was concluded that plasticization and hydrolysis of the matrix resulted in a reduced composite stiffness; however, the swelling of the matrix after moisture saturation induced compressive stresses in the material, leading to matrix crack closure, and thus an increase in the functional failure load for certain bi-axial loading conditions, specifically those in which the hoop stress was greater than the axial stress. Karbhari et al. [19] studied the durability of uni-axial, bi-axial and tri-axial nonwoven E-glass fabrics reinforced vinyl ester composites in an aqueous environment. A resin infusion process was used for the fabrication of composites. Samples are immersed in de-ionized water at 23 °C and 60 °C, respectively. Reduction in both tensile strength and tensile modulus were observed. The tensile strength retention was about 30-60% after 50 weeks immersion in water at 23 °C and 60 °C, while the tensile modulus decreased 10-30% of the control value. From the above literature review, it can be said that tensile strength and tensile stiffness reduction due to environmental ageing depends on type of glass fibre, resin, manufacturing method, ageing temperature and applied stress during ageing. Although in many cases, tensile modulus of GFRP rebar has been not reported along with tensile strength, however, in general, tensile strength and tensile stiffness reduction in GFRP rebar due to environmental ageing range between 0 to 75% and 0 to 20%, respectively, as registered in many cases [20].

The modulus of elasticity of GFRP materials exposed in controlled environment has not been assessed thoroughly yet as said earlier, although modulus of elasticity is paramount important for reliable design of reinforced concrete (RC) structures to protect against cracking. The relationship between crack width and elastic modulus of fibre reinforced polymer (FRP) rebar is shown in Eq. (i) [6].

$$w = 2\frac{f_f}{E_f}\beta \times k_b \times \sqrt{d_c^2 + \left(\frac{s}{2}\right)^2} \qquad \text{(i)}$$

where w = maximum crack width (mm), E_f = modulus of elasticity of FRP rebar (MPa), f_f = stress in FRP reinforcement in tension (MPa), k_b = bond-dependent coefficient, β = ratio of distance from neutral axis to extreme tension fibre to distance from neutral axis to centre of tensile reinforcement, d_c = thickness of concrete cover measured from extreme tension fibre to centre of bar (mm) and s = longitudinal FRP bar spacing (mm).

As seen earlier, only a limited amount of literature focused on the effect of controlled environment on tensile modulus of GFRP rebar, which highlighted the need for a systematic investigation of the elastic modulus of GFRP rebar. In addition, the literature review affirmed that no systematic investigation has been conducted on dynamic mechanical modulus of GFRP rebar to compare directly with that of quasi-static tensile and flexural moduli. It would seem appropriate, therefore, that in any work undertaken, long term influence of controlled environmental ageing on the quasi-static modulus should be compared with dynamic modulus of GFRP rebar.

2. EXPERIMENTAL

2.1. Material

The glass fibre reinforced polymer (GFRP) composite rebars were received from Pultron Composites Limited, New Zealand. These threaded rebars of 14 mm diameter (outer diameter) were made up of unidirectional roving of ECR glass (i.e. "Electrical" grade "Corrosion Resistant" glass) and epoxy vinyl ester resin as manufactured through pultrusion process. The minimum glass content of the rebar was about 63% by volume.

2.2. Methods

Alkali resistance test: A temperature controlled stainless steel tank was used to immerse the GFRP rebars in alkaline solution for 1, 2, 3, 4, 6, 14 and 24 months at 60 °C. Typical alkaline solution was made up of 118.5g of $Ca(OH)_2$, 0.9g of NaOH, and 4.2g of KOH in 1l of deionised water [21]. The alkaline solution had a pH value of about 13, a representative pH value of mature concrete pore solution. Photograph of alkaline aged rebar samples is presented in Figure 1.

Tensile test: Specimen preparation and tensile testing were conducted according to the guidelines as specified in ASTM D7205/D7205M [22] and ACI 440.3R [21]. Tensile specimen for testing was prepared by anchoring two ends of the rebar in steel plugs filled with epoxy resin (see Figure 2). The free length between the steel plugs was about 40 times of root diameter of the rebar. The destructive quasi-static tensile test (a time dependent test but is "slow" enough such that inertial effects can be ignored) was performed on 4 replicates of each batch using a universal testing machine (WDW-200E) fitted with 200 kN load cell (see Figure 3). Load ramp rate was set at 667 N/s, and an extensometer of 50mm gauge length was mounted with clips in the centre of test specimen. Load versus extension data were acquired by computer, which was also facilitated machine control.

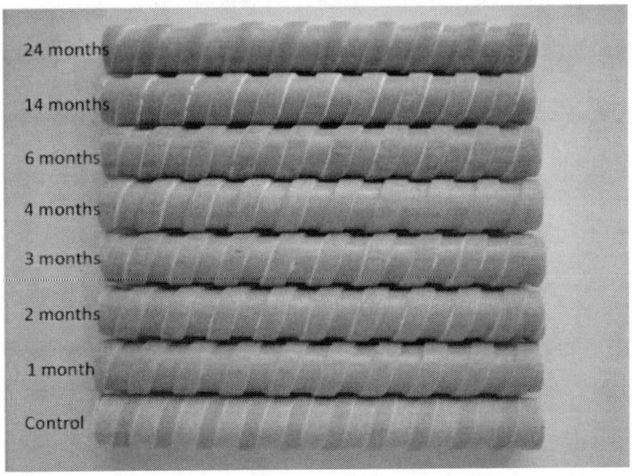

Figure 1. Photograph of control and alkaline aged GFRP rebar samples.

Tensile modulus was evaluated using Eq. (ii):

$$E_L = (F_1 - F_2)/\{(\varepsilon_1 - \varepsilon_2) * A\} \quad \text{(ii)}$$

where E_L = tensile modulus; F_1 and ε_1 = load and corresponding strain, respectively, at approximately 50% of the ultimate tensile capacity; F_2 and ε_2 = load and corresponding strain, respectively, at approximately 20% of the ultimate tensile capacity; and A = cross-sectional area of rebar sample.

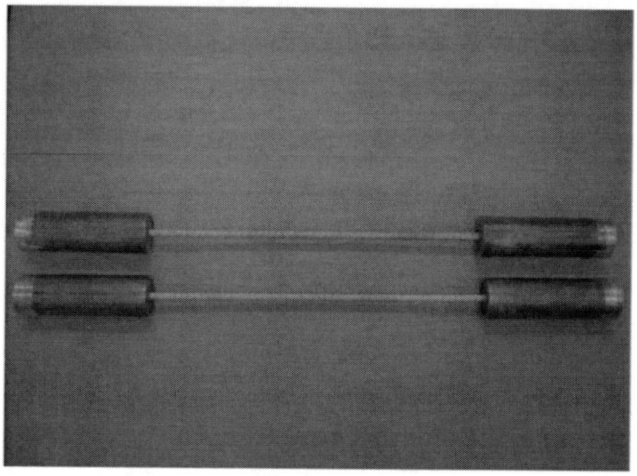

Figure 2. Typical GFRP rebar samples for tensile test.

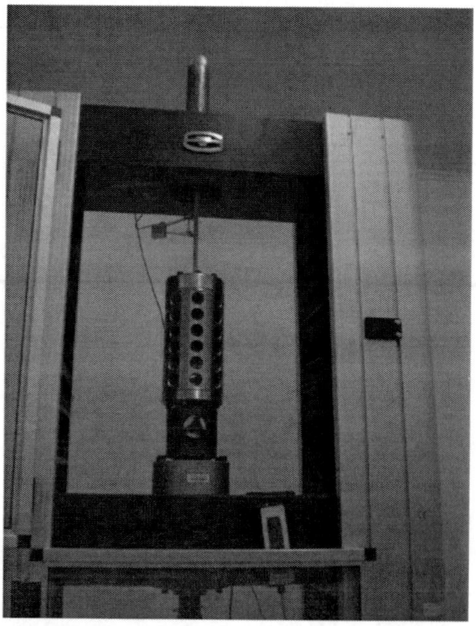

Figure 3. Quasi-static tensile testing machine setup.

Flexural test: Quasi-static flexural testing was performed similar to the method ASTM D790 [23] in three point bending mode using a universal testing machine as described previously. The sample span-to-depth ratio was 16:1 and the load ramp rate was set at 667 N/s. Four specimens were tested from each batch of samples.

Flexural modulus of circular cross section rod sample was calculated using Eq. (iii):

$$E_f = \left(L^3\, m\right)/\left(12\,\pi\, R^4\right) \tag{iii}$$

where L = distance between support points, m = slope, and R = radius of sample.

Dynamic mechanical thermal analysis (DMTA): Dynamic mechanical thermal analysis is a powerful tool for developing insight into structure, morphology and viscoelastic behaviour of polymer while simultaneously enabling a high level of sensitivity in the detection of changes in mobility of the polymer chain segments [24, 25]. One of the most common uses of DMTA data is the evaluation of the various transitions (e.g. gamma transition, T_γ; beta transition T_β; and alpha transition, T_α or glass transition, T_g) in a polymer [26]. Also DMTA provides a direct link between a material's chemical makeup (e.g. a polymer blend or alloy, semicrystalline or amorphous thermoplastic and thermoset) and its mechanical behaviour.

Viscoelastic material shows time-dependent deformation under stress. Upon release of the stress, the proportion of strain recovered relates to the energy stored in the material, while the proportion of unrecovered strain relates to the energy dissipated as heat [27]. DMTA technique applies an oscillating force to test a sample and the sinusoidal stress and strain curves are recorded as a function of time (see Figure 4).

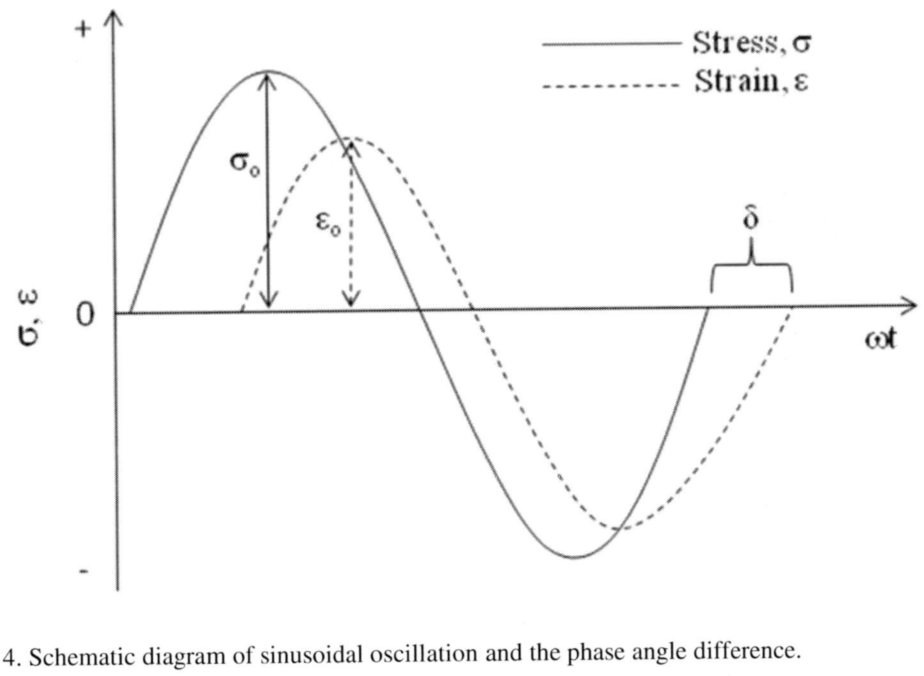

Figure 4. Schematic diagram of sinusoidal oscillation and the phase angle difference.

An oscillating stress (σ_A) can be expressed as:

$$\sigma_A = \sigma_0 \, e^{iwt} \tag{iv}$$

Due to the loss of energy during the loading cycle the strain (ε_A) in the sample will exhibit a phase shift relative to the applied stress, thus

$$\varepsilon_A = \varepsilon_0 \, e^{i(wt-\delta)} \tag{v}$$

where ω is the angular frequency, and δ is the phase shift angle between dynamic stress and dynamic strain in a viscoelastic material that is exposed to a sinusoidal load.

The modulus of the material can therefore be expressed as:

$$E = \frac{\sigma_A}{\varepsilon_A} = \frac{\sigma_0 \, e^{iwt}}{\varepsilon_0 \, e^{i(wt-\delta)}} = \frac{\sigma_0 \, e^{i\delta}}{\varepsilon_0} \tag{vi}$$

Again,

$$e^{i\delta} = \cos\delta + i\sin\delta \tag{vii}$$

From the Eq. (vi) it can be written as:

$$E = \frac{\sigma_0}{\varepsilon_0}\cos\delta + i\frac{\sigma_0}{\varepsilon_0}\sin\delta \tag{viii}$$

or,

$$E^* = E^{/} + iE^{//} \tag{ix}$$

where E^*, $E^{/}$ and $E^{//}$ are the complex modulus, storage modulus and loss modulus, respectively [26-29].

Therefore, the important parameters that can be obtained from a DMTA test are: (a) storage modulus ($E^{/}$), which is a measure of the maximum energy, stored in the material during one cycle of oscillation, (b) loss modulus ($E^{//}$), which is proportional to the amount of energy that has been dissipated as heat by the sample and (c) mechanical damping factor ($\tan\delta$), which is the ratio of the loss modulus to storage modulus and is related to the degree of molecular mobility in the polymer material.

In this work, dynamic mechanical properties of the rebars were measured using a dynamic mechanical thermal analyser (DMTA V: Rheometric Scientific Inc.). The analyser was operated at 1 Hz from 25 to 190 °C at a heating rate of 3 °C/min. The strain amplitude was 0.05%, which was well within the linear viscoelastic region. The samples were rectangular coupons with dimensions of about 40 x 5 x 0.5 (± 3%) mm^3 and the tests were conducted in three point bending mode (see Figure 5). The span length was 28 mm. Four specimens of each type were tested.

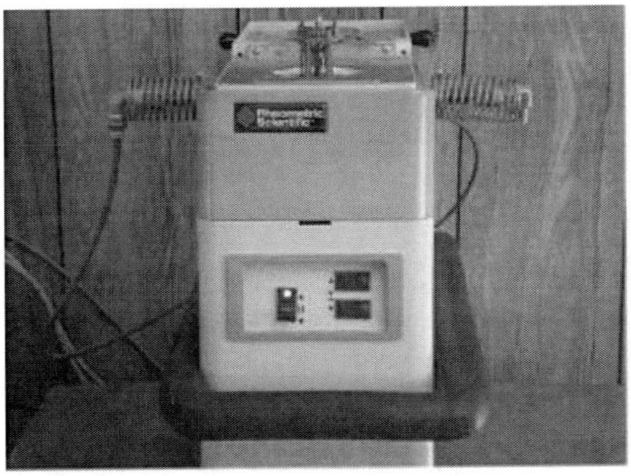

(a)

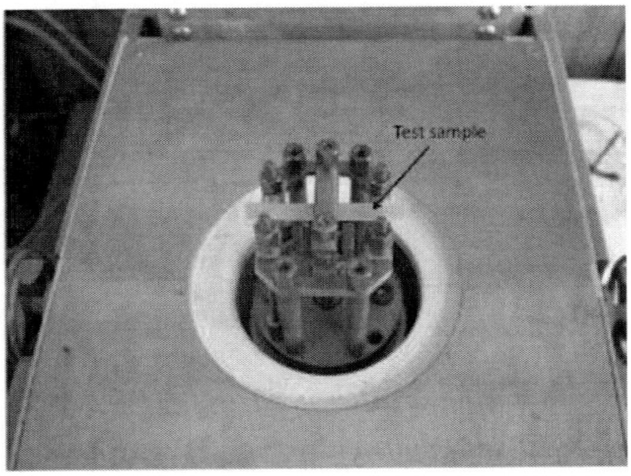

(b)

Figure 5. Photographs of (a) DMTA machine and (b) test sample in 3-point bend fixture.

Glass transition temperature (T_g): Glass transition temperature is one of the key parameters to analyse the durability behaviour of GFRP composites. DMTA is considered a sensitive way to measure the T_g of a composite with a highly crosslinked thermoset matrix and a high fibre volume fraction [29]. In this work, T_g of the alkaline aged and control GFRP samples were evaluated according to the method ASTM E1640 [30] using a DMTA machine as described previously.

Statistical analysis: In this study, moduli of the GFRP rebars of alkaline aged and control samples as obtained by different methods (i.e. quasi-static tensile, quasi-static flexural and dynamic mechanical) were statistically analysed using OriginPro 7.5 software from OriginLab Corporation, USA. The data was analysed using two sample t-test at a confidence level of 95% ($P<0.05$) to determine the significance of the difference in the modulus of alkaline aged sample in relation to that of the control sample.

Scanning electron microscope (SEM) and Energy dispersive X-ray (EDX): The effect of alkaline ageing on the morphology of fibre/matrix interface of GFPR rebar samples was investigated using a scanning electron microscope (Model: MV2300) from CamScan Electron Optics. Elemental analysis of glass fibre of GFRP rebar was conducted by EDX. Samples were sputter coated with gold to make them conductive prior to SEM and EDX analysis.

3. Results and Discussion

3.1. Quasi-static Tensile Deformation and Tensile Modulus

Figure 6 is showing typical tensile load versus displacement curves of control (0 month) and high temperature alkaline environment aged rebar samples. It may be observed that the curves initially showed linear deformation, followed by a slight non-linear deformation prior to attainment of the maximum load. Once the load reached maximum, the recorded load dropped suddenly. In general, it also appeared that load-displacement curves of the alkaline aged rebars did not change significantly compared to control rebar. However, load-displacement curves of 14 and 24 months aged rebar samples showed knee (indicated by arrow) at relatively lower load level, which indicated the onset of microcracking (see Figure 7) as a result of increase in matrix brittleness with ageing time. This increase in matrix brittleness was believed to be due to leaching of non-crosslinked residue (e.g. mould release agent) during high temperature alkaline ageing.

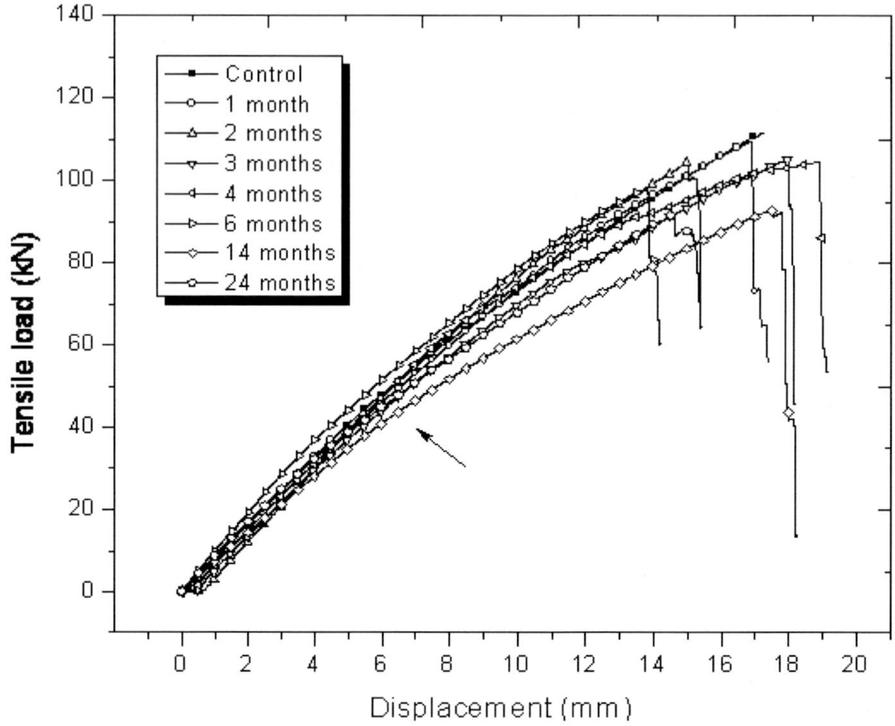

Figure 6. Typical tensile load versus displacement curves of GFRP rebar.

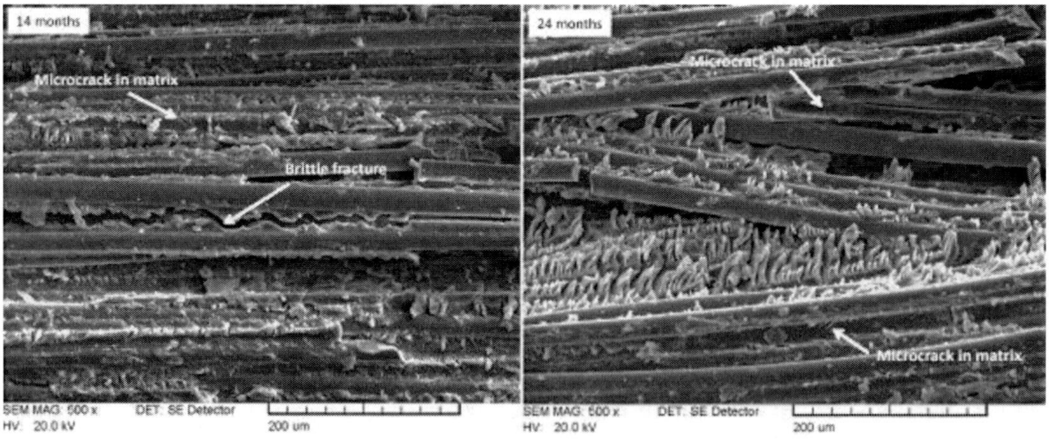

Figure 7. SEM micrographs of tensile fracture surface of GFRP rebar showing microcrack and brittle fracture.

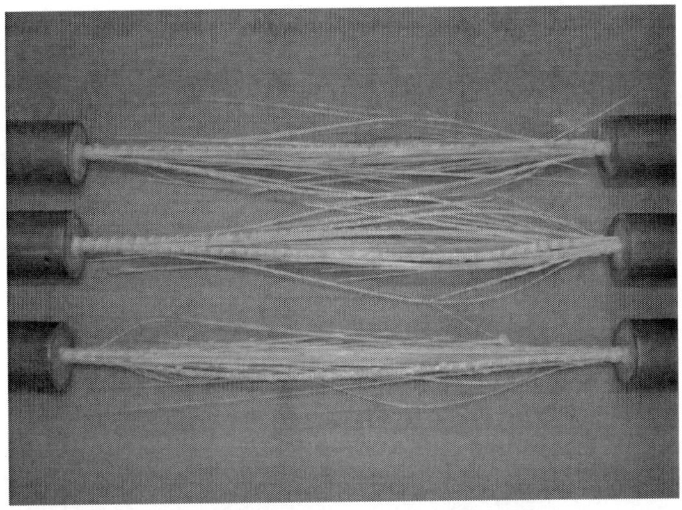

Figure 8. Photograph of typical tensile failure behaviour of GFRP rebar.

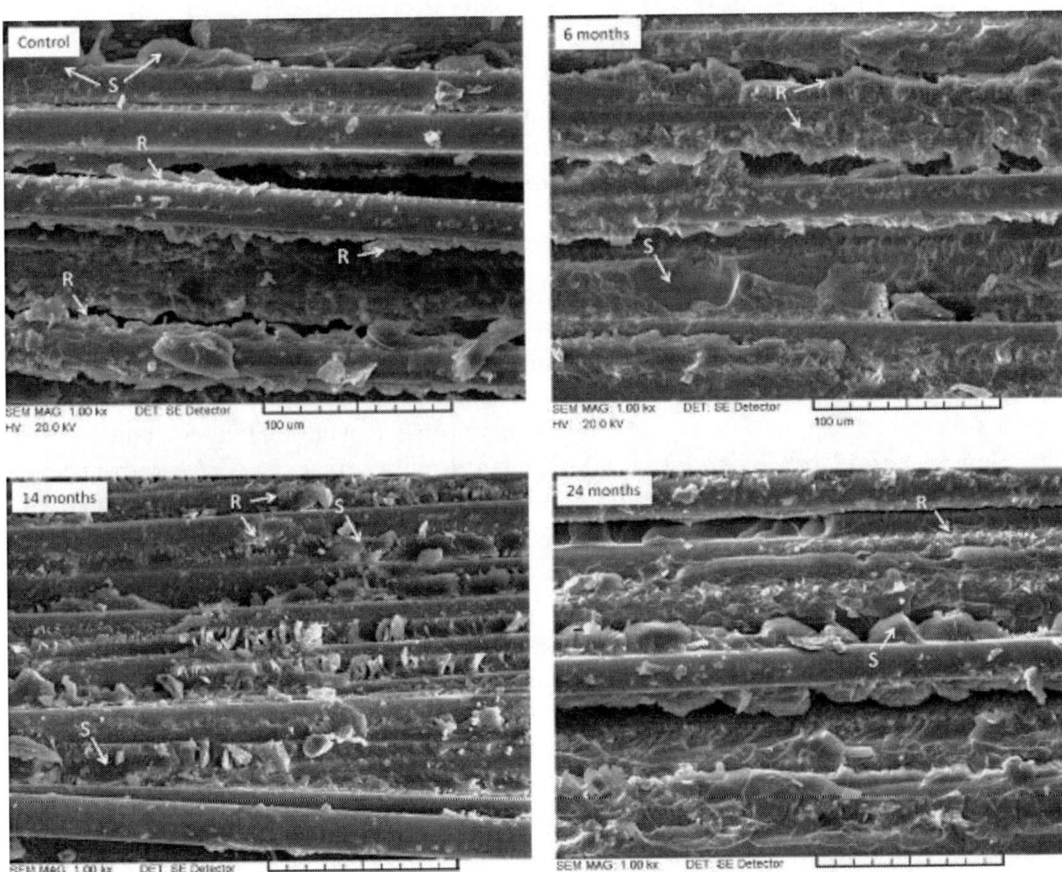

Figure 9. SEM micrographs of tensile fracture surface of control and alkaline aged GFRP rebars.

Table 1. Two sample paired t-test tensile elastic modulus statistics summary

Sample	Mean	Standard deviation	Standard error	t - distribution	P - value	Remarks
Control	55.025	1.394	0.697	-	-	P-value is greater than the significance level (0.05). At the 0.05 level, the difference of the population means is not significantly different than the test difference.
1 month	56.475	0.830	0.415	2.339	0.101	
2 months	54.750	1.603	0.802	0.468	0.672	
3 months	53.375	1.767	0.884	1.108	0.348	
4 months	54.350	1.533	0.766	1.157	0.331	
6 months	54.125	1.415	0.708	2.183	0.117	
14 months	54.750	1.905	0.953	0.230	0.832	
24 months	54.805	0.688	0.344	0.218	0.840	

During the tensile test, all the rebar samples were failed by splintering along the specimen free length (see Figure 8) as a result of combined transverse compression and longitudinal extension in the composites. SEM micrographs (see Figure 9) of tensile fracture surfaces indicated that fibres were well connected with the matrix and the fibre/matrix interface was not affected remarkably due to exposure in alkaline environment for long time. In all the cases, there were signs of matrix adhering to the fibres which indicated a good bond in the fibre/matrix interface. Besides, it can be seen that the fracture surface was moderately flat and smooth (S), however, matrix tearing and/or roughness (R) was also noticed to some extent. Thus, tensile failure of the GFRP rebar was neither completely brittle nor completely ductile, but rather mixed. This was also supported by the load versus displacement traces (see Figure 6).

The average tensile modulus of GFRP rebar samples as a function of ageing time is presented in Figure 10. It was found that average tensile modulus ranged from 53.4 to 56.5 GPa irrespective to ageing time. Although tensile modulus showed a general declining trend with respect to ageing time, however, statistical analysis (*t*-test at a confidence level of 95%) did not support a significant difference in aged samples in relation to that of the control sample (see Table 1). Bakis et al. [31] also reported that tensile modulus of pultruded E-glass/vinylester resin composite rebar did not significantly affected after ageing up to 12 months in alkaline environment at 60 °C. Tensile modulus of the unconditioned rebar was around 40 GPa. In another work, Micelli and Nanni [32] reported tensile modulus of alkaline aged E-glass and polyester resin based pultruded GFRP rebar. They found a slight increase in tensile modulus from 29.9 to 30.4 GPa after 42 days ageing at 60 °C. Davalos et al. [33] reported that tensile moduli of pultruded GFRP rebars (unconditioned samples) with E-glass and vinyl ester resin from different sources varied from 40.8 to 49.4 GPa. Similarly, Benmokrane et al. [5] reported tensile modulus of various GFRP rebars (e.g. helically wrapped, ribbed and sand-coated) as manufactured by different combinations of E-glass and AR-glass (alkali resistant glass) with vinyl ester and polyester resins. Tensile moduli of the unconditioned rebars were found to be varied from 31.6 to 51.8 GPa. In a different manufacturing technique (i.e. braiding process), tensile modulus of E-glass and vinyl ester resin based GFRP rebar was reported to be 42.8 GPa [3]. Thus it appears that use of various/combinations of commercially available fibre/polymer materials and manufacturing methods resulted in widely diverse tensile modulus of GFRP rebar. Nevertheless, tensile modulus of unconditioned GFRP rebar (i.e. control sample) as obtained from this work was found to be higher than the other literature data as mentioned earlier. This high tensile modulus result of the rebar was believed to be due to (a) good fibre/matrix bonding and (b) improved load transfer efficiency from the matrix to the fibre.

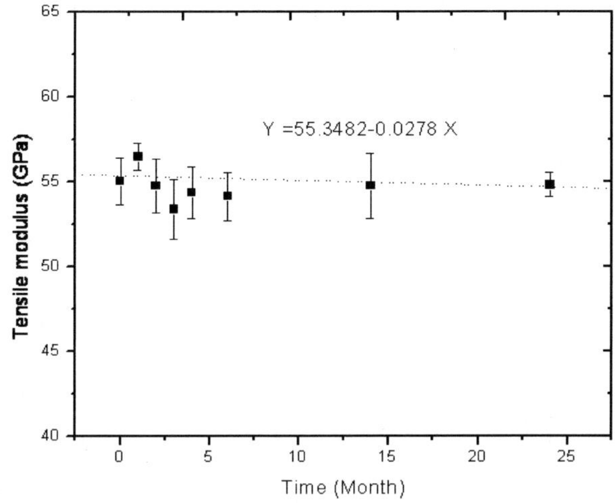

Figure 10. Quasi-static tensile elastic modulus of GFRP rebar as a function of alkaline ageing time.

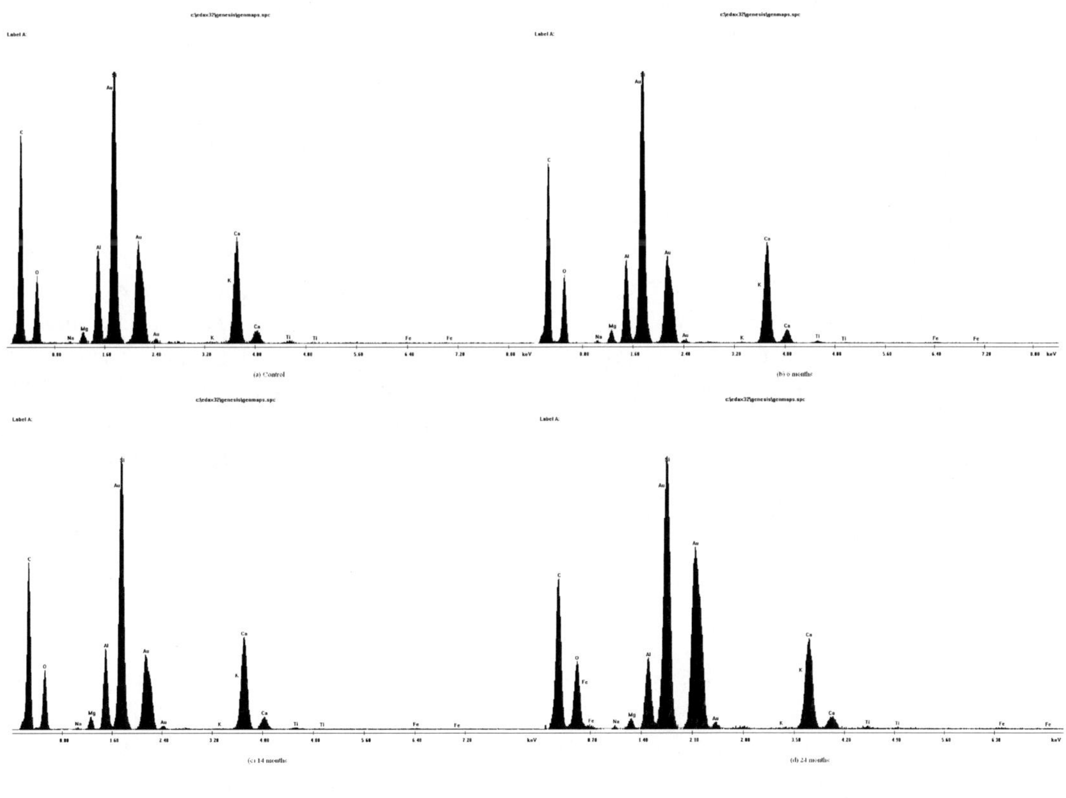

Figure 11. EDX spectra of control and alkaline aged GFRP rebar.

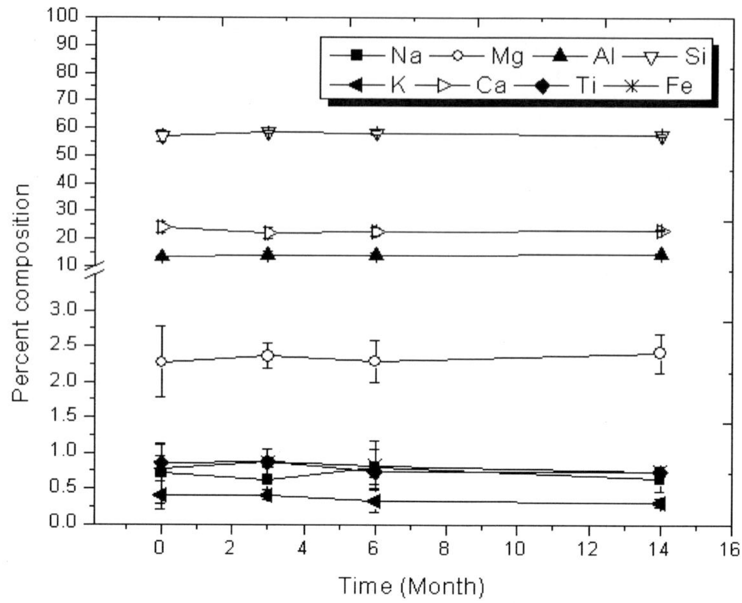

Figure 12. EDX results versus alkaline ageing time of GFRP rebar.

The elastic modulus of a composite is largely a property of fibres. In FRP composites, it is generally expected that polymer matrix will provide the needed chemical protection for the fibres. Moisture absorption or chemical diffusion through weaker crosslinked polymer structure towards fibre/matrix interface can damage the interfacial bond strength, which in turn reduce the load transfer efficiency. The tensile modulus results as discussed earlier (see Figure 10) indicated that fibre/matrix interface and fibre themselves were not deteriorated considerably due to ageing in high temperature alkaline environment for up to 24 months as supported by the SEM micrographs (see Figure 9). Additional investigation regarding alkaline attack on the glass fibre (i.e. corrosion of glass fibre) of tensile tested GPRP rebar was conducted using energy dispersive X-ray (EDX) as shown in Figure 11. As can be seen, the glass fibre mainly consists of SiO_2, Al_2O_3, CaO, MgO, Na_2O, K_2O, TiO_2 and Fe_2O_3. If the glass fibre degraded during ageing due to dissolution, leaching and electrochemical reaction of metal ions of glass and OH^- ions of the alkaline solution, then it could be expected that the composition of glass fibre was also changed quantitatively. Figure 12 is showing that fibre glass compositions of alkaline aged rebars were not changed significantly when compared with that of control rebar sample. The variations in glass fibre compositions for different ageing period were within the standard deviation limit.

Characterizing long term performance of materials is a critical aspect of understanding the reliability of structures. The required service life of concrete reinforcement is usually in the order of 50 to 100 years. However, long term durability data of GFRP composite is not readily available. Short term accelerated ageing has often been extrapolated to make long term mechanical property estimates [34, 35]. In this work, average tensile modulus retention results as a function of extrapolated time in logarithmic scale was drawn to show the trend in tensile modulus of GFRP rebar for up to 100 years (see Figure 13). As can be seen in figure, the trend in tensile modulus with respect to long term alkaline accelerated ageing indicated that tensile modulus of GFRP rebar will be retained by about 96% after 100 years. It is worth to mention here that properties of GFRP composite deteriorate much faster in alkaline solution at high temperature (i.e. accelerated ageing) than in concrete due to easy mobility and diffusion of moisture and ions of the solution through the matrix to fibre/matrix interface.

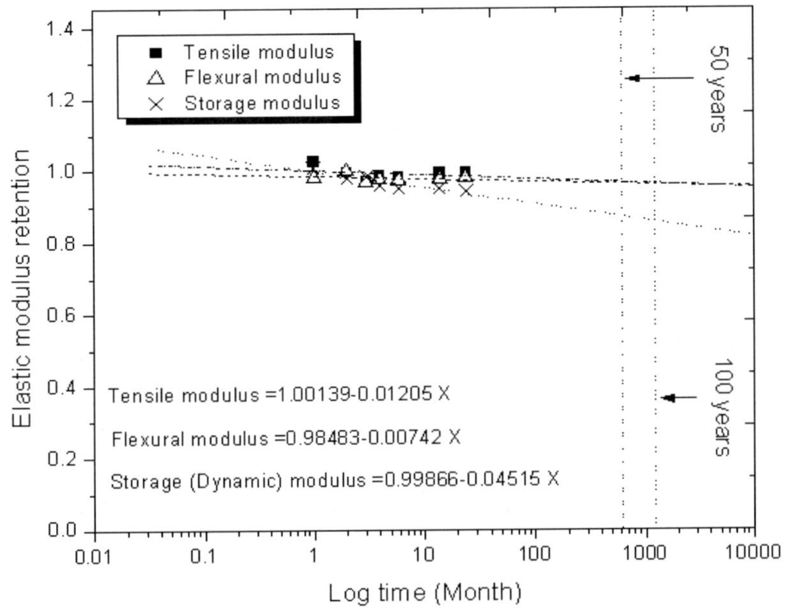

Figure 13. Tensile, flexural and dynamic mechanical elastic moduli retention with respect to alkaline ageing time in logarithmic scale.

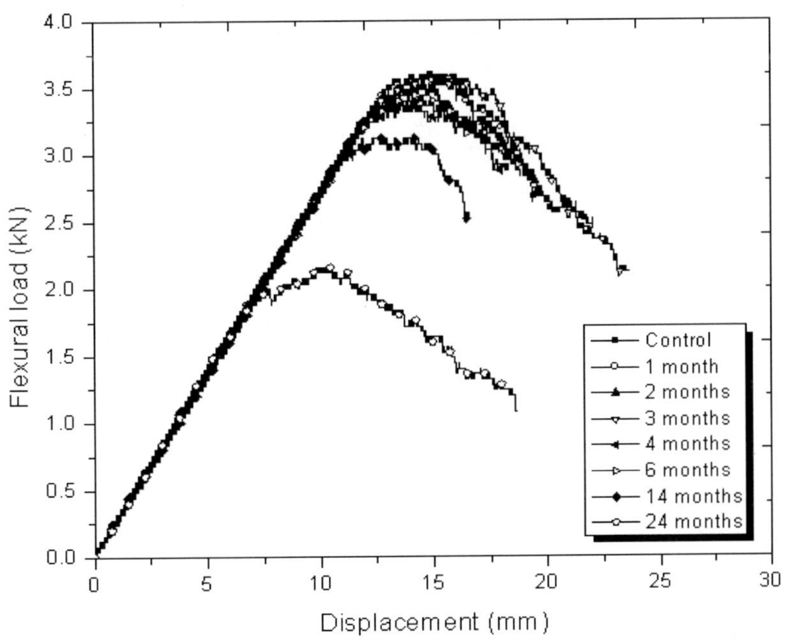

Figure 14. Typical flexural load versus displacement curves of GFRP rebar.

3.2. Quasi-Static Flexural Deformation and Flexural Modulus

Flexural load versus displacement curves are presented in Figure 14. It may be observed that, like tensile deformation, flexural curves showed initially linear deformation, followed by an amount of non-linear deformation before reaching maximum load. Once the maximum load was attained, the recorded load diminished gradually which was a result of shear failure along with limited plastic deformation of the rebars. It was observed that flexural deformation curves was quite linear and there was no appearance of knee as seen for the 14 or 24 months aged samples in the tensile deformation curves. This was because (a) ultimate flexural failure load was significantly lower than the load at which knee appeared in tensile deformation curves, and (b) flexural deformation mechanism was controlled generally by the properties of fibre/matrix interface. It is also evident that area under the curve of 24 months aged sample decreased reasonably compared to other samples. This observation again could be attributed to the increase of matrix brittleness as observed for tensile deformation behaviour. Typical flexural failure behaviour of the rebars is shown in Figure 15 where it can be observed that maximum axial fibre stresses occurred in the outer fibres and the failure was dominated by the properties of fibre/matrix interface as said earlier.

Figure 15. Photograph of typical flexural failure behaviour of GFRP rebar. The labelled regions are compressive (C) and tension (T) sites of the tested specimen.

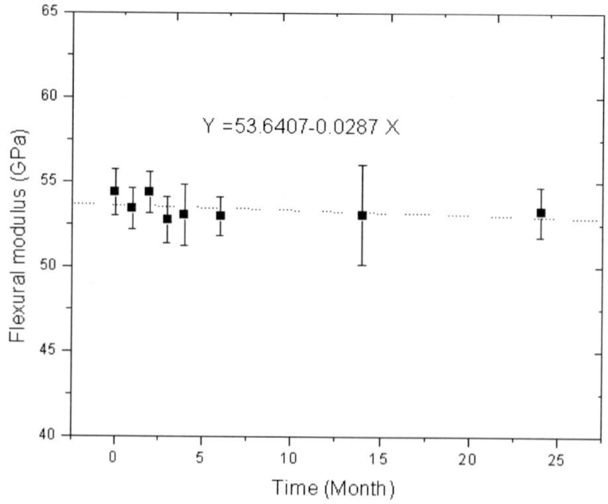

Figure 16. Quasi-static flexural modulus of GFRP rebar as a function of alkaline ageing time.

Table 2. Two sample paired t-test flexural elastic modulus statistics summary

Sample	Mean	Standard deviation	Standard error	t - distribution	P - value	Remarks
Control	54.370	1.369	0.684	-	-	P-value is greater than the significance level (0.05). At the 0.05 level, the difference of the population means is not significantly different than the test difference.
1 month	53.423	1.245	0.623	2.340	0.101	
2 months	54.398	1.202	0.601	0.031	0.977	
3 months	52.750	1.346	0.673	2.140	0.122	
4 months	53.055	1.779	0.889	1.140	0.337	
6 months	52.980	1.159	0.580	2.587	0.081	
14 months	53.068	2.939	1.470	0.761	0.502	
24 months	53.242	1.442	0.721	1.165	0.328	

Figure 16 is showing the average flexural modulus of alkaline aged and control GFRP rebar samples with respect to ageing time. Flexural modulus was found to be in the range of about 52.7 to 54.4 GPa irrespective to ageing time, which was very close to the tensile modulus results. It was noted that the average flexural modulus of rebar samples in all the cases (control and aged samples) was slightly lower than the average tensile modulus. The slight lower flexural modulus compared to tensile modulus for the same sample could be explained as follows. In a flexural test, there is a combination of tensile and compressive loads (see Figure 15). The compression of fibres in a polymer matrix can be subjected to a compression wave form (i.e. buckling) in the fibres. As said earlier, the modulus of a composite is mainly fibre dominated property; therefore, lower flexural modulus would be expected as a result of reduced load bearing efficiency due to wave form in the fibres on the compression site of composite. Nevertheless, the small difference in flexural and tensile moduli was within the range of the standard deviation, indicating that accelerated alkaline environmental conditions as studied here have no considerable affect on the tensile and flexural moduli of GFRP rebar. Again like tensile modulus, flexural modulus showed a general declining trend with respect to ageing time; and the flexural modulus of aged samples was found to be changed insignificantly when compared with that of control samples (see Table 2). The trend in flexural modulus as extrapolated in logarithmic time plot (see Figure 13) indicated that flexural modulus will be retained by about 95% after 100 years, a finding quite close to that for tensile modulus. However, it was not possible to compare the above flexural modulus results with other literatures because flexural modulus of GFRP rebar is not well documented yet.

3.3. Dynamic Mechanical Modulus

Typical dynamic mechanical or storage modulus *(E´)* curves of the rebars as a function of temperature is presented in Figure 17. A general declining trend in *E´* of the rebars can be observed when the specimens went through higher temperatures. This trend was more prominent above the glass transition region (around 115-125 ^{o}C), which was due to increased mobility of polymer networks.

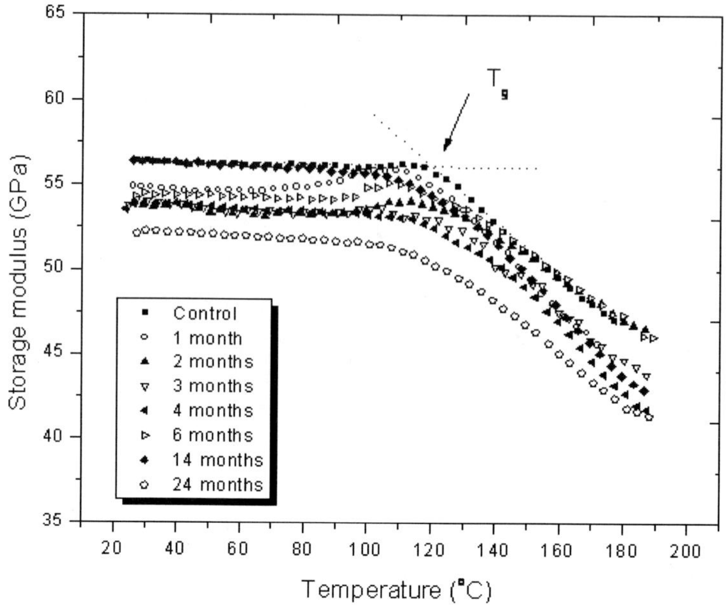

Figure 17. Typical storage modulus curves as a function of temperature.

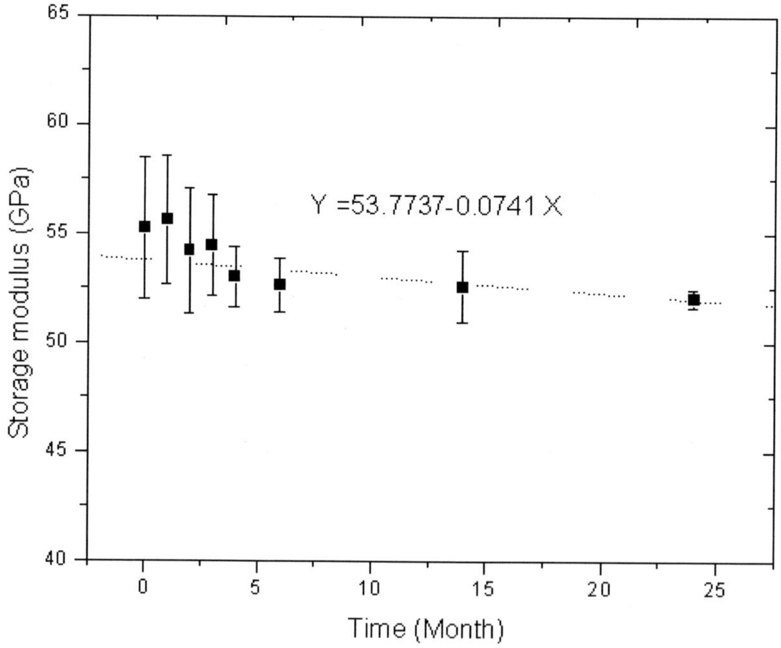

Figure 18. Storage (dynamic) modulus of GFRP rebar as a function of ageing time.

The average E' at 25 °C of the rebar samples with respect to ageing time is presented in Figure 18. The E' is a measure of how elastic the material is and ideally is equivalent to Young's modulus. However, according to the literature [26], for many reasons E' is not considered equivalent to Young's modulus. First, Young's modulus is estimated over a range of stresses and strains as it is the slope of a linear line while E' comes from what can be considered a point on the line. Second, the tests are very different as in the stress-strain test, one material is constantly stretched, whereas it is oscillated in the dynamic test. Nevertheless, in the present work, *E´* of the samples was found to be in the range of 52.6 to 55.3 GPa irrespective to ageing time, which was very similar to the quasi-static tensile and flexural moduli results. The above similarity in quasi-static tensile and flexural moduli with dynamic modulus results could be attributed to high degree of orientation (i.e. straightness) in fibres along specimen length, good distribution of fibres across specimen thickness, reasonably high fibre content and good curing. Although dynamic test was conducted in 3-point bend mode but the effect of compression wave form in the fibres would not be prominent when compared with quasi-static 3-point bend flexural test as discussed earlier. This was because thickness of the dynamic test sample was very small compared with that for quasi-static flexural test.

Like tensile and flexural moduli, *E´* also showed a general declining trend with respect to ageing time; and *E´* of alkaline aged samples changed insignificantly when compared with that of control samples (see Table 3), which again indicated that fibre/matrix interface and fibres were not deteriorated notably. Like tensile and flexural moduli, *E´* of the rebar was extrapolated in logarithmic time plot (see Figure 13) and found that *E´* will be retained by about 85% after 100 years. This finding indicated that dynamic mechanical modulus of GFRP rebar could be lesser than the quasi-static tensile and flexural moduli in long term service.

3.4. Glass Transition Temperature

According to the standard ASTM E1640 [30], glass transition temperature (T_g) should be taken as the onset glass transition of *E´* curve. The temperature of onset glass transition of *E´* is useful in design as it describes the absolute temperature limit for structural properties to remain stable. T_g of the rebar samples was evaluated as the extrapolated onset to the sigmoidal change in the *E´* curve observed in going from the glassy region to the soft rubbery region. Figure 19 shows T_g of the rebar samples as a function of ageing time. T_g of the samples was found to be in the range of about 115 to 120 °C irrespective to ageing time. Also T_g of the alkaline aged rebar did not change significantly compared with control rebar. This finding suggested that majority of the specimen was not degraded. There was no sign of considerable post curing of the rebar during alkaline ageing as confirmed from the T_g results.

Table 3. Two sample paired t-test storage (dynamic) modulus statistics summary

Sample	Mean	Standard deviation	Standard error	t - distribution	P - value	Remarks
Control	55.250	3.236	1.618	-	-	P-value is greater than the signify-cance level (0.05). At the 0.05 level, the difference of the population means is not significantly different than the test difference.
1 month	55.625	2.974	1.487	0.321	0.769	
2 months	54.200	2.893	1.446	0.478	0.665	
3 months	54.450	2.316	1.158	0.397	0.718	
4 months	53.025	1.391	0.696	1.078	0.360	
6 months	52.650	1.223	0.612	1.312	0.281	
14 months	52.575	1.634	0.817	2.199	0.115	
24 months	52.025	0.411	0.205	2.121	0.124	

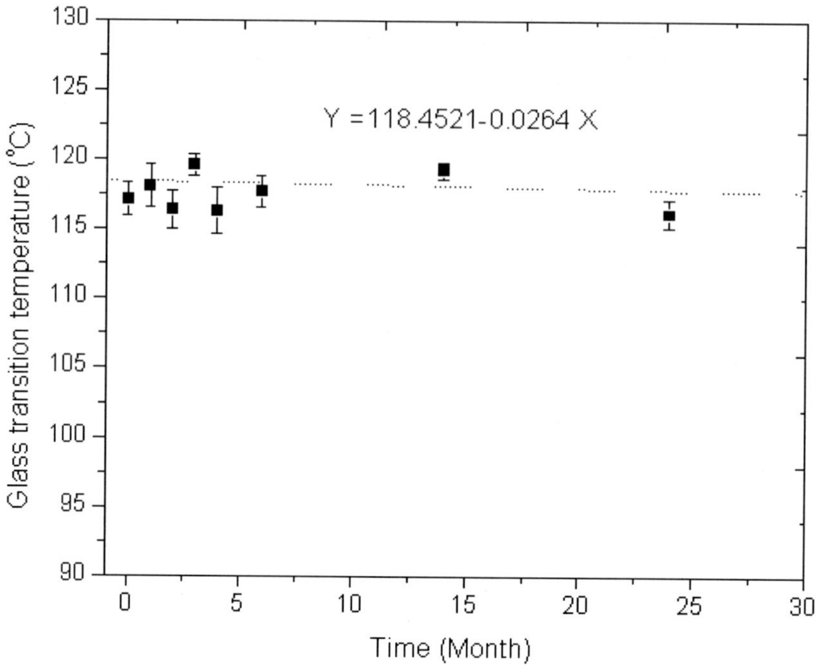

Figure 19. Glass transition temperature of GFRP rebar as a function of alkaline ageing time.

Conclusion

To date many research works have been devoted mainly to analyse the effect of alkaline ageing on tensile strength of GFRP rebar. However, alkaline ageing affect on the elastic modulus of GFRP rebar was not well documented. This paper presented long term alkaline ageing affect on the quasi-static and dynamic moduli of GFRP rebar. It was revealed that quasi-static tensile, quasi-static flexural and dynamic mechanical moduli of alkaline aged samples changed insignificantly with ageing time when compared with those of control samples; and the average elastic modulus results were ranged from 52 to 57 GPa regardless of testing method and ageing time. Furthermore, degradation of fibres and fibre/matrix interface due to exposure in alkaline environment was not apparent in the SEM micrographs and there was no evidence of resin degradation as supported by T_g results.

Preparation of GFRP rebar test sample according to the standards ASTM D7205/D7205M and ACI 440.3R for tensile modulus evaluation is a lengthy process. In particular, the sample preparation is quite difficult for large diameter GFRP rebar and a large scale machine is required for the test. On the other hand, dynamic mechanical modulus assessment by DMTA is relatively a quick method and very easy to prepare test sample from GFRP rebar of any diameter. In this work, quasi-static tensile modulus was in good agreement with dynamic mechanical modulus for the same material and the modulus results were varied within the range of standard deviation. So it can be suggested that dynamic mechanical test could be useful as an alternative of quasi-static tensile test to evaluate the elastic modulus (i.e. Young's modulus) of a well cured unidirectional GFRP rebar with high glass content.

ACKNOWLEDGMENT

Staff members of research and development of Pultron Composites Limited are acknowledged for contributing to the activities in this project work. Thanks to Dr. Abdullah A. Mamun, University of Kassel, for SEM-EDX test.

REFERENCES

[1] Matthews, F. L. & Rawlings, R. D. (1994). Composite Materials: Engineering and Science. London: Chapman and Hall.

[2] Karbhari, V. M. (2007). Fabrication, quality and service-life issues for composites in civil engineering. In V. M. Karbhari (Ed.), Durability of composites for civil structural applications (23-24). Boca raton, FL: CRC Press.

[3] Won, J. P. & Park, C. G. (2006). Effect of Environmental Exposure on the Mechanical and Bonding Properties of Hybrid FRP Reinforcing Bars for Concrete Structures. *Journal of Composite Materials*, *40*, 1063.

[4] American Association of State Highway and Transportation Officials, (2009). Design Guide Specifications for GFRP-Reinforced Concrete Bridge Decks and Traffic Railings.

[5] Benmokrane, B., Wang, P., Ton-That, T. M., Rahman, H. & Robert, J. F. (2002). Durability of Glass Fiber-Reinforced Polymer Reinforcing Bars in Concrete Environment. *Journal of Composites for Construction*, *6*, 143.

[6] American Concrete Institute, (2006). Guide for the Design and Construction of Structural Concrete Reinforced with FRP Bars. ACI 440.1R.

[7] Chu, W, Wu, L. & Karbhari, V. M. (2004). Durability evaluation of moderate temperature cured E-glass/vinylester systems. *Composite Structures*, *66*, 367.

[8] Dejke, V. (2002). Durability and Service Life Prediction of GFRP for Concrete Reinforcement. Chalmers University of Technology.

[9] Apicella, A., Migliaresi, C., Nicodemo, L., Nicolais, L., Iaccarino, L. & Roccotelli, S. (1982). Water sorption and mechanical properties of a glass-reinforced polyester resin. *Composites*, *13*, 406.

[10] Apicella, A., Migliaresi, C., Nicolais, L., Iaccarino, L. & Roccotelli, S. (1983). The water ageing of unsaturated polyester-based composites: influence of resin chemical structure. *Composites*, *14*, 387.

[11] Pantuso, A., Spadea, G. & Swamy, R. N. (1998). An Experimental Study on the Durability of GFRP Bars. *In Proceedings of Second International Conference on Composites in Infrastructure* (476-487).

[12] Liao, K., Schultheisz, C. R. & Hunston, D. L. (1999). Effects of environmental aging on the properties of pultruded GFRP. *Composites Part B: Engineering*, *30*, 485.

[13] Abbasi, A. & Hogg, P. J. (2005). Temperature and environmental effects on glass fibre rebar: modulus, strength and interfacial bond strength with concrete. *Composites Part B: Engineering*, *36*, 394.

[14] Chen, Y., Davalos, J. F., Ray, I. & Kim, H. Y. (2007). Accelerated aging tests for evaluations of durability performance of FRP reinforcing bars for concrete structures. *Composite Structures*, *78*, 101.

[15] Masmoudi, R., Nkurunziza, G., Benmokrane, B. & Cousin, P. (2003). Durability of glass FRP composite bars for concrete structure reinforcement under tensile sustained load in wet and alkaline environments. Annual Conference of the Canadian Society for Civil Engineering, Moncton, Canada.

[16] Sen, R., Mullins, G. & Salem, T. (2002). Durability of E-Glass/Vinylester Reinforcement in Alkaline Solution. *ACI Structural Journal*, *99*, 369.

[17] Ellyin, F. (2003). The Influence of Aqueous Environment, Temperature and Cyclic Loading on Glass-Fibre/Epoxy Composite Laminates. *Journal of Reinforced Plastics and Composites*, *22*, 615.

[18] Ellyin, F. & Maser, R. (2004). Environmental effects on the mechanical properties of glass-fiber epoxy composite tubular specimens. *Compos. Sci. Techn.*, *64*, 1863.

[19] Karbhari, V. M. (2003). Durability of FRP composites for civil infrastructure - Myth, mystery and reality. *Advances in Structural Engineering*, *6*, 243.

[20] Ceroni, F., Cosenza, E., Gaetano, M. & Pecce, M., (2006). Durability issues of FRP rebars in reinforced concrete members. *Cement and Concrete Composites*, *28*, 857.

[21] American Concrete Institute, (2004). Guide Test Methods for Fiber-Reinforced Polymers (FRPs) for Reinforcing or Strengthening Concrete Structures. ACI 440.3R.

[22] ASTM International, (2006). Standard Test Method for Tensile Properties of Fiber Reinforced Polymer Matrix Composite Bar. ASTM D7205/D7205M.

[23] ASTM International, (2003). Standard Test Methods for Flexural Properties of Unreinforced and Reinforced Plastics and Electrical Insulating Materials. ASTM D790.

[24] Karbhari, V. M. & Wang, Q., (2004). Multi-frequency dynamic mechanical thermal analysis of moisture uptake in E-glass/vinylester composites. *Composites Part B: Engineering*, *35*, 299.

[25] Akay, M. (1993). Aspects of dynamic mechanical analysis in polymeric composites. *Compos. Sci. Techn.*, *47*, 419.

[26] Menard, K. P. (2008). Dynamic mechanical analysis A Practical Introduction. Boca raton, FL: CRC Press.

[27] Earl, J. S. & Shenoi, R. A. (2004). Hygrothermal ageing effects on FRP laminate and structural foam materials. *Composites Part A: Applied Science and Manufacturing*, *35*, 1237.

[28] Kelly, A. & Zweben, C. (2000). Comprehensive Composite Materials. Oxford: Pergamon.

[29] Goertzen, W. K. & Kessler, M. R., (2007). Dynamic mechanical analysis of carbon/epoxy composites for structural pipeline repair. *Composites Part B: Engineering*, *38*, 1.

[30] ASTM International, (2004). Standard Test Method for Assignment of Glass Transition Temperature By Dynamic Mechanical Analysis. ASTM E1640.

[31] Bakis, C. E., Boothby, T. E., Schaut, R. A. & Pantano, C. G. (2005). Tensile Strength of GFRP Bars Under Sustained Loading in Concrete Beams. *In Proceedings of 7th International Symposium on Fiber-Reinforced (FRP) Polymer Reinforcement for Concrete Structures* SP-230 (1429-1446).

[32] Micelli, F. & Nanni, A. (2004). Durability of FRP rods for concrete structures. *Construction and Building Materials*, *18*, 491.

[33] Davalos, J. F., Chen, Y. & Ray, I. (2008). Effect of FRP bar degradation on interface bond with high strength concrete. *Cement and Concrete Composites*, *30*, 722.

[34] Weber, A. & Baquero, C. W. (2010). New durability concept for FRP reinforcing bars Combined durability and creep rupture testing allows for realistic design values. *Concrete International*, July 49.

[35] Karbhari, V. M. & Abanilla, M. A. (2007). Design factors, reliability, and durability prediction of wet layup carbon/epoxy used in external strengthening. *Composites Part B: Engineering*, *38*, 10.

In: Polymer-Matrix Composites
Editor: Rakesh Kumar

ISBN: 978-1-61122-005-6

Chapter 8

Conducting Polymer-Metal Oxide Nanocomposite and Their Property Towards Energy Storage: An Overview

***Bhavana Gupta*[1*], *Ambrose A. Melvin*[2], *P. Revathi*[1,4], *T. Matthews*[1], *S. Dash*[1], *A. K. Tyagi*[1] and *Rajiv Prakash*[3]**

[1] Surface and Nanoscience Division, Materials Science Group, Indira Gandhi Centre of Atomic Research, Kalpakkam, Tamilnadu, India

[2]Catalysis Division, CSIR-National Chemical Laboratory, Dr. Homi Bhabha, Pune, India

[3] School of Materials Science and Technology, Indian Institute of Technology, Banaras Hindu University, Varanasi, India

[4]Vellore Institute of Technology,Vellore, Tamil Nadu

Abstract

Supercapacitors combine the unique properties of two type of energy storage devices *i.e.* battery and capacitor. Thus supercapacitor reveals higher energy and power densities than conventional batteries and capacitors. Energy storage in supercapacitor takes place by two distinct phenomena *i.e.* surface charging/discharging as well as redox reaction. For further improvement in the current state of art of supercapacitors, it is required to develop new and efficient conductive and electroactive materials to meet ever increasing energy storage demand. In this context, conducting polymer and transition metal oxides are the versatile classes of materials for further modification and several combinations. Moreover, the above materials in nano-scale dimensions exhibit unique properties or combinations of properties as needed for supercapacitor electrode application. Nanotechnology enabled material synthesis strategies, have pushed the specific energy storage capacity to ~ 1000 F/g. Supercapacitor design and construction are likely to reap these benefits. The present article attempts to bring forward specific advantages of nanoscale materials deposited by chemical means towards device applications.

[*]Corresponding author: Tel.: 044-2458500-22171; Email address: bgupta1206@gmail.com, bhavanag@igcar.gov.in.

Keywords: Supercapacitor; Nanocomposite; Conducting polymer; Metal oxide

1. INTRODUCTION

With the advancing swing towards renewable energy, it is required to carry out research in the area of energy storage methods so that it is available whenever desired. Basically energy can be stored by two types of basic devices i.e. battery and capacitor. In battery, chemical energy gets converted to electrical energy at a particular applied voltage. However, in capacitors, there is nonfaradic energy storage which helps in the formation of an electrostatic double layer of charge over the surface of electrode. Hence, it derives the name electric double layer capacitor (EDLC). It has relatively fast charge – discharge characteristic. Slow energy storage mechanism of battery reduces its power density and charging/discharging rates. In contrast, a capacitor can store small amount of charge, resulting low energy density. This magnitude is normally scaled up with increase in the size of the electrode. With progressive enhancement in specific surface, there has been substantial increase in energy density. A typical value for carbon powder is 100 F/g with 1000-2000 m^2/g of specific surface. Practical realization of above technicality has led to construction of new type of capacitor termed as supercapacitor [1-4]. Charge and discharge profiles of supercapacitor and battery are shown in Figure 1.

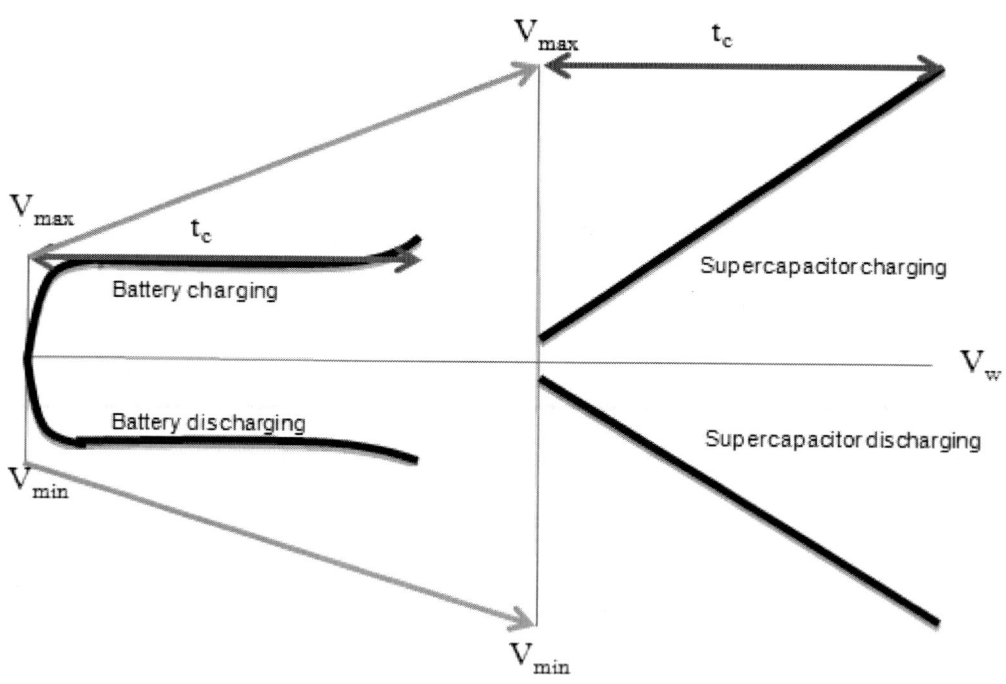

Figure 1. Charge-discharge profiles of battery and supercapacitor with respect to time. V_w is the required voltage for supercapacitor operation similar to that of open circuit voltage. V_{max} and V_{min} are highest charging and lowest discharging voltages repectively.

Recently, supercapacitors have received growing interest due to the origin of pseudocapacitive properties of redox materials. When the charge storage mechanism is based on both faradic and nonfaradic processes, specific capacitance gets enhanced profoundly. Non faradic process leads to the formation of charged interface of porous electrode and the electrolyte which is termed as electrochemical double layer.

Nonetheless, the faradic capacitance arises due to peculiar pseudocapacitive characteristics of an electroactive material. Redox materials allow the involvement of sub-surface layers to store energy. In consequence to this, huge specific capacitance results. Conductive polymers and oxides of transition metals are specifically known to be appropriate materials for electrochemical or redox supercapacitor usage [5].

Due to the availability of π-conjugated backbone, conducting polymer can readily be transform into oxidized and reduced states simply by varying the applied potentials which leads to the doping and dedoping respectively. On the other hand, the metal ion in metal oxide shows variable oxidation state under different applied potentials. The above class of materials are redox active and show pseudocapacitive response despite certain inherent advantages and disadvantages.

Conducting polymers have multiple advantages such as high pseudocapacitance, conductivity and feasibility of fabrication process. However, these suffer from low cycle life, stability and poor mechanical property [6]. Apart from these deleterious aspects, the volumetric shrinkage after ion ejection and low conductivity in de-doped state cause high ohmic polarization of the supercapacitor. Similarly, metal oxides/hydroxides also have pseudocapacitive properties along with large charge/discharge potential windows. Here too, the low conductivity and cycle life limit their further application. Combination of such dissimilar materials in nanoscale is a unique approach to optimize their structure and properties in thin film form for enhancing supercapacitive performance. These composites posses advantages of polymers (flexibility, toughness and coat ability) and metal oxide (hardness and durability) simultaneously. In addition, certain other properties are also manifested. Resulting properties of the electrode materials depend on the three features:

1) Composition
2) Combined morphology
3) Interfacial characteristics

The last one is very important as it controls interaction among components [7, 8]. Extensive work has been done in the area of the synthesis and fabrication of such composite materials for use in supercapacitors. Some methods are expensive and their limited feasiblity leads to the non-suitability for making thin film materials of complex composition. Such ultrafine composite architecture promotes interaction among nano-sized components. The morphology control of such exquisite thin films can be achieved by simply altering the experimental conditions.

This chapter describes the literature survey of conducting polymer/metal oxide nanocomposite thin films with respect to their synthesis technique (chemical or electrochemical) for supercapacitor application [7]. The investigations on these materials attempt to correlate role of surface morphology and other measured physico-chemical properties on the capacitive performance.

2. Synthesis/Fabrication Methods

Microstructural and morphological parameters specific to a fabrication method vastly influence properties of a nanocomposite thin film. Broadly two different methods are known for the fabrication of these composites *i.e.* chemical and electrochemical. Next sub-division describes the variants in chemical and electrochemical methods adopted by various research groups to obtain nanocomposites in various combinations.

2.1. Chemical Fabrication

Chemical synthesis is carried out in the presence of an oxidant or catalyst. Chemical synthesis method is advantageous due to easy process condition and cost-effectiveness [9]. Chemical polymerization is amenable to processing purposes and it is easily scalable. It also results, in the formation of powder and colloidal dispersions. With chemical polymerization, insulating surfaces can also be coated. Generally the chemical route is preferred to electrochemical one as it allows formation of conducting polymer films with required morphology. The chemical procedure has much less number of experimental variables compared to electrochemical methods involving cell assemblies and electrolytes. For making nanocomposites by chemical methods, two different techniques are adopted. In one method, metal oxide nanoparticles are prepared separately and this is followed by polymerization over a surface as shown in Figure 2A. Emulsion and interfacial methods of chemical synthesis are also applicable in this case. Usually here the polymerization process is guided by the metal oxide surface, which enables nanocomposite films to acquire the morphology of the metal oxide. Such type of synthesis is termed as *ex-situ* synthesis or polymerization [10]. Other methods involve both metal oxides and polymers to form simultaneously as shown in Figure 2 B.

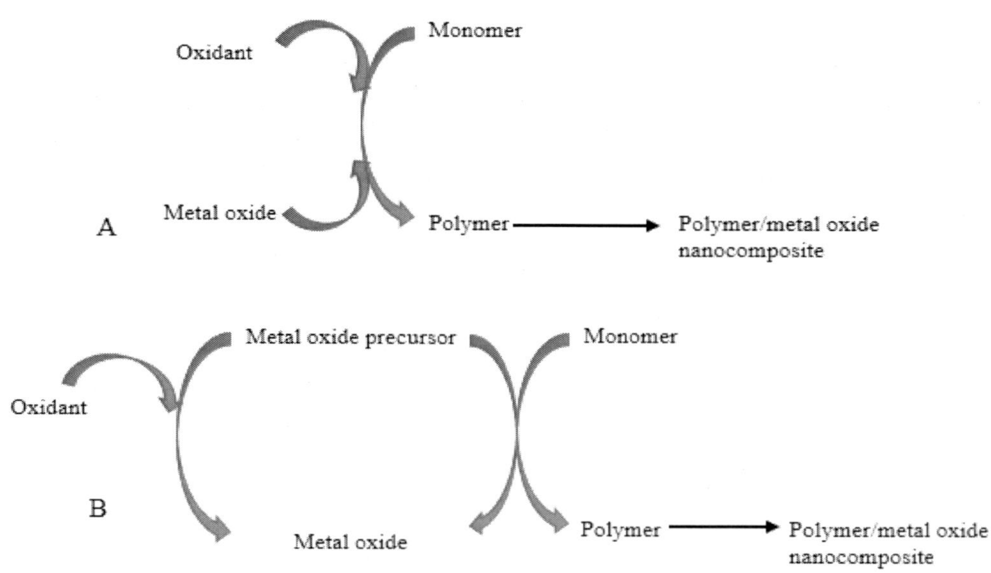

Figure 2. Possible ways of chemical synthesis of conducting polymer/metal oxide nancomposites.

This method is known as in-situ synthesis [11-13]. Here, control of morphology of the nanocomposite is difficult. The resulting nanocomposites are formed mostly in the form of powders. Such powders are further used for fabrication of both conducting and insulating surfaces which is amenable to both electrochemical and other suitable characterizations. Ni et al. synthesized MnO_2-PAni by using first methods of synthesis. Their synthesis procedure used 1:2 wt. % of MnO_2 in aniline with excess oxidizing agent (ammonium persulphate) prepared in acidulated water (19.6 ml water+0.4 ml H_2SO_4). This mixture was kept in an ice bath. Time provided for polymerization was 1, 2 and 5 hrs to get distinct changes in properties [14].

Qunting et al. also adopted similar procedure, where the solvent used is sodium dodecyl sulfonate dissolved in 1M HCl. Molar ratio of aniline to oxidant is 1:1 (5 mM). The wt. % ratio between aniline to metal oxide was kept at 1:1. For the synthesis of V_2O_5 with polypyrrole, first V_2O_5 nanosheet surface was modified with sodium dodecyl sulfonate for uniform dispersion followed by polymerization using $FeCl_3$ in acidic medium to get the monomer protonated at $0^{o}C$ [15]. Liu et al. modified the surface of samarium oxide (Sm_2O_3) with amino group using γ-amino propyl triethoxysilane (APTES) for carrying out controlled deposition of polypyrrole over the surface. Here, surface functionalized Sm_2O_3 is synthesized separately. This synthesis was followed by polymerization in acidic medium (4.16 g p-toluene sulphonate dissolved in 120 ml water) using ammonium persulphate (~0.5 M in 1:1 wt. % ratio to monomer) in ice cold condition. Sm_2O_3 wt. % was varied from 0 to 50 with respect to monomer [16]. Zhang et al. carried out in-situ synthesis of metal oxide precursor using Molybdenum (Mo) and H_2O_2 in an ice bath. The resultant peroxomolybdate was used as an oxidizing agent with and without ultrasonic treatment for polyaniline synthesis in 2.5 M HCl medium at 0 ^{o}C and 40 ^{o}C. The aniline concentration was maintained at 20 mM [17]. Rao et al. in-situ synthesized SnO_2-PAni nancomposite using $SnCl_2.2H_2O$ and aniline monomer (about 1:1 wt. %) in an acidic medium (150 ml) followed by polymerization process which used ammonium persulphate (equal wt. % ratio to monomer). The duration of reaction was 6 h at room temperature [18]. We have synthesized PEDOT-SnO_2 nanocomposite by first methods (ex-situ) directly on the solid substrate (both conducting and insulating). SnO_2 synthesis was achieved by electrospinning in thin film form in nanofibrous morphology using $SnCl_2$ and polypyrrolidone reaction mixture (1:1) in DMF and ethanol solvents (1:1) followed by annealing at $600^{o}C$ to remove polymer and transform $SnCl_2$ to SnO_2 [19]. Annealing temperature was identified by thermal gravimetric analysis (TGA) of PVP+$SnCl_2$ nanofibre which revealed massive degradation in the temperature range of $400^{o}C$ to $600^{o}C$. Confirmation of SnO_2 formation came from XRD and FT-IR analysis. XRD shows all characteristic planes of SnO_2 and FT-IR shows the presence of Sn-O and O-H bands at 1639 and 669 cm^{-1}, respectively, [20-22] as shown in Figure 3 (a,b).

The SnO_2 film on solid support was directly used for *in-situ* oxidative polymerization of EDOT. For *in-situ* oxidative polymerization, reactant mixture (EDOT: Fe(III)OTs: Imidazole: 1:4.2.5) in n-butanol [23] was spread on SnO_2 coated solid substrates like glass and ITO *etc.* After spreading the mixture substrate was kept for baking at $80^{o}C$ for achieving polymerization as shown in Figure 4. Once substrate color changed to green, it was cleaned with butanol for removal of extra reactant and dried in air. This sample was subjected to further characterization. Here the requirement of post purification and fabrication could be easily avoided.

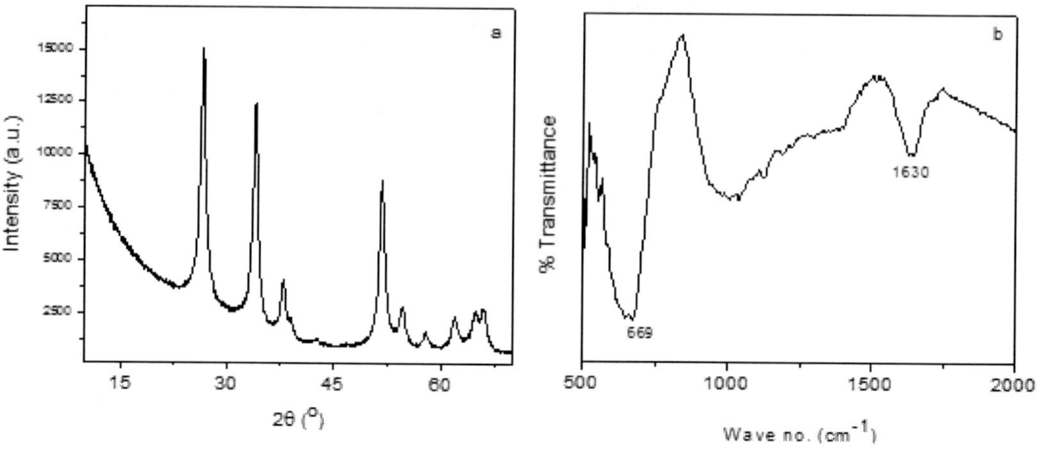

Figure 3. (a) X-ray diffractogram and, (b) FT-IR spectrum of SnO_2 nanofibre.

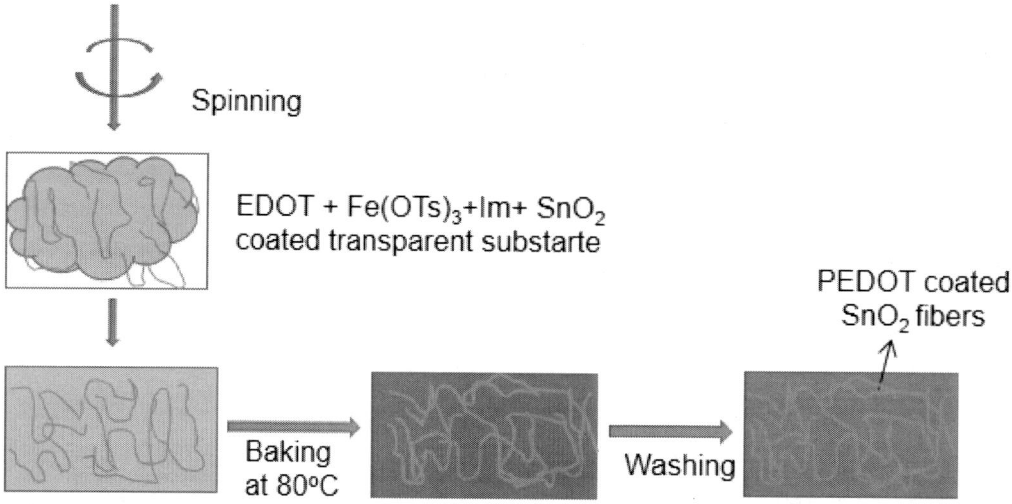

Figure 4. In-situ oxidative polymerization of EDOT on SnO_2 coated transparent substrates like glass and ITO.

2.2. Electrochemical Fabrication

The electrochemical synthesis is relatively clean and requires a three electrode assembly. There are several advantages in the synthesis of conducting polymer/metal oxide nanocomposites through electrochemical approach [9]. Since electrochemical process takes place directly on the electrode surface, it forms a free standing film. Physical properties of such composite films can be easily tailored according to the experimental conditions followed during electropolymerization [24]. Moreoever, electrochemical polymerization is more suitable for *in-situ* spectroscopic studies. However, these methods suffers from the yield problem of the resultant materials. The amount of the material deposited, entirely depends on the electrode surface area. In addition to this an electrode-electrolyte system is also required which varies with the required electroactive material. The electrode-electrolyte system helps the movement of ionic species (monomer and electrolyte) towards electrodes under potential or current biased condition [9].

The range of dopant ion incorporated with the polymer is limited in comparison to the ions present in the oxidant. For making conducting polymer/metal oxide nanocomposite two types of approaches are agenerally followed. In the first approach, metal oxide is synthesized and separated by electrochemical method. The modified electrode is then further used as a working electrode for electrochemical coating of polymer in a defined nanostructure shape. The nanostructure is guided by metal oxide morphology as shown in Figure 5A. In second case, a monomer and metal oxide dispersion are used for electrochemical polymerization on a bare electrode. In this case, the resultant morphology is the intrinsic property of nanocomposite which can be controlled by applied bias as shown in Figure 3B. Cuentas-Gallegos et al. synthesized polyoxometalates like $H_4SiW_{12}O_{40}$, $H_4PW_{12}O_{40}$ and $H_4MoW_{12}O_{40}$ (similar to metal oxide) with PAni by electrochemical (potentiodynamic) technique where metal oxide and polymer depositions take place simultaneously. Silicotungstanic acid, phosphotungstanic acid and molybdotungstanic acid along with aniline in 100 ml water were used as reaction mixture for electrochemical synthesis. Nanocomposite synthesis was optimized by control of scan rate, potential range and no of cycles. Best condition for $H_4SiW_{12}O_{40}$/PAni and $H_4PW_{12}O_{40}$/Pani synthesis are 2.5 mV/sec (-0.25-0.9V vs Ag/AgCl) and 10 mV/sec for 40 cycle (0.35 to 0.9V vs Ag/AgCl), respectively. Cycles more than 40 resulted in decrease in current possibly due to loss of conductivity of working electrode. In case of $H_4MoW_{12}O_{40}$/PAni, the applied potential was kept between -0.1-0.9 V vs Ag/AgCl. Two scan rates were required *i.e.*, 0.5mV/sec for PAni deposition and latter 1.5 mV/sec for metal oxide deposition. This procedure was followed as polyanion creates passivation of the electrode surface and retards the polymer deposition process [25].

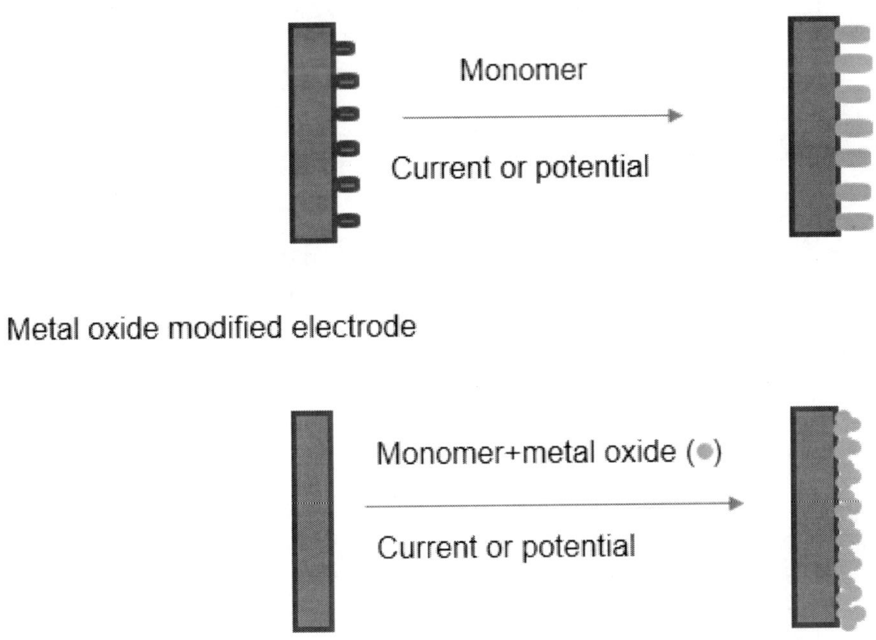

Figure 5. Possible ways of electrochemical synthesis of conducting polymer/metal oxide nancomposites.

PAni/WO_3 composite is synthesized by electrochemical method where metal oxide coated substrate was used as a working electrode for polymerization. Here, WO_3 (synthesized by sol-gel methods) was used for coating on ITO substrate. This was in turn used as a working electrode for electrochemical polymerization. This technique known as PAni by potentiodynamic method uses 0-1.3 V vs Ag/AgCl at 50mV/sec scan rate in 0.5 M H_2SO_4 [26]. Xei et al. first modified Ti foil by oxidation to TiO_2 by applying an anodization process and further used this material as a working electrode for polymerization of pyrrole into polypyrrole. Pulse potential of 0.7 to 1.1 V vs. Hg/$HgCl_2$ in 0.1 M $LiClO_4$ electrolytic solution containing 0.15 M pyrrole with pulse width and pulse period of 0.04 and 20 s, were applied respectively, for pyrrole polymerization. Actual enhancement in current occurred in the range 0.85 to 1.1 V which is an indication of the growth of highly conducting polypyrrole on TiO_2 surface [27].

3. Physicochemical Properties of Conducting Polymer/Metal Oxide Nanocomposite

Multiple physicochemical tools have been utilized to characterize such nanocomposites for their structural, morphological and electrical properties [28]. The experimental techniques include TGA, FT-IR, XRD, SEM and TEM. Basically such studies are used to evaluate interaction, interfacial characteristics and morphology of the nanocomposite. Assessment of material properties by several authors who have made use of these experimental tools will be highlighted. Different compositions were tried out for these characterizations. The results of these experimental investigations were critically examined to benchmark an optimized composition for supercapacitor application. Electroanalytical measurements were employed for elucidation of capacitive response from these materials.

3.1. Structural Characterization

Structural characterization involves evaluation and investigation of change in chemical backbone or formula molecular unit of the component and change in charge concentration upon nanocomposite formation. Generally XRD and FT-IR were used for structural characterization.

3.1.1. FT-IR

To get the composition of metal oxide, FT-IR spectroscopy was performed in transmission mode [29]. Ni et al. synthesized PAni-MnO_2 nanocomposite by in-situ polymerization and the time duration was varied for this purpose. Characteristics peak or transmission rate of MnO_2 decreased with increase in polymerization time with concomitant reduction in wt. % of MnO_2 [14]. Xei et al. reported the appearance of all PPy peak in the TiO_2/PPy nanocomposite where TiO_2 was present in the form of nanotubes. From this feature, it is concluded that the resultant polymer is present uniformly over the nanotube surface and covering the entire surface. Presence of vibrational frequency is related to both components (*i.e.* PPy and TiO_2).These facts corroborates to mixing of both the components in the composite [27]. Wei et al. reported the shifting in peak position in PAni-WO_3 nanocomposite towards lower wave number side. Once again this peak is related to both PAni and WO_3. Such shifting occurs only when N atom of polymer tend to co-ordinate with metallic site of the metal oxide [26]. Liu et al. stated that interaction between Sm_2O_3 and PPy is rather insignificant when 5 wt. % Sm_2O_3 is taken. This insignificant change in PAni-Sm_2O_3 nanocomposite implies that the Sm_2O_3 nanofibre is wrapped by PAni chain [16]. However, when Sm_2O_3 fraction increases to 20 wt. %, pronounced shift in PPy frequency takes place towards low wave number. This significant shift is attributed to charge transfer from PPy to Sm_2O_3. Rao et al. studied the effect of in-situ and ex-situ PAni-SnO_2 through FT-IR spectra. Here, the peak, related to polymer shifted towards lower wave number side due to de-doping or charge transfer effect. This shifting is comparatively higher in nanocomposite prepared by in-situ methods than that of ex-situ prepared nanocomposite [18]. In case of PEDOT/SnO_2 nanocomposite, FT-IR studies revealed structural property of PEDOT as shown in Figure 6. The peak at 1325 and 1513 cm^{-1} belongs to C-C and C=C stretching of benzenoid and quonenoid ring structures respectively. Similarly' peaks at 1201, 1125 and 1035 cm^{-1} correspond to C-O-C stretching frequencies. The vibrations belonging to C-S bond appears at 973, 915, 799 and 683 cm^{-1} wave numbers.

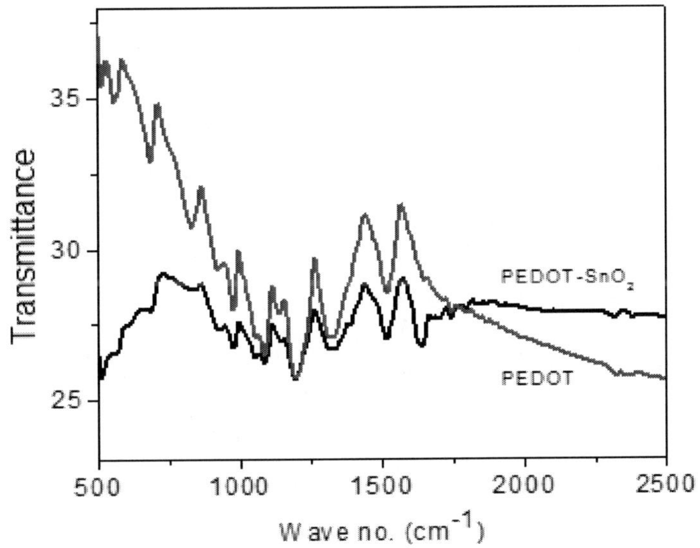

Figure 6. FT-IR of PEDOT and PEDOT-SnO_2 composite.

3.1.2. XRD

XRD data give an idea about the effect of polymer amorphous pocket on metal oxide crystallinity [28]. The diffractograms obtained by Wei et al. depicted superior crystalline features belonging to nanocomposite as compared to the polymer. Authors attributed this observation to occurrence of slow polymerization of PAni over WO_3 modified ITO surface. Furthermore, PAni peaks are shifted out to higher angle side because of electron transfer from PAni to WO_3 which possibly brings in formation of nanocomposite [26]. In case of PAni-SnO_2, nanocomposite peaks belonging to SnO_2 broaden and intensity declines due to its low mass fraction in composites. Zheng et al. described XRD results in terms of interlayer spacing of layered metal oxide structure [17]. Metal oxide interlayer spacing remains intact and sustains intercalation of single chain of PAni, where aniline ring is oriented in a perpendicular direction. In this condition, PAni has got fixed conformation and its orientation inside metal oxide interlayer space is supported by HRTEM analysis. Liu et al. studied effect of polymer wt. % on metal oxide by XRD. It was noticed that as the polymer quantity decreases in the composite matrix, the XRD pattern of the composite resembles that of the metal oxide. This supports the fact that polymer growth takes place on the surface of metal oxide [16]. In our case of PEDOT-SnO_2 nanocomposite, diffraction pattern reveals the features of SnO_2 in form of diffraction planes belonging to 101, 110, 200, 111, 210, 211, 220, 002, 310, 112 and 301 at 2θ 33.87, 26.53, 37.89, 38.93, 42.58, 51.8, 54.68, 57.96, 61.89, 64.73 and 65.98, respectively (Figure 7). After PEDOT incorporation, the peak that belongs to SnO_2 becomes sharper (low FWHM). This observation supports the increase in particle size. Crystallite size calculation based on Scherrer formula gives the value of 10 and 12 nm for SnO_2 and PEDOT-SnO_2 nanocomposites respectively. All these establish the fact that there is polymer over SnO_2 surface which is uniform.

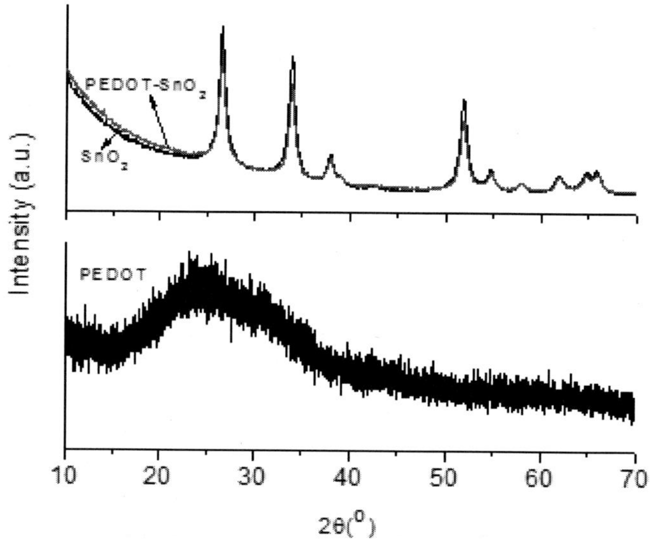

Figure 7. XRD pattern ofof PEDOT, SnO_2 and PEDOT-SnO_2 nanocomposite.

3.2. Thermal Characterization

Several authors have attempted to figure out degree of stability, composition and stoichiometry of composites from thermograms [30, 31]. Ni et al. studied the thermal degradation of PAni-MnO_2 nanocomposite where the polymerization process was carried out for 1, 2 and 5 hrs. The thermal degradation study on 5 hrs polymerized nanocomposite samples carried out upto 800 °C depicted higher residual weight compared to other nanocomposites formed by short duration of polymerization. This observation supports the fact that higher the polymerization time, better is the polymer yield [14]. Zheng et al. found about 35.7 wt. % loss at 500°C and this was related to evolution of organic species. Based on the TGA results, authors tried to formulate mechanism for nanocomposite formation and established a stoichiometry relation as shown in the following equation [17]:

Based on this formula, inorganic component is found to be 61.0 % which is well co-related with the TGA result. Liu et al. observed that weight loss of PPy is significantly reduced in the presence of Sm_2O_3. Apart from this, the degradation of PPy continued even after 700°C in the presence of Sm_2O_3. In comparison, pure PPy degradation almost reaches completion at 700°C [16]. We studied the effect of SnO_2 on thermal property of PEDOT as shown in Figure 8. First, TGA showed an initial weight loss at 60 °C followed by another minor loss at 100 °C. This could be related to vaporization of moisture and other gaseous species. Degradation at around 225°C corresponds to PEDOT chain decomposition. PEDOT-SnO_2 is found to be more stable in terms of degradation temperature; a scenario that supports strong material interaction. At 500°C, the total weight loss is about 2 % which supports the fact that SnO_2 in the matrix constitutes around 95-98 wt. % and remaining is polymer.

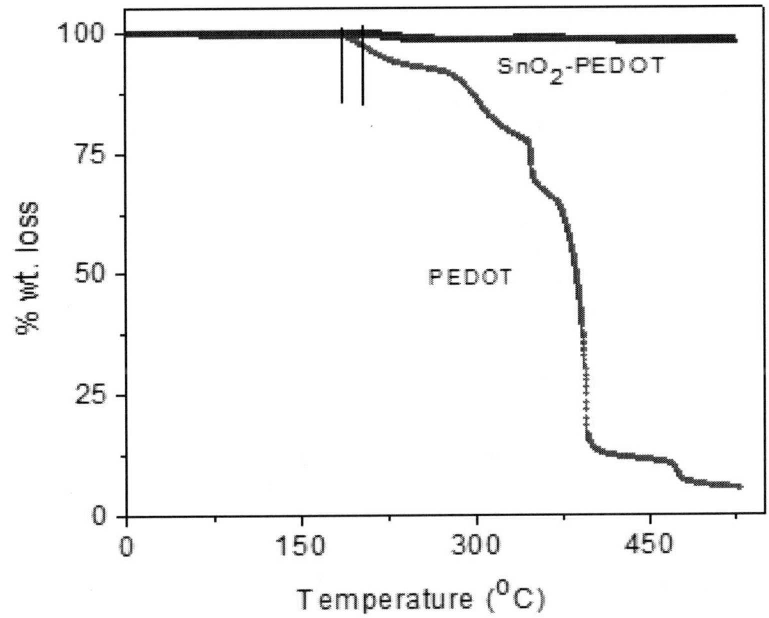

Figure 8. TGA of PEDOT and PEDOT-SnO_2 nanocomposite. Shift in degradation temperature is shown in the form of two parallel lines.

3.3. Electrical Characterization

Electrical characteristics of a nanocomposite is a crucial property for accessing its use in electrochemical applications like supercapacitor. A highly conducting material bestowed with redox property is a promising candidate for applications requiring high specific capacitance [32, 33]. Four probe conductivity measurements are routinely employed to carry out conductivity measurements. Liu et al. carried out conductivity measurement on PPy coated Sm_2O_3 and found that PPy-Sm_2O_3 (5 wt. %) possesses superior conductivity to that of pristine

PPy. This happened due to formation of crystalline pockets of PPy in the presence of Sm_2O_3, a scenario which is responsible for high charge carrer mobility. Further increase in Sm_2O_3 fraction in the nanocomposite brought in opposite trend. This phenomenon is corroborated by the disappearance of polaronic transition as seen in absorption spectra which is caused by de-doping or electron transfer from polymer to metal oxide [16]. In our system (PEDOT-SnO_2) conductivity reduces after SnO_2 incorporation in comparison with pristine PEDOT. Pristine polymer containing glass shows a resistance of 7 kΩ/cm^2. In composite form the resistivity is raised to 18-20 kΩ/cm^2 which is nearly three and a half time more than that of pure PEDOT (Figure 9). Such a rise supports formation of a thin coating of polymer over SnO_2 surface. This occurs while maintaining the molar concentration of constituents as EDOT: OTS: Im:1:4:2.5 during spin coating process. As the number of coated layers increases, the trend in conductivity ascends towards positive side where S1, S2 and S3 refer to mono, di and tri layer coatings respectively. The conductivity shows ascendancy with increasing layer number which establishes the suitability of a coated composite for electrochemical supercapacitor application.

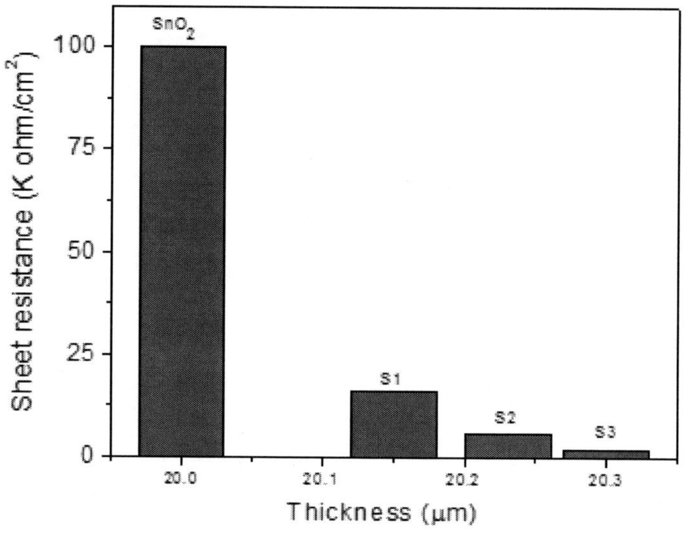

Figure 9. Four probe conductivity of PEDOT, PEDOT-SnO_2-S1, PEDOT-SnO_2-S2 and PEDOT-SnO_2-S3 with respect to the thickness.

3.4. Contact Angle/Surface Property

For an electrolyte to enter into the electrode, presence of an optimum porosity is desirable. Although a porous material facilitates electrolyte movement in/out from the surface [34], large pore volume can cause hindrance to the electrolyte/ion movement as the gas held within the pores blocks the electrolyte. Surface area required for the electrochemical reaction gets pruned. An optimized interconnected pore network brings in efficient ion transport [35]. Flow blockage and wetting properties of a material surface can be correlated to contact angle which is a measurable property. Blocked electrolyte movement is revealed in the form of high contact angle and vice versa. From this idea, we carried out contact angle measurements on

PEDOT, SnO_2 and PEDOT-SnO_2nanocomposite. From the contact angle measurements, it appears that the highest electrolyte accessibility occurs in case of SnO_2 (almost zero contact angle) which is followed by PEDOT-SnO_2 (45 °C) nanocomposite and PEDOT film (90 °C) (Figure 10). SnO_2 formed by electrospining probably forms nanofibres which provides a high surface area. Due to high pore volume, it also provides the highest accessibility to the water droplets into the porous network. A pristine polymer is a dense packed structure along with hydrophobic organic chain which repels water. Ingress of a hydrophilic liquid into porous network is ensured by high contact angle. But when the same polymer is coated over SnO_2 coated solid substrates, contact angle assumes an intermediate value. This optimized porous architecture not only facilitates ion transport but also provides enhanced conductivity. Such a condition is very suitable for the fabrication of an electrochemical device such as super capacitor. This data is further supported by roughness measurement which shows highest roughness occurs in case of SnO_2 nanofibre followed by SnO_2-PEDOT nancomposite and thin PEDOT film.

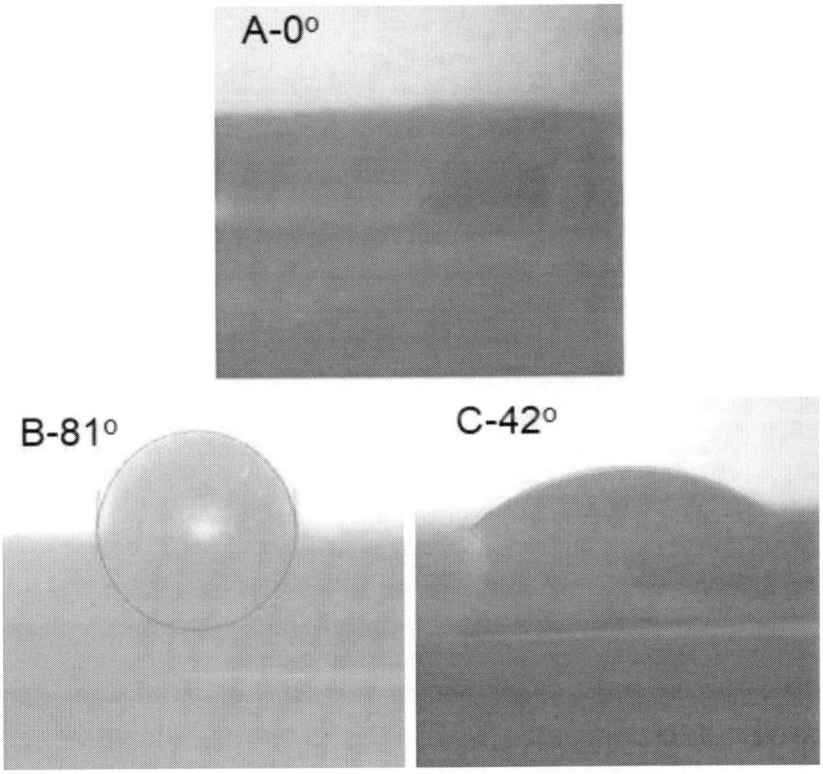

Figure 10. Contact angle measurements of A) SnO_2 nanofibre, B) PEDOT and C) PEDOT-SnO_2 nancomposite films using falling water droplet method.

3.5. Morphological Characterization

Surface morphology is a key parameter which vastly influences ionic mobility. A surface laden with porosity has significant influence on ensuing electrochemical processes [36-40]. An optimized porous network and the distribution achieved by judicious combination of metal oxide and polymer constituents cause considerable reduction in both ion and electron diffusion lengths resulting in enhanced transport. Nanocomposites are usually based on microporous structure where metal oxide is embedded in the polymer matrix. Nei et al. evaluated particle size of MnO_2 in MnO_2-PAni nancomposite. Size of nancomposite grains are larger (80-120 nm) than embedded MnO_2 (5-15 nm) particles. The metal oxide gets wrapped by polymer and this prevents subsequent particle agglomeration and growth [14]. Xei et al. electrochemically coated PPy over TiO_2 nanotubes. The coating was confirmed to be an inner liner with dimensions ranging from 120-150 nm to 50-80 nm. It has also increased the outer wall thickness from 10-20 nm to 30-90 nm [27]. Presence of metal oxide filler was also evaluated by morphological characterizations. Wei et al. succeeded in getting clear diffraction fringes of WO_3 from the polymer matrix, which could not be captured in XRD pattern. This further establishes the possibility of metal oxide wrapping by polymeric layer. Later, it was further confirmed by thickness measurement of the WO_3, PAni and WO_3-PAni nanocomposite films by atomic force topography. Nanocomposite roughness is less (30 nm) than that of pure PAni (80 nm) and nearly equal to WO_3 film (20 nm). Thus there is a clear indication of polymer coating over WO_3 particles [26]. Zheng et al. got flower type structure of PAni-MoO_3 nanocomposite without any sharp edge. HRTEM studies showed lamellar architecture with interlamellar space of 1.5 nm. When the metal oxide acquires lamellar structure, the existence of polymer is expected to be within the interlamellar space. This happens when polymer is present in small quantities. Mono molecular layers of PAni and MoO_3 have nearly same thickness of 1.5 nm. This aspect supports polymer intercalation into interlamellar space. Value of interlayer space also matches well with the d-spacing measured form XRD studies [17]. Effect of metal oxide feed ratio on degree of nancomposite dispersion was studied by Liu et al. These authors stated that with increase in feed ratio of metal oxide, binding of growing polymer with surface absorbed monomer weakens. In such a situation, the polymer layer over metal oxide (NH_2-Sm_2O_3) becomes thinner leading to formation of smaller particles as observed in microscopic analysis. Since particle size significantly affects the degree of dispersion, the feed ratio of Sm_2O_3 is related to degree of dispersion [16]. Effect of surfactant (DBS) on resultant morphology was observed by Qunting et al. in the form of protuberance on the edges. According to their studies, the nanoribbon morphology of V_2O_5 with layered structure remains intact even after polymerization. The protuberance is observed in the form of leaf like particles decorated on the surface of V_2O_5 [15]. Hou et al. studied the effect of PEDOT chain on the dispersion and stability of MnO_2 grown over the surface of Few Walled Carbon Nanotube (FWCNT). From TEM studies, it appears that there is no off-FWCNT scaffold of MnO_2 formed under control deposition along with "crumped paper ball" mesoporous morphology. Such a morphology forms interaction with surface oxygen and it contains groups like OH and COOH present on FWCNTs. Without PEDOT, the resultant material having high sheet resistance peeled off from the current collector. Presence of PEDOT in such hybrid materials not only facilitates electron transfer but also work as a binder [41]. Cuentas-Gallegos et al. optimized the synthesis parameters for obtaining $H_4SiW_{12}O_{40}$, $H_4PW_{12}O_{40}$ and $H_4MoW_{12}O_{40}$ in combination with PAni. This optimization involved control of metallic concentration and polymer growth. In case of $H_4SiW_{12}O_{40}$-PAni nanocomposite, the high tungsten (W) concentration was obtained at a scan rate of 2.5 mV/sec over 40 cycles and 5 mV/sec over 60 cycles. Optimum synthesis parameter was found to be 5 mV/sec scan rate carried out over 40 cycles. Similarly, $H_4PW_{12}O_{40}$-PAni was optimized in terms of scan rate and number of cycles where 10 mV/sec was impressed over

40 cycles. This resulted in greenish dark blue thin layer polymer coating as a major constituent. Later, $H_4PW_{12}O_{40}$-PAni was allowed to grow further by increasing beyond 40 cycles. Cycle numbers 60and 80 showed formation of yellow particles on the electrode surface, which was absent when the synthesis was carried out over 40 cycles. Elemental analysis revealed localized pockets where concentration of tungsten (W), phosphorus (P) and carbon (C) were enhanced. Thus, a 10 mV/sec carried out over 40 cycles turned out to be an optimum condition for material synthesis. In a similar manner, synthesis of $H_4MoW_{12}O_{40}$-PAni nanocomposite was found to be requiring two sets of scan rates *i.e.*, 0.5 mV/sec for PAni and 1.5 mV/sec for PM12 clusters. The resultant morphology showed the uniform distribution of P and W throughout the surface [25]. We have made nanofibres of SnO_2 followed by spin coating of the PEDOT over it. The as-spun PVP/$SnCl_2$ nanofibers have a variable diameters which ranges between 65 and 100 nm. These fibers constituted a self-supporting network, as shown in Figure 11 A. After calcination, the resultant SnO_2 fibre depicted a smooth surface with an outer diameter of 65 nm. The polymer deposition process induced the roughness over the surface which is a clear indication of the formation of polymer coating on SnO_2 as shown in Figure 11 C.

Figure 11. SEM of A) as synthesized electrospun $SnCl_2$ + PVP, B) $SnCl_2$+PVP transformation into SnO_2 nanofibre after 600°C annealing and C) PEDOT+SnO_2 nanofibrous nanocomposite: Zoomed area of the PEDOT+SnO_2 (Inset).

Evaluation of the core of nanofibre and thickness of the coating was carried out with the help of TEM measurement. TEM imaging revealed that the resultant nanofiber of SnO_2 possesses a solid core. Coating of polymer over SnO_2 surface is seen in the form of a differing contrast amounting to 2 nm thickness as shown in Figure 12. Hybrid nanofibrous morphology is supposed to give high surface area for electrochemical reactions. In addition to this, the porosity will provide the pathway for the ion transport through the nanocomposite film surface.

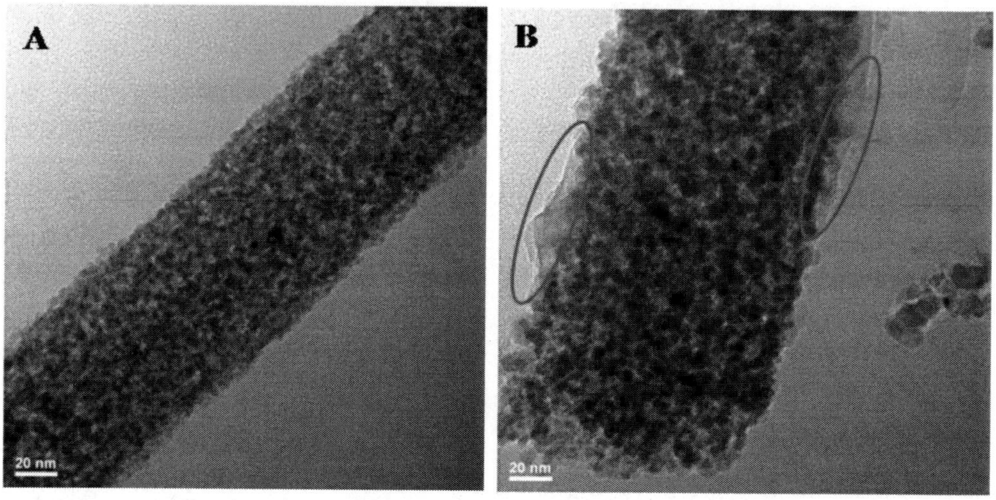

Figure 12. TEM of A) as synthesized SnO_2 nanofibre, B) PEDOT+SnO_2 nanofibrous nancomposite (red circle represents the polymer deposited area).

5.6. Electrochemical Property

Evaluation of electrochemical property of metal oxide-conducting polymer nanocomposite is essential as it provides a basis for assessing the suitability of materials for supercapacitor fabrication. Basically three methods are in vogue for the estimation of specific capacitance of the materials. First methods are based on cyclic voltammetry where average current is used for different materials at same scan rate. Following formula was applied for specific capacitance measurement [42, 43]:

$$C = \frac{I}{Scan\ rate}$$

Impedance and galvanostatic charge/discharge sweeps are used to obtain capacitive response of materials. Following formulae were used [44, 45]:

$$C = \frac{1}{2\pi f Z''}$$

$$C = \frac{I\ \Delta t}{\Delta V}$$

In above equation, "f" is the lowest frequency and Z'' is imaginary impedance at same frequency. ΔV and Δt are potential window and discharge time, respectively. Amongst above three methods, any one can be used for the estimation of specific capacitance of a material. Ni et al. evaluated specific capacitance of MnO_2/PAni nanocomposite where PAni was grown for different time durations like 1 hr, 2 hrs and 5 hrs. The nanocomposites where PAni grown for 5 hrs exhibited higher capacitance value i.e. 510 F/g. Presence of PAni not only provides conducting channels for electron but also contains micro-environmental water which facilitates movement of ion. Such a scenario is suitable for getting high capacitance value [14]. Xei et al. studied the effect of PPy on TiO_2 capacitance. In the presence of PPy, capacitance value is enhanced manifold. The material also sustains high current densities. Material remains stable even after 200 cycles while retaining 85 % of initial capacity [27]. Corresponding electrical characteristics of PAni, WO_3 and their nanocomposite were evaluated by Wei et al. They have stated that capacitance of PAni is more than the nanocomposite probably due to larger extent of polymerization on bare ITO than WO_3 coated ITO. However, in terms of stability, nanocomposite is superior as it remains stable upto 500 cycles with only 4.5 % degradation in the capacity [26]. In case of PAni, crack formation takes place even at low current densities during chrge/discharge cycles. In comparison, pure WO_3 only works under very small current densities. Hu et al. have carried out detail electrochemical property measurements on SnO_2/PAni by Cyclic Voltammetry (CV) and impedance spectroscopic techniques. CV reveals the presence of a redox peak which supports incorporation of PAni with SnO_2. In addition to this, the charge transfer resistance also reduces in case of nanocomposite compared to that of metal oxide [46]. Nanosheets and nanotubular architectures of PAni-MoO_3 nanocomposite were also electrochemically investigated. It was found that nanosheet structure of the material showed enhancement in the capacitance after certain number of cycles. This happened due to crack formation after certain cycles which helped in the ion movement. Apart from this, one chain of PAni sandwiched between the layers prevented the layer from collapsing [17]. Liu et al. studied specific capacitance of PPy-SnO_2 nanocomposite with varying wt. % of metal oxide embedded in polymer matrix. It was found that nanocomposite specific capacitance is higher than that of PPy. Small amount of SnO_2 (5 wt. %) is sufficient for getting high specific capacitance. Above this content, the specific capacitance declined significantly. Large amount of SnO_2 creates blockages in charge carrier channels which retard the ion transport causing reduction in capacitance [16]. Nanocomposite retained its capacity with only 47 % degradation. However, pure polymers degrade upto 73 % after 800 cycles. Effect of ternary architecture (PEDOT:PSS, FWCNT and MnO_2) was also evaluated and considered as an effective material even under very high metal oxide loadings (PEDOT: PSS, FWCNT and MnO_2 = 10, 30 and 60 wt. %). The electrical performance of this ternary composite was far superior to both the individual components and the double mix combinations. In addition, the ternary mix displayed extraordinary electrochemical stability. This material remains stable after 1000 cycles with only 1 % capacity degradation. Effect of current density on capacitive property was also evaluated [41]. Effect of variation in current density from 5mA to 25mA/cm^2 is least in case of nanocomposite. In other case, the capacity declined significantly with an increase in current density from 5mA to 25mA/cm^2. In our group, we have made thin films of SnO_2 and SnO_2-PEDOT by electrospinning and electrospinning plus spin coating for electrochemical characterization in both basic (0.5 M KOH) and acidic (0.5 M H_2SO_4) electrolyte media. After PEDOT incorporation, current density increased and charge transfer resistance decreased in both the electrolytes as shown in Figure 13 and 14. Values of capacitance of SnO_2 and SnO_2-PEDOT nanocomposite are 10, 14 and 96 and 102 F/g in 0.5 M KOH and 0.5 M H_2SO_4 electrolyte media, respectively. Comparative values of capacitance are given in Table 1.

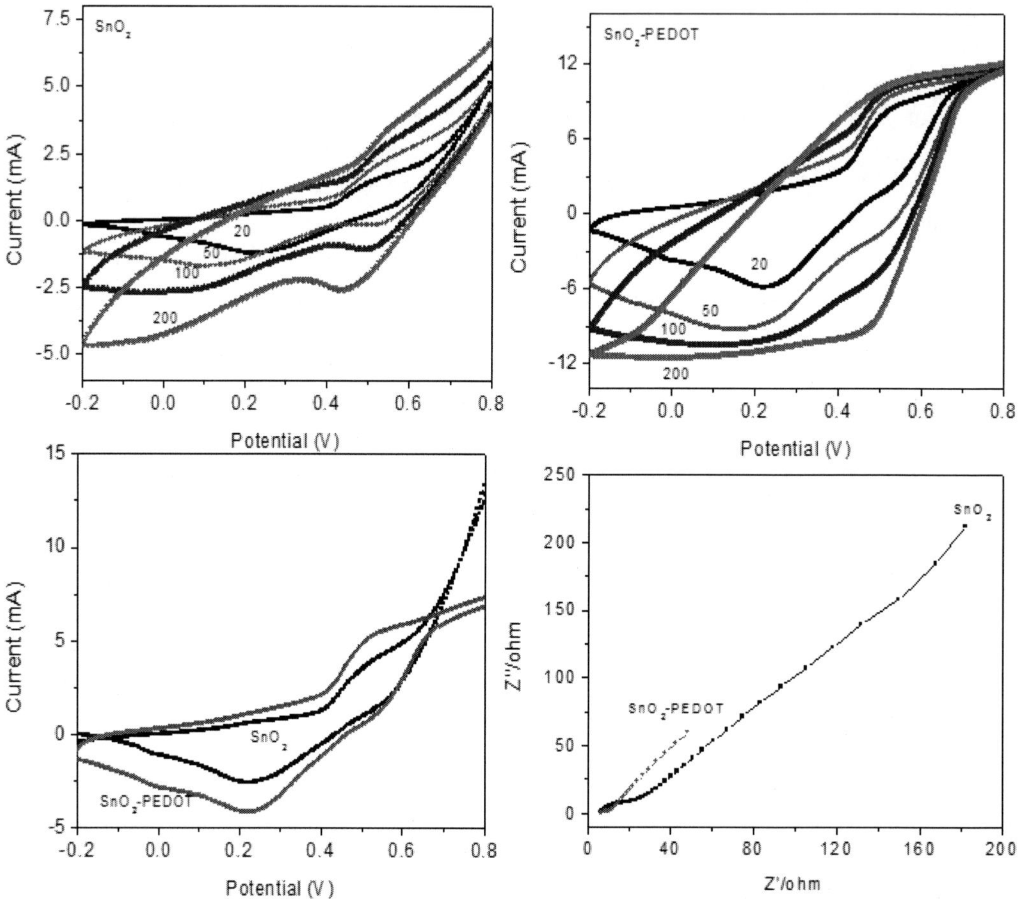

Figure 13. CV of SnO_2 and SnO_2-PEDOT at different scan rates, comparative CV of SnO_2 and SnO_2-PEDOT and comparative impedance of SnO_2 and SnO_2-PEDOT in 0.5 M H_2SO_4.

Enhancement in capacitance with the addition of very small amount of conducting polymer (2-3 wt. %) is the main highlight of this work. Further enhancement in current density was observed while increasing the amount of polymer. However, electrochemical stability declined. Thus, the single layer deposition of PEDOT (2-3 wt. %) on SnO_2 fibre is the best composition to get highest stability even after impressing 1000 redox cycles.

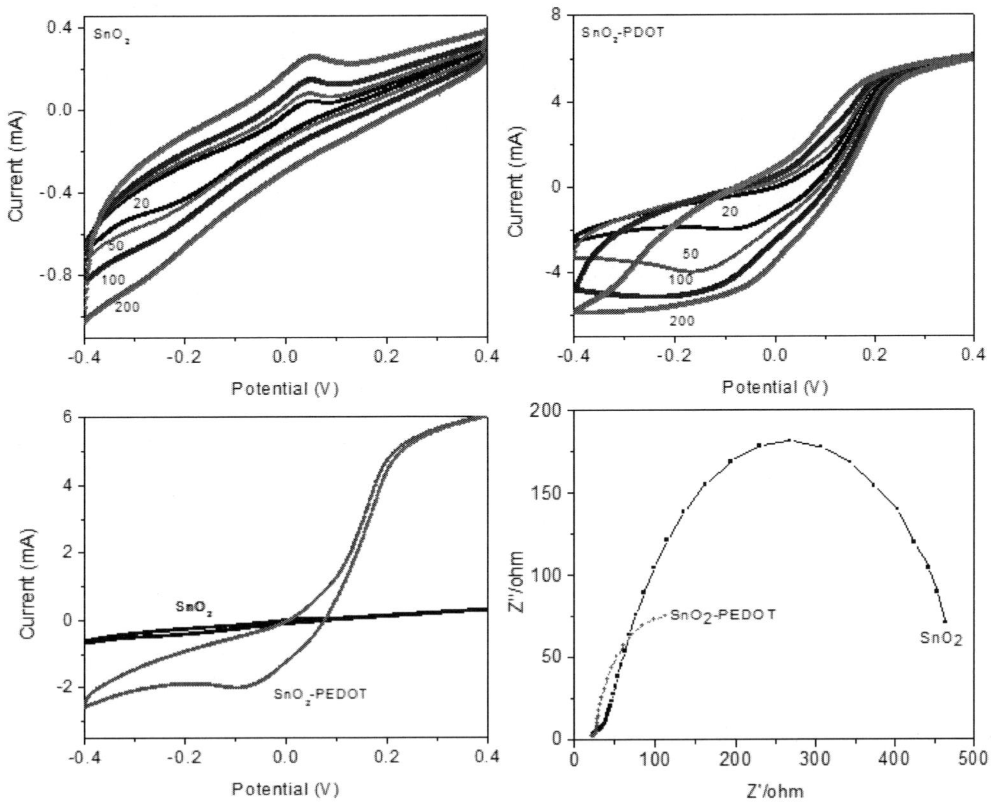

Figure 14. CV of SnO_2 and SnO_2-PEDOT at different scan rates, comparative CV of SnO_2 and SnO_2-PEDOT and comparative impedance of SnO_2 and SnO_2-PEDOT in 0.5 M KOH.

Table 1. Electrochemical property of metal oxide-conducting polymer nanocomposite under different experimental conditions

S. No.	Electrode	Electrolyte	Specific capacitance	Ref.
1.	15 wt. % acetylene black and 5 wt. % Polyetrafluoroethylene (PTFE) in Ni foam	0.5 M Na_2SO_4	F/g (1.0 A/g)	[14]
2.	Ti/TiO_2/PPy	1.0 M H_2SO_4	179F/g (1.85mA/cm^2)	[27]
3.	ITO/WO_3/PAni	0.5 M H_2SO_4	0.012 F/cm^2 (0.008mA/cm^2)	[26]
4.	75wt. % PAni-SnO_2 10 wt. % acetylene black and 10 wt. % conducting graphite and 5 wt. % Polyetrafluoroethylene (PTFE) on graphite	1.0 M H_2SO_4	305 F/g (5mA/cm^2)	[46]
5.	MoO_3-PAni (nanotube and nanosheet), Acetylene black and Poly (vinylidene fluoride) (PVDF) 80:10:10 on Ni plate	1.0 M Na_2SO_4	140 F/g (300mA/g)-Nanotube 60F/g (300mA/g)-nanosheet	[17]

Table 1. (Continued)

S. No.	Electrode	Electrolyte	Specific capacitance	Ref.
6.	Sm_2O_3-PAni, Carbon black and Poly (vinylidene fluoride) (PVDF) 80:15:5 in DMF on Ni foam	1.0 $NaNO_3$	771 F/g (0.02 A/cm^2)	[16]
7.	V_2O_5/PPy, PTFE on Ni mesh	0.5 M K_2SO_4	308 F/g (308mA/g)	[15]
8.	Au/MnO_2/fWCNT/PEDOT	1.0 Na_2SO_4	200F/g (5mA/cm^2)	[41]
9.	Pt/SnO_2-PAni (80 wt.%), Carbon (10 wt.%), PVDF (10 wt. %) in NMP	1M $HClO_4$	219 F/g (1mA/cm^2)	[18]
10.	PAni/PMo12/ (52 wt. %), Kinar flex (22 wt. %), carbon super P (24 wt. %) in 2ml acetone containing 2 drops of DBP	Nefion 117 (12 mm diameter and 0.1778 mm thick	110-120 F/g (400 mA/g)	[25]
11.	FTO/SnO_2-PEDOT	H_2SO_4 KOH	102 F/g 96 F/g	Our result

CONCLUSION

Several material architectures based on metal oxide polymer nanocomposites have been synthesized and experimentally investigated. Extensive studies undertaken by several leading researchers in the field including our own experiments have indicated amenability of these high quality materials for supercapacitor applications. Conducting polymers in different combinations with metal oxides like PAni-SnO_2, PAni-MoO_3, PPy-SnO_2, and PPy-Sm_2O_3 were evaluated for their electrochemical properties directed towards supercapacitor application. Conventional methods based on chemical and electrochemical synthesis techniques have been adopted to obtain these materials. The procedure adopted by researchers, only differ slightly from each other. However, such variation often induced interesting features linked to enhanced surface area, conductivity, energy, density and columbic efficiency. Most of the electrochemical studies are based on asymmetric three electrode assembly where active materials reveal pseudocapacitive property. In this geometry the other electrode is based on a double layer capacitive material.

Although nanocomposite films exhibit excellent electrochemical characteristics with very high stability due to synergistic effects, control of various physical parameters like micro/nanostructure, surface area, interfacial characteristics and chemical composition still remains challenging. With minor variation in experimental conditions, behavior of material changes drastically. This scenario renders reproducibility of structural and condensed phase transport parameters and other associated issues. Contamination of reactants and generation of byproduct in the nanocomposite also need to be resolved before venturing for large scale applications and production. Lastly and most importantly cost-effectiveness of material and the synthesis process also merits careful consideration. The impending energy crisis warranting development of storage devices is all set to benefit from these collective developmental endeavors on nanocomposite material research. In addition, iterative efforts, reproducibility of benchmarked parameters, process and implication will bring down cost to benefit ratio significantly.

ACKNOWLEDGMENT

Dr. Bhavana Gupta would like to thank DST for Inspire-Faculty Fellowship. Encouragements received from Dr. C. S. Sunder, Director, MSG is sincerely acknowledged.

REFERENCES

[1] Conway, B.G., (1999). Electrochemical supercapacitor: Scientific fundamentals and Technological application, *Kluwer Academic/Plenum Publisher,* NY.

[2] Jarvis, L.P., Atwater, T.B., Cygan, P.J., (1999). Fuel cell/electrochemical capacitor hybrid for intermittent high power applications. *Journal of Power Sources* 79, 60.

[3] Sukla, AK., Sampath, S., Vijayamohanan, K., (2000) Electrochemical supercapacitor: Energy storage beyond battery. *Current Science* 79, 1656.

[4] Ghosh, S., Inganäs O., (1999). Conducting polymer hydrogels as 3D Electrodes: Applications for supercapacitors. *Advanced Materials* 11, 1214.

[5] Liu, R., Cho, S., Lee S.B., (2008). Poly(3,4-ethylenedioxythiophene) nanotubes as electrode materials for a high-powered supercapacitor. *Nanotechnology* 19, 215710

[6] Lokhande, C.D., Dubal, D.P., Joo O.S., (2011). Metal oxide thin film based supercapacitors. *Current Applied Physics* 11, 255.

[7] Lu, X., Zhang W., Wang, C., Wen T.C., Wei. Y., (2011). One-dimensional conducting polymer nanocomposites: Synthesis, properties and applications. *Progress in Polymer Science* 36, 671.

[8] Yang, D., Nanotechnology and Nanomaterials "Nanocomposite-New Trends and Development" DOI 10.5772/50409 (Chapter 12).

[9] Anand, J., Palaniappan, S., Sathyanarayana, D.N., (1998) Conducting polyaniline blends and composites. *Progress in Polymer Science* 23, 993

[10] Khanna, P.K., Singh, N., Charan, S., Subbarao, V.V.V.S., Gokhale, R., Mulik, U.P., (2005). Synthesis and characterization of Ag/PVA. *Mater. Chem. Phys.* 93, 117.

[11] Abargues, R., Marque´s-Hueso, J., Canet-Ferrer, J., Pedrueza, E., Valdes, J. L., Jime´nez, E., Martı´nez-Pastor, J.P., (2008). High-resolution electron-beam patternablenanocomposite containing metal nanoparticles for plasmonics. *Nanotechnology* 19, 355308.

[12] Fritzche, W., Porwol, H., Wiegand, A., Bornmann, S., Kohler, J. M., (1998). In-situ formation. *Nanostruct. Mater.* 10, 89.

[13] Porel S., Singh S., Harsha, S. S., Rao, D. N., Radhakrishnan, T. P., (2005). Nanoparticle-embedded polymer: In situ synthesis, free-standing films with highly monodisperse silver nanoparticles and optical limiting. *Chem. Mater.* 17, 9.

[14] Ni, W., Wang, D., Huang, Z., Zhao, J., Cui, G., (2010). Fabrication of nanocomposite electrode with MnO2 nanoparticles distributed in polyaniline for electrochemical capacitors. *Materials Chemistry and Physics* 124, 1151.

[15] Qunting Q., Zhu, Y., Xiangwen G., Wu, Y., (2012). Core-shell structure of polypyrrole grown on V_2O_5 nanoribbon as high performance anode material for supercapacitors. *Adv. Energy Mater.* 2, 950.

[16] Liu, P., Wang, Y., Wang, X., Yang, C., Yi, Y., (2012). Polypyrrole-coated samarium oxide nanobelts: fabrication, characterization, and application in supercapacitors. *J. Nanopart. Res.* 14, 1232.

[17] Zheng, L., Xu, Y., Jin, D., Xie, Y., (2011). Polyaniline-intercalated molybdenum oxide nanocomposites: simultaneous synthesis and their enhanced application for supercapacitor. *Chem. Asian J.* 6, 1505.

[18] Rao, C.R.K., Vijayan, M., Anwar, S., Jeyakumar D., (2012). Hydrolysis of $SnCl_2$ on polyaniline: Formation of conducting PAni-SnO2 composite with enhanced electrochemical properties. *Journal of Applied Polymer Science* 124, 4819.

[19] Xia, X., Dong, X.J., Wei, Q.F., Cai, Y.B., Lu, K.Y., (2011). Formation mechanism of porous hollow SnO2nanofibers prepared by one-step electrospinning, *eXPRESS Polymer Letters* 6, 169.

[20] Yang, Z., Du, G., Guoa, Z., Yu, X., Chen, Z., Zhang, P., Chen, G., Liu H., (2010). Easy preparation of SnO2@carbon composite nanofibers with improved lithium ion storage properties. J. *Mater. Res.* 25, 8.

[21] Wang, H., Lib, B., Gao, J., Tang, M., Feng, H., Li, J., Guo, L., (2012). SnO_2 hollow nanospheres enclosed by single crystalline nanoparticles for highly efficient dye sensitized solar cells. *Cryst. Eng. Comm.* 14, 5177.

[22] Ayeshamariam, A., Sanjeeviraja, C., Jayachandran, M., (2011). Synthesization, characterization and gas sensing properties of SnO_2 nanoparticles, International Journal of Chemical and Analytical Science 2(6), 54.

[23] Ha, Y.H., Nikolov, N., Pollack, S. K., Mastrangelo, J., Martin, B. D., Shashidhar, R., (2004). Towards a transparent, highly conductive poly (3,4-ethylenedioxythiophene). *Adv. Funct. Mat.* 14, 615.

[24] Gupta, B., Singh, A.K., Melvin, A., Prakash, R., (2014). Influence of monomer concentration on Polycarbazole-Polyindole (PCz-PIn) Copolymer Properties: Application for Schottky Diode. *Solid State Sciences* DOI: 10.1016/j.solidstate-sciences.2014.05.015

[25] Cuentas-Gallegos, A. K., Lira-Cantu, M., Casan-Pastor, N., Gomez-Romero, P., (2005). Nanocomposite hybrid molecular material for application in solid state electrochemical supercapacitor. *Adv. Func. Mat.* 15, 1125.

[26] Wei, H., Yan, X., Wu, S., Luo, Z., Wei, S., Guo, Z., (2012). Electropolymerized polyaniline stabilized tungsten oxide nanocomposite films: Electrochromic behavior and electrochemical energy Storage. *J. Phys. Chem.* C 116, 25052.

[27] Xie, Y., Du, H., (2012). Electrochemical capacitance performance of polypyrrole–titania nanotube hybrid. *J. Solid State Electrochem.* 16, 2683.

[28] Yang, P., Zhao, D., Margolese, D.I., Chmelka, B.F., Stucky, G. D., (1998). Generalized syntheses of large-pore mesoporous metal oxides with semicrystalline frameworks. *Nature* 396, 12.

[29] Stuart, B., (2004). Infrared spectroscopy: Fundamentals and applications, *Wiley Publisher.*

[30] Choudhury, A., Bhowmick, A.K., Ong, C., Soddemann, M., (2010). Effect of various nanofillers on thermal stability and degradation kinetics of polymer nanocomposites. *J Nanosci. Nanotechnol.* 10(8), 5056.

[31] Gao F., (2012). Advances in polymer nanocomposites: Types and applications, *Wood Head Publisher.*

[32] Simon P., Gogotsi Y., (2008). Materials for electrochemical capacitors. *Nature Materials* 7, 845.

[33] Lu, W., Hartman, R., (2011). Nanocomposite electrodes for high-performance supercapacitors. *J. Phys. Chem. Lett.* 2, 655.

[34] Huang, F.L, Wang, Q.Q., Wei, Q.F., Gao, W.D., Shou, H.Y., Jiang, S. D., (2010). Dynamic wettability and contact angles of poly(vinylidene fluoride) nanofiber membranes grafted with acrylic acid. eXPRESS Polymer Letters 4, 551.

[35] Mittal K.L., (2013). Advances in contact angle, wettability and adhesion, *Wiley, Scrivener Publisher.*

[36] McCann, J.T., Li, D., Xia. Y., (2005). Electrospinning of nanofibers with core-sheath, hollow, or porous structures. *J. Mater. Chem.* 15, 735.

[37] Duc, Q., Murukanahally, T., Devaraju, K., Ganbe, Y., Tomai, T., Honma, I., (2014). Controlling the shape of LiCoPO4 nanocrystals by supercritical fluid process for enhanced energy storage properties. *Nature/ Scientific Report* 4, 3975.

[38] Dai L., Chang D.W., Baek J.B., Lu W., (2012). Carbon nanomaterials for advanced energy conversion and storage. *Small* 8, 1130.

[39] Dar, F.I., Moonooswamy, K.R., Es-Souni, M. (2013). Morphology and property control of NiO nanostructures for supercapacitor applications Morphology and property control of NiO nanostructure for supercapacitor application. *Nanoscale Res. Lett.* 8(1), 363. Morphology and property control of NiO nanostructures for supercapacitor applications.

[40] Wu, J., Krishna, G., Dathar, P., Sun, C., Theivanayagam, M.G., Applestone, D., Dylla1, A.G., Manthiram, A., Henkelman, G., Goodenough, J.B, Stevenson, K.J, (2013). In situ Raman spectroscopy of $LiFePO_4$: Size and morphology dependence during charge and self-discharge. *Nanotechnology* 24, 424009.

[41] Hou, Y., Cheng, Y., Hobson, T., Liu, J., (2010). Design and synthesis of hierarchical MnO_2 nanospheres/carbon nanotubes/conducting polymer ternary composite for high performance electrochemical electrodes *Nano Lett.*10, 2727.

[42] Ravi, S., Prabhin, V.S., (2013). High specific capacitance of electrochemically synthesized nano MnO2-gold electrodes for supercapacitors. *Adv. Mat. Lett.* 4(4), 296.

[43] Gupta, B., Prakash, R., (2009). Processible polyacid doped polyaniline composites: Application for charge storage devices. *Materials Science and Engineering* C 29, 1746.

[44] Lufrano, F., Staiti, P., Minutoli M., (2003). Evaluation of nafion based double layer capacitors by electrochemical impedance spectroscopy. *Journal of Power Sources* 124, 314.

[45] Luan, F., Wang, G., Ling, Y., Lu, X., Wang, H., Tong, Y., Liu, X.X., Li, Y., (2013). High energy density asymmetric supercapacitors with a nickel oxide nanoflake cathode and a 3D reduced graphene oxide anode. *Nanoscale* 5, 7984.

[46] Hu, Z.A., Xie, Y.L., Wang, Y.X., Mo, L.P., Yang, Y.Y., Zhang, Z.Y., (2009). Polyaniline/SnO2 nanocomposite for supercapacitor applications. *Materials Chemistry and Physics* 114, 990.

In: Polymer-Matrix Composites
Editor: Rakesh Kumar

ISBN: 978-1-61122-005-6

Chapter 9

BEHAVIOR OF SULFUR-POLYMER MATRIX COMPOSITES WITH DIFFERENT FILLERS UNDER ACID AND SALT INFLUENCE

Milica M. Vlahović [1*], ***Predrag B. Jovanić*** [2], ***Sanja P. Martinović*** [1], ***and Tatjana D. Volkov Husović*** [3]

[1]University of Belgrade, Institute for Chemistry, Technology and Metallurgy, Belgrade, Serbia
[2]University of Belgrade, Institute for Multidisciplinary Research, Belgrade, Serbia
[3]University of Belgrade, Faculty of Technology and Metallurgy, Belgrade, Serbia

ABSTRACT

Based on the nature of the matrix material, modern composites can be generally classified into three categories: polymer matrix composites, metal matrix composites, and ceramic matrix composites. Polymer matrix composites are not as strong or heat-resistant as metal matrix and ceramic matrix composites and therefore are restricted to secondary structures in which operating temperatures are lower than 300° C while for higher temperatures, metal matrix and ceramic matrix composites are required.

Polymeric materials based on inorganic components can play an important role in various technological processes but their production on an industrial scale is still problematic. However, inorganic polymers, in combination with other materials can generate composite materials of very high quality and possibilities of applications. Among plenty of inorganic materials that can generate polymer chains, special attention should be paid to sulfur. One of the most popular applications of waste sulfur from the oil refining process is to produce a modified sulfur binder, which is predominantly implemented in different compositions of sulfur- polymer matrix composite materials used for diverse applications. Using sulfur as a binder is based on its physical and

* Corresponding Author address University of Belgrade, Institute for Chemistry, Technology and Metallurgy, 12 Njegoseva St., Belgrade, Serbia Email: milica.vlahovic@yahoo.com.

chemical characteristics: chemical passivity, excellent resistance to aggressive agents (mainly acids and salt solutions, but not bases) and hydrophobic properties.

Sulfur implementation for composites production has started with using unmodified sulfur as a binder. However, despite excellent mechanical properties after preparation, samples exhibited low stability, so spalling and failure occurred after a short period. The development of modified sulfur binder contributed to better endurance of sulfur- polymer matrix composites, which focused its application for roads construction and repairing and as a building material. Sulfur itself tends to polymerize to a large extent while chemical modification increases this tendency or prolongs the time required for the polymerization. Except the prevention of sulfur transformation from monoclinic to orthorhombic form, the degree of sulfur polymerization is increased and long chains are created due to modification.

Sulfur- polymer matrix composites are thermoplastic materials made of mineral aggregate and filler, with sulfur as a binder (instead of cement and water) at temperature above the hardening point of sulfur (120 °C). The proportion of aggregate, filler and binder for the preparation of sulfur- polymer matrix composite mixture may vary depending on the application. The choice of filler is important because it forms with sulfur paste that coats and binds the aggregate particles. In this chapter, various fillers were used for the production of sulfur- polymer matrix composite samples: talc, alumina, microsilica and fly ash. Durability of the obtained sulfur- polymer matrix composite samples was investigated under influence of acids (hydrochloric and sulfuric acids) and salt (sodium chloride) by monitoring their physico-mechanical characteristics. It was concluded that the used filler had influence on the behavior of the sulfur- polymer composite samples in the investigated environments.

Keywords: Sulfur- polymer matrix composite, Fillers, Durability, Physico-mechanical characteristics

1. INTRODUCTION

Modern trends of obtaining materials with desired properties are based on combining ingredients with different properties in different proportions, as well as on the application of various manufacturing procedures and additional material processing. Having in mind that while in service use, all materials are exposed to diverse external stresses and environmental influences that provoke some type of material response, the idea in this research was to analyze the level of material properties changes by imposed combination of the mechanical and chemical influence.

Sulfur- polymer matrix composites are high performance thermoplastic materials made of mineral aggregate, filler, and modified sulfur binder, instead of cement and water as in ordinary Portland cement- based composites at temperatures above the hardening point of sulfur (120°C). Using sulfur to obtain modified sulfur binder is based on its physico-chemical characteristics. According to our terminology, modified sulfur binder considers a mixture of elemental sulfur and modified sulfur- sulfur polymer. Contemporary experience all over the world shows that composite materials with modified sulfur binder instead of cement and water have significant chemical and physico- mechanical advantages comparing with Portland cement- based composites [1, 2].

A wide range of materials that had a great impact in the development of practical applications of materials were polymers. From the perspective of nowadays-used technologies and process solutions, polymeric materials based on inorganic components can play an important role in various technological processes. The difficulty lies in the fact that their production on an industrial scale is still problematic. However, inorganic polymers, in combination with other materials can generate composite materials of very high quality and possibilities of applications.

Beside plenty of inorganic materials that can in some circumstances generate polymer chains, special attention should be paid to sulfur.

In the last decades, the availability of sulfur has considerably grown mainly due to the current environmental restrictions regarding the petroleum and gas refining processes, which limit the maximum quantity of sulfur present at combustibles. Sulfur from oil processing was stored leading to more or less controlled disposal. Extremely large quantities of sulfur are thus obtained as a by-product of these processes.

Therefore, the development of new applications for sulfur becomes fundamental.

Building materials certainly represent media that should be examined as potential acceptors for large quantities of wastes from various sources. The fact is that wide range of hazardous waste can be inverted by their incorporation into usable building materials.

During the 1960s, there was a remarkable investment in environmental protection against discharge of sulfur into the atmosphere, thus making sulfur a surplus commodity on the market, particularly in the United States and Canada. This was a crucial point, that made the interest for usage of sulfur as a structural binder to grow further, and initiated extensive research programs which became very active in the 1970s and focusing on various properties of the material, including durability [3].

Development in sulfur implementation for concrete production has started with using unmodified sulfur as a binder. However, experiments pointed out that despite excellent mechanical properties after preparation, samples exhibited low stability, so spalling and failure occurred after a short period of time. Namely, exposure of such material to repeated cycles of freezing and defrosting in terms of high humidity or immersion into water caused its degradation and failure. This can be explained by sulfur transformation that is occurring during the preparation of sulfur concrete with unmodified sulfur. On cooling of casted liquid mixture, sulfur crystallizes as monoclinic S_β at 114 °C with volume contraction of 7%. Below 95.5°C, it transforms completely into rhombic S_α within 20 h. Since S_α is more dense than S_β high stresses and micro-cracking within the material are induced. This exposes the material to moisture penetration and it fails prematurely. The development of modified sulfur binder contributed to better endurance of sulfur concrete, which focused its application for roads construction and repairing and as a building material. Sulfur itself tends to polymerize to a large extent while chemical modification increases this tendency or prolongs the time required for the polymerization. Except the prevention of sulfur transformation from monoclinic to orthorhombic form, the degree of sulfur polymerization is increased and long chains are created due to modification. Modified sulfur has much lower thermal expansion coefficient compared with unmodified one, therefore shrinkage and residual stresses upon cooling are lower. The polymer prevents the growth of macro- sulfur crystals. Long-term durability of modified sulfur concrete lies in the stability of microcrystalline sulfur [4].

On the other side, Portland cement may be partially or completely substituted with other binders such as modified heated sulfur to form a stable, hard product. Unlike material made from Portland cement (which can be cold mixed), composite made with modified heated sulfur needs to be heated during production. Although modified sulfur, together with aggregates, needs to be mixed at 135°C, this process uses less energy than conventional cold mixing processes because it avoids the energy needed to heat limestone to 1450°C to manufacture Portland cement [5, 6].

Construction materials such as sulfur concrete and sulfur modified asphalt continue to receive more attention since they are environmentally friendly and cost-effective.

The use of sulfur to produce sulfur- polymer matrix composites- sulfur concrete, sulfur mortar and sulfur- modified asphalt is a relatively new technology that has to be proven in practice. The fact is that these materials are already in use, but refining their application quality is still certainly needed. Sulfur- polymer matrix composite has a relatively simple composition and manufacture, and very interesting characteristics and properties. The recent research programs on sulfur- polymer matrix composite have been continuing to study different properties related to the material performance, most of which are reported to be excellent. Its extremely high corrosion resistance, mechanical strength and fast hardening, make it a high performance material suitable for several applications, especially the ones in which other materials, fail [3]. Furthermore, sulfur- polymer matrix composites can be used for solidification and encapsulation of different waste materials (fly ash, cement kiln dust, phosphogypsum, mercury), thus obtaining the sustainable development of construction and industrial sectors and for corrosion protection of reinforcing steel and concrete [7-11]. Sulfur-polymer matrix composite should also find widespread application in chemical and fertilizer replacement for materials in acid and salt environments as well as in metallurgical operations since they fail in acid and salt environments.

In the characterization of new materials, two main problems can be pointed out. The first problem is to understand the technology for material production and the other is to characterize solely the material properties. In the case of material properties characterization the basic thing is the adequate choice of the methodologies. The decision between the characterization of material properties and the characterization of functional properties of material due to its possible application has to be made.

The data obtained by materials testing, that is by measuring characteristics and behavior of such substances as metals, ceramics, or plastics under various conditions, can be used in specifying the suitability of materials for various applications- e.g., building construction, machinery or packaging. A full- or small- scale model of a proposed machine or structure may be tested. Alternatively, investigators may construct mathematical models that utilize known material characteristics and behavior to predict capabilities of the structure.

Materials testing breaks down into five major categories:

- Mechanical testing;
- Thermal properties testing;
- Electrical properties testing;
- Testing for resistance to corrosion, radiation, and biological deterioration;
- Nondestructive testing.

Nondestructive testing (NDT) is a wide group of analysis techniques used in science and industry to evaluate the properties of a material, component or system without causing damage. The terms nondestructive examination, nondestructive inspection, and nondestructive evaluation are also commonly used to describe this technology. Because nondestructive testing does not permanently alter the article being inspected, it is a highly valuable technique that can save both money and time in product evaluation, troubleshooting, and research.

In this case, beside the classical material testing procedures the quantification of visual information was used in order to characterize the behavior of the material under chemical influence. The characterization of long-term responses of materials and structures to complex or cyclic environmental conditions presents a difficult challenge to those involved in materials usage, selection or structural design. The problem is particularly challenging when the required service life is inordinately long compared with the time available for development and evaluation of materials and components. In most cases, service failures due to inaccurate characterization of fatigue and aging responses result in potential repair or premature component replacement.

Simulated service testing to characterize materials aging is fraught with difficulties that come from the inability to reproduce complex service conditions in the laboratory. Important aspects of the environmental conditions encountered in service cannot be accurately simulated. These environments are influenced by changing loads, temperature, humidity, pH, radiation, and other effects that are interactive and cannot be consistently recreated and controlled. Moreover, critical degradation mechanisms and failure modes in materials are not always well known and understood prior to testing. While most service simulation testing has provided insights concerning degradation mechanisms and potential failure modes, it has yielded little engineering- quality data useful in the systematic characterization of some materials aging. Thus, alternative approaches are required to develop such data.

Microscope image processing dates back a half century when it was realized that some of the techniques of image capture and manipulation, first developed for television, could also be applied to images captured through the microscope. Initial approaches were dependent on the application: automatic screening for cancerous cells in Papanicolaou smears; automatic classification of crystal size in metal alloys; automation of white cell differential count; measurement of DNA content in tumor cells; analysis of chromosomes; etc. In each case, the solution lay in the development of hardware (often analog) and algorithms highly specific to the needs of the application. Thus, there existed an unbridgeable gap between the theory of how microscope images could be processed and what was practically attainable.

The modern programmable digital computer has removed all the old constraints—incredible processing power, speed, memory and storage come with any consumer computer. The challenge lies not in processing images, but in processing them correctly and effectively. Microscope Image Processing provides the tools to meet this challenge. Analytic results can only be as reliable as the processes used to obtain them. Spurious results can be avoided when users understand the limitations imposed by diffraction optics, empty magnification, noise, sampling errors, etc. as a quantitative tool in their research.

The greatest impact of the new technologies is on a stronger understanding of the structure- function relationship in materials, and on the ability to translate data gathered from microstructures into better process control. Traditionally, much image analysis has been based on the geometric information (area, perimeter, projection, etc.) derived from the feature- specific analyses in morphometry. While the histograms of population distributions generated by this approach can be valuable, all too often they fall short of truly describing the bulk properties that influence mechanical strength, thermal sensitivity, and other properties.

Stereology provides the necessary field- oriented analyses, but it requires a large number of fields to be statistically valid. Currently, relatively few automated image analysis systems include any stereology, necessitating hours of tedious manual work. To complicate further matters, the mathematical proofs behind this method are extremely challenging. Texts written on the subject are aimed at communication within that field of mathematics rather than for the lay user, making entry daunting, and masking relevance to everyday applications.

Ironically, the actual measurements needed for stereology are extremely simple and available in the digitized images common to all image analysis packages, and the actual math is trivial and easily inserted. The result is a body of parameters that provide strong relationships between structure and function. They can be readily translated into process control of factors such as time at temperature, moisture content, and concentration of constituent or catalyst.

2. EXPERIMENTAL

2.1. Materials

The initial materials for technological procedure of sulfur- polymer matrix composite obtaining were aggregate, modified sulfur binder and filler.

2.1.1. Aggregate

Sand with maximum grains size of 2 mm, obtained by sieving of locally available classical building mixture of sand and gravel, was used as an aggregate. Chemical analysis indicated that the aggregate mainly consisted of oxides of silica (89.98 %), aluminium (3.61 %), calcium (0.84 %), iron (0.62 %), potassium (0.59 %), sodium (0.57 %), and magnesium (0.19 %).

Grain size distribution, determined by sieve analysis, before and after sieving, is presented in Figure 1.

2.1.2. Sulfur

Sulfur, the basic component for a modified sulfur binder, originates from oil refining process by Claus's procedure in the Oil Refinery Pancevo, and its purity is 99.9%. Dicyclopentadiene (DCPD), cyclic hydrocarbon, was used for sulfur modification. The procedure performed according to the literature [12-15] consisted of mixing DCPD and melted sulfur in the temperature range from 120 to 140 °C and ambient pressure for 30 min, and then rapid cooling and solidification of the obtained sulfur polymer.

Because of the importance of preparing the sulfur- polymer matrix composite with modified sulfur, both elementary and the obtained modified sulfur were investigated by scanning electron microscope (SEM), type JEOL JSM-5800 with EDX, Figure 2, and their microstructures were analyzed according to the literature [16]. The results showed that the elementary sulfur was composed of dense orthorhombic crystals of alpha form (S_α), Figure 2a, while modified sulfur consists of plate monoclinic crystals of beta form (S_β), partially polymerized in zigzag chains, Figure 2b. Therefore, it was proved that modification of sulfur was achieved.

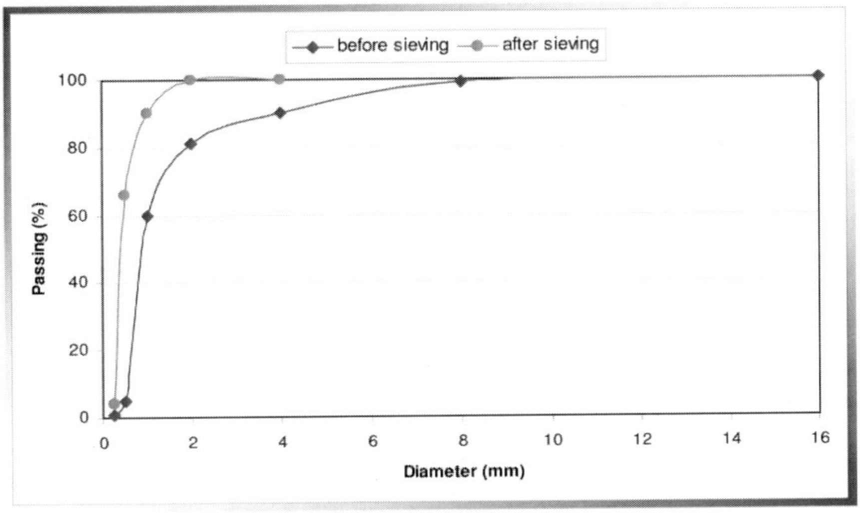

Figure 1. Grain size distribution of aggregate.

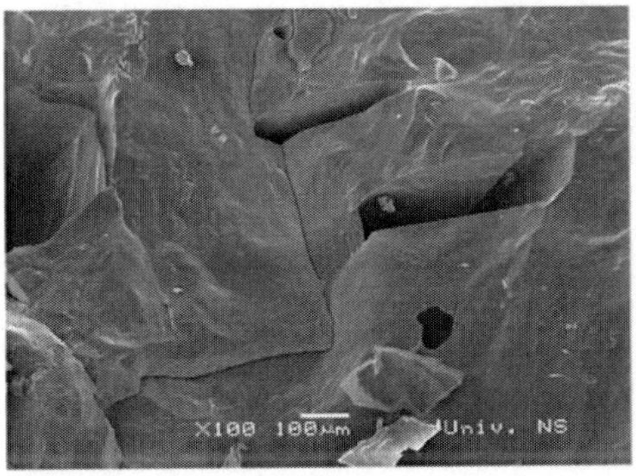

(a)

(b)

Figure 2. Structure of sulfur: (a) elementary; (b) modified.

2.1.3. Fillers

Fillers used in this production were: talc (technical quality, China), alumina (Almatis, Germany), microsilica (Sika, Switzerland) and fly ash (Power Plant "Nikola Tesla- A", Obrenovac).

The choice of filler is important because it forms with sulfur paste that coats and binds the aggregate particles. The main functions of fillers are to [17]: control the viscosity of the sulfur- filler paste and provide nucleation sites for crystal formation and growth, fill the voids in the aggregate and, hence, reduce the required amount of sulfur, reduce shrinkage during hardening, reinforce the matrix and therefore increase the strength. Talc, microsilica and fly ash are fillers that fulfill these requirements and therefore are recommended for the sulfur-polymer matrix composite production [18]. Fine fractions of calcined alumina are used as fillers for refractory concretes [19].

Common characteristic of all selected fillers is their particles size below 75 µm. Chemical composition and densities of applied fillers are given in Table 1.

Table 1. Chemical composition and densities of used fillers

Filler	Chemical composition (wt. %)										Density, ρ (g/cm³)
	SiO_2	Al_2O_3	Fe_2O_3	CaO	MgO	Na_2O	K_2O	TiO_2	P_2O_5	SO_3	
T	57	0.3	0.4	0.7	31	-	-	-	-	-	2.52
A	0.15	99.3	0.02	0.03	0.004	0.40	0.03	0.02	-	-	3.86
MS	95.1	0.5	0.5	0.4	0.6	0.2	0.7	-	1.1	0.3	2.32
FA	56.7	3.78	5.85	5.89	2.08	0.54	1.05	0.66	0.06	1.61	2

T-talc; A-alumina; MS-microsilica; FA-fly ash.

Particle size distribution, determined by the laser diffraction using device type HELOS (H1597) & SUCELL, Sympatec, Germany, for four fillers is represented in Figure 3.

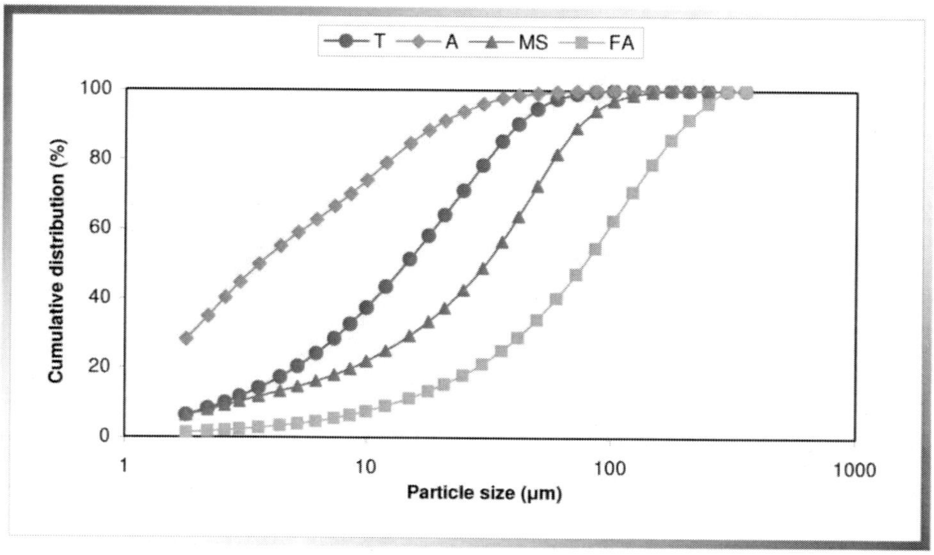

Figure 3. Particle size distribution of used fillers: talc (T); alumina (A); microsilica (MS); fly ash (FA).

2.2. Preparation and Characterization of Sulfur- Polymer Matrix Composite Samples

Sulfur- polymer matrix composite was prepared according to the manufacturing technological procedure described in literature [20- 22]. Preheated aggregate and filler (up to 160 ºC) were stirred for about 15 min in a mixer, then melted sulfur and modified sulfur were mixed into homogenized dry mixture of aggregate and filler at sulfur melting temperature, 132-141°C. Preheating is desirable to avoid solidification of the molten sulfur by contact with aggregate at a low temperature and to reduce the mixing time. The heated aggregate and filler were then properly mixed with the molten modified sulfur until a homogenous viscous mixture was obtained. After homogenization and mixing that lasted for 2 minutes, the sulfur-polymer matrix composite mixture was casted into molds preheated at 120ºC and vibrated for 10 seconds. Any extra material was removed to get a well-finish at top surface and left inside the molds at room temperature for hardening. A very important characteristic of sulfur-polymer matrix composite is rapid hardening (from 15 minutes to several hours, depending on size and shape of the sample), which allows removal from the mold and curing in a relatively short period of time (only 24 h at room temperature). After 3 h of hardening at room temperature, samples were demolded and kept at room temperature of 20°C for 24 h. Mechanical properties measurements were made after 75 h.

A schematic representation of the mixing process for sulfur- polymer matrix composite preparation is shown in Figure 4.

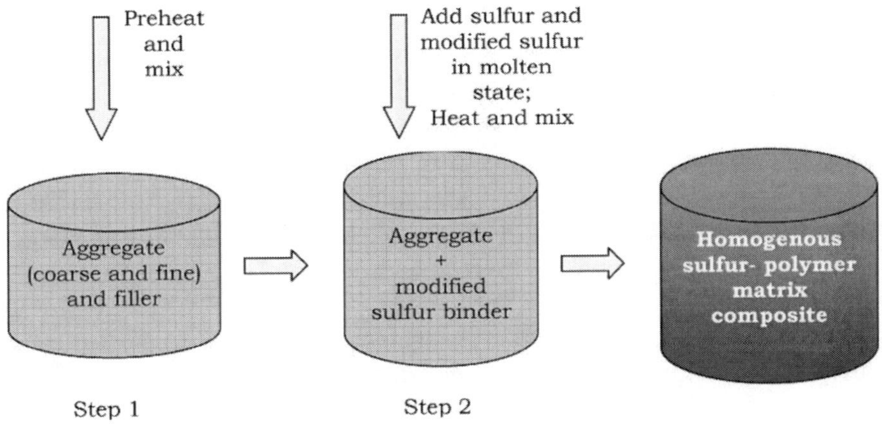

Figure 4. Mixing scheme for sulfur- polymer matrix composite preparation.

For investigating, prism samples with dimensions (4 x 4 x 16) cm were prepared.

Each reported value is the average of three readings obtained from three different samples to ensure the reliability of the test results.

2.3. Methods

The experimental program consisted in performing the following tests on concrete samples.

2.3.1. Acid Resistance

For sulfur- polymer matrix composite corrosion resistance and durability testing the standard testing method [20] was used. The prismatic samples with dimensions of 40 x 40 x 160 mm were immersed in three various aggressive environments: 10 % HCl solution, 20 % H_2SO_4 solution and 3 % NaCl solution.

The destruction of the material during 360 days of immersion was observed by periodical mass measuring and comparing with an initial mass, as well as variations of mechanical strength and apparent porosity with time.

In order to determine mass and mechanical strength changes and apparent porosity, the samples were removed from the acid solution, washed and dried in the oven at 105±5 °C to constant mass.

2.3.2. Mass Change

Masses of the samples before and after immersion in acid aggressive environment were measured by digital laboratory balance, type "Shollex", capacity of 5 kg, and accuracy of 0.01 g. The obtained mass values were the average of three samples. Mass change was recorded after 7, 14, 21, 60 and 180 days exposure to the aggressive solution.

2.3.3. Mechanical Strength Testing

The mechanical characterization of the concrete samples was carried out using measurements of compressive and flexural strength.

Mechanical strength of the concrete samples before and after immersion tests was conducted using the "Amsler" press with maximum load of 200 kN, and method for testing the strength of concrete according to the standard [23]. Three replicate samples were tested to obtain strength values. Mechanical strength change was recorded after 7, 14, 21, 60 and 180 days exposure to the aggressive solution.

2.3.4. Apparent Porosity Testing

Apparent porosity of the samples before and after immersion tests was determined using boiling water saturation technique [24]. The samples were boiled for 5 h and then cooled for 19 h to a final temperature of 20-25 °C. The apparent porosity was calculated by measuring dry, saturated and buoyant masses of the samples and using the following equation [Eq. (1)]:

$$\text{Apparent porosity} = \frac{m_1 - m_2}{m_1 - m_3} \cdot 100\% \tag{1}$$

where: m_1 – saturated surface- dry mass of the sample in air, m_2 – oven- dry mass of the sample in air, m_3 – mass of the saturated sample in water.

3. Results and Discussion

3.1. Referent Sulfur- Polymer Matrix Composite Samples Properties

Physico-mechanical properties of referent sulfur- polymer matrix composite samples were determined after removing it from the mold after 24 h and curing at room temperature. The obtained values of bulk density, mechanical strength, both compressive and flexural, and apparent porosity are given in Table 2.

It can be noticed that mechanical properties of all referent samples of sulfur- polymer matrix composite are similar which can be connected with approximately the same values of their bulk densities.

By comparing properties of prepared samples with sulfur- polymer matrix composite according to literature [20], it can be concluded that, regarding mechanical strength, the samples with various fillers have completely satisfying quality.

Mutual differences in mechanical strength and apparent porosity values of referent samples of sulfur- polymer matrix composite probably originate from the physical and chemical properties of applied fillers, since the other components were the same in all samples.

Table 2. Physico-mechanical properties of referent sulfur- polymer matrix composite samples after 24 h of curing at room temperature

Sample	Bulk density (g/cm^3)	Mechanical strength (MPa)		Apparent porosity (%)
		Compressive	Flexural	
SPMC-T	2.33	55.4	8.3	3.14
SPMC-A	2.34	49.2	8.4	1.38
SPMC-MS	2.31	50.3	7.2	3.21
SPMC-FA	2.25	48.9	7.8	4.93

SPMC-T = sulfur- polymer matrix composite with talc, SPMC-A = sulfur- polymer matrix composite with alumina,

SPMC-MS = sulfur- polymer matrix composite with microsilica, SPMC-FA = sulfur- polymer matrix composite with fly ash.

3.2. Durability Testing Results

The problem is the choice of the representative parameter for the system response, which will present the material response to the input stress. In the presented research, porosity is accepted as the main property of the investigated material, which will present the material behavior during life circle.

Sulfur- polymer matrix composite samples in acid and salt environments after 12 months did not show any deterioration while mass and compressive strength changes were limited.

Appearance of sulfur- polymer matrix composite samples with various fillers after 12 months of immersion in acid and salt environments is shown in Figure 5.

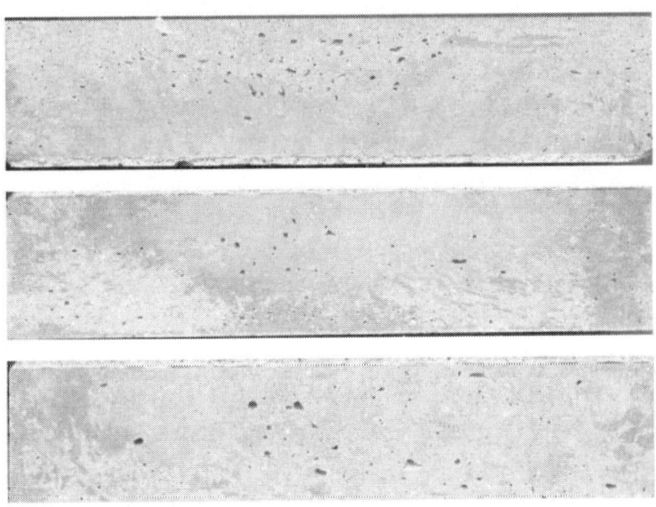

Figure 5. Examples of samples surfaces of sulfur- polymer matrix composite: a) with talc in 10 % HCl; b) with microsilica in 20 % H_2SO_4; c) with fly ash in 3 % NaCl.

As addition to the classical porosity test in this research the porosity changes due to chemical influence were followed by the microscopic image analysis. In this case, the samples were analyzed with the minimum resolution in order to characterize only primary interactions between material and the induced stress agents.

Figure 6 presents schematic diagram of the induced chemical stress.

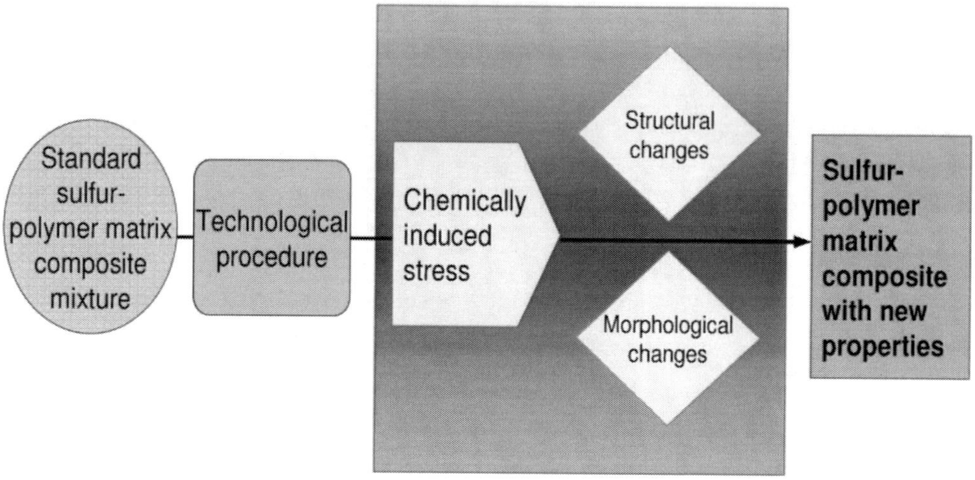

Figure 6. Schematic diagram of the induced chemical stress.

The ability to characterize cell sizes, and separated features within a microstructure falls under the generic term "Sizing". Since pores and micro-pores are often irregular in size, shape, and orientation, no single parameter can be used to characterize their true form. However, for the population distributions typical of morphometry, if the shapes of features are known, then size distributions can be derived, and vice versa. For regular polygonal grain structures, individual components can be assumed to have a pseudo- spherical shape and a simple intercept length and area measurement on individual features, recorded from the plane section, can be used to define the grain size distributions.

For example, to evaluate pores size in sulfur- polymer matrix composite, sizing can be done by the linear intercept method. In this approach, a test pattern of lines of known length is randomly laid over a plane section, and the individual intercept lengths across features of interest are measured. The pattern intercepts the grains and produces chords of varying lengths, with a maximum length equal to the diameter of the sphere.

This approach is acceptable for the analysis of the material behavior during life circle or under influence of environmental stress agents.

3.2.1. Behavior of Sulfur- Polymer Matrix Composite in Acid and Salt Environments

Behavior of sulfur- polymer matrix composite was investigated for 12 months in acid and salt aggressive environments i.e., in medium concentrated solutions of hydrochloric acid, sulfuric acid and sodium chloride for different intervals during a period of 360 days.

The obtained results represent durability of sulfur- polymer matrix composite samples depending on the type of filler and aggressive agent expressed through:

- Mass change as a function of time;
- Compressive strength degradation during the treatment time, and
- Apparent porosity change during the time.

It is known that dilute hydrochloric and sulfuric acids do not affect sulfur. The aggregate and fillers used in the manufacture of sulfur- polymer matrix composite are constituted by mineral oxides, which are mainly basic. The attack of sulfur- polymer matrix composite by hydrochloric or sulfuric acid solution is based on the reactions of basic and amphoteric oxides with acids and resulted in the formation of metal chlorides or sulfates. Most bases dissolve in water, and thereby release hydroxide ions (OH^-). Hydroxide ions react with acids and form salts. Hydrogen ions are accepted by a calcium oxide base, hence a base is a proton acceptor [Eq. (2) – (5)]:

$$\text{Acid} + \text{Basic Oxide} \rightarrow \text{Salt} + \text{Water} \quad (2)$$

$$H_2SO_4\,(aq) + CaO\,(s) \rightarrow Ca\,SO_4\,(aq) + H_2O\,(l) \quad (3)$$

Amphoteric aluminium oxide reacts in the same way:

$$\text{Acid} + \text{Amphoteric Oxide} \rightarrow \text{Salt} + \text{Water} \quad (4)$$

$$6HCl\,(aq) + Al_2O_3\,(s) \rightarrow 2AlCl_3\,(aq) + 3H_2O\,(l) \quad (5)$$

Sodium chloride does not react with sulfur or with the oxides present in the aggregate and fillers. Hence, physical processes on the contact regions sulfur/ aggregate and filler are responsible for the attack in the saline media. The growth of sodium chloride crystals leads to a partial detaching between sulfur and aggregate and filler [18].

a) Mass Change of Sulfur- Polymer Matrix Composite under Acid and Salt Influence

The obtained results, given in the following diagrams in Figures 7-9, represent durability of sulfur- polymer matrix composite samples depending on the type of filler and aggressive agent expressed through mass change as a function of time.

The results show that the mass changes of sulfur- polymer matrix composite samples with different fillers investigated in three aggressive environments were divers and dependent on the filler type. The lowest mass change, without taking into account whether it is positive or negative, was for the sulfur- polymer matrix composite samples with talc and very similar for the samples with fly ash. Samples with alumina and microsilica showed remarkably higher mass change. As filler is the only component by which the samples mutually differ, mentioned diversity can evidently be attributed to the applied filler, its influence on the structure of sulfur- polymer matrix composite and behavior of such structure in relation to the physico-chemical properties of the solution.

Figure 7 presents behavior of sulfur- polymer matrix composite with various fillers in medium concentrated solution of hydrochloric acid during the period of 360 days.

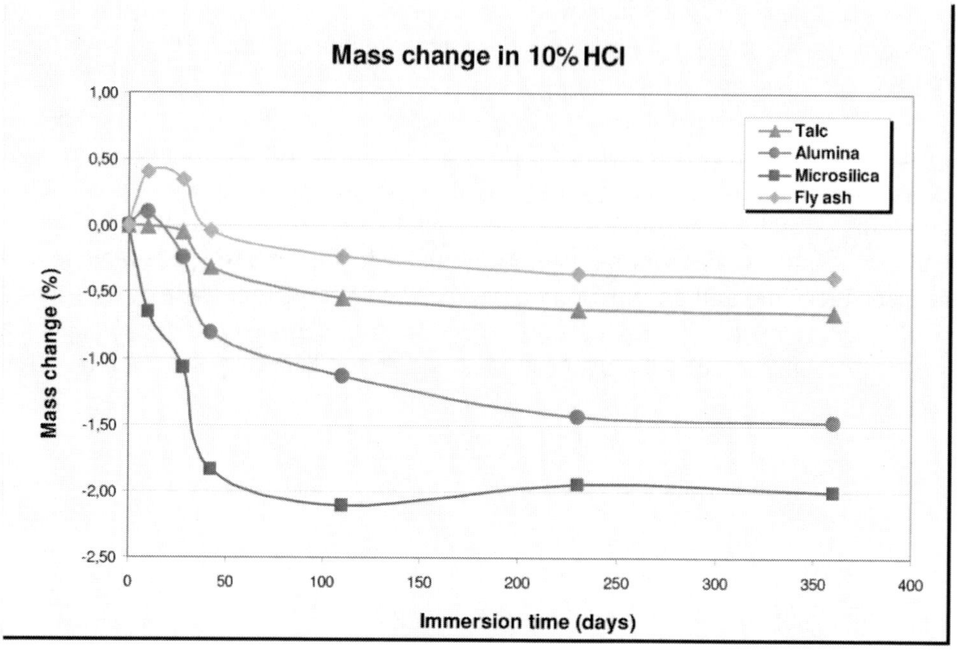

Figure 7. Mass change of sulfur- polymer matrix composite samples with different fillers in 10 % HCl.

All sulfur- polymer matrix composite samples exhibited loss of mass in hydrochloric acid solution where the lowest was for the sample with fly ash (~ 0.4 %) and the highest for the sample with microsilica (~ 2 %). At the beginning of the acid treatment, in the first ten days, mass increase for the samples with fly ash (~ 0.4 %) and alumina (~ 0.1 %) can be noticed. During the period of about 100 days, mass loss for all investigated samples is obvious, whereby the lowest for the samples with fly ash (~ 0.35 %), and the highest for the samples with microsilica (~ 2 %). After the period of 200 days, masses of all samples remained constant. The fact that time is a limitation factor for mass loss, and that mass does not change after 200 days of treatment, points to the conclusion that during this period, samples corroded in varying degrees depending on the type of filler. Corrosion probably took place on the surface in open pores that were not wrapped by sulfur and lasted until such surfaces existed. The amount of those surfaces and open pores obviously depends on the applied filler since the other components were the same in all samples.

Figure 8 presents behavior of sulfur- polymer matrix composite with various fillers in medium concentrated solution of sulfuric acid during the period of 360 days.

All sulfur- polymer matrix composite samples exhibited mass increase in sulfuric acid solution- the highest was for the sample with alumina (~ 8 %), which after 100 days began to crumble) and for the sample with microsilica (~ 4 %). This behavior is probably the result of increasing porosity owing to corrosion and significant absorption of the liquid phase, whereby the absolute amount of sample mass change is actually the material balance between weight loss due to corrosion and absorption of the liquid phase in the pores and voids caused by corrosion. On the other hand, for the other samples, the increase in mass was significantly lower (~ 0.1 % for the sample with talc, ~ 1 % for the sample with fly ash). After 200 days, for these two samples there weren't any mass changes or they were insignificant. It can be concluded that the mass increase is a result of absorption of the liquid phase due to surface porosity.

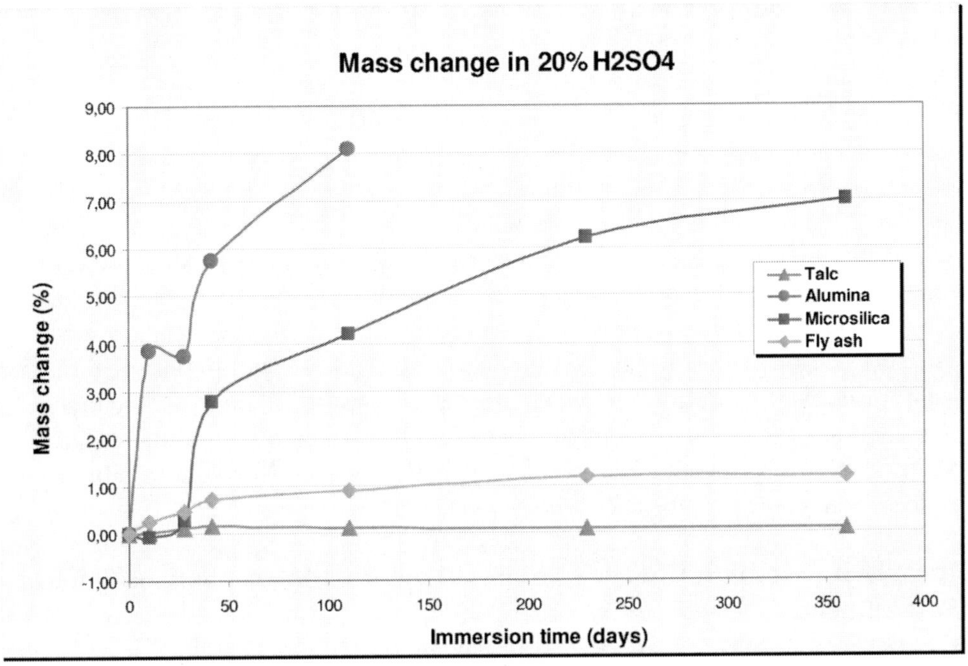

Figure 8. Mass change of sulfur- polymer matrix composite samples with different fillers in 20 % H2SO4.

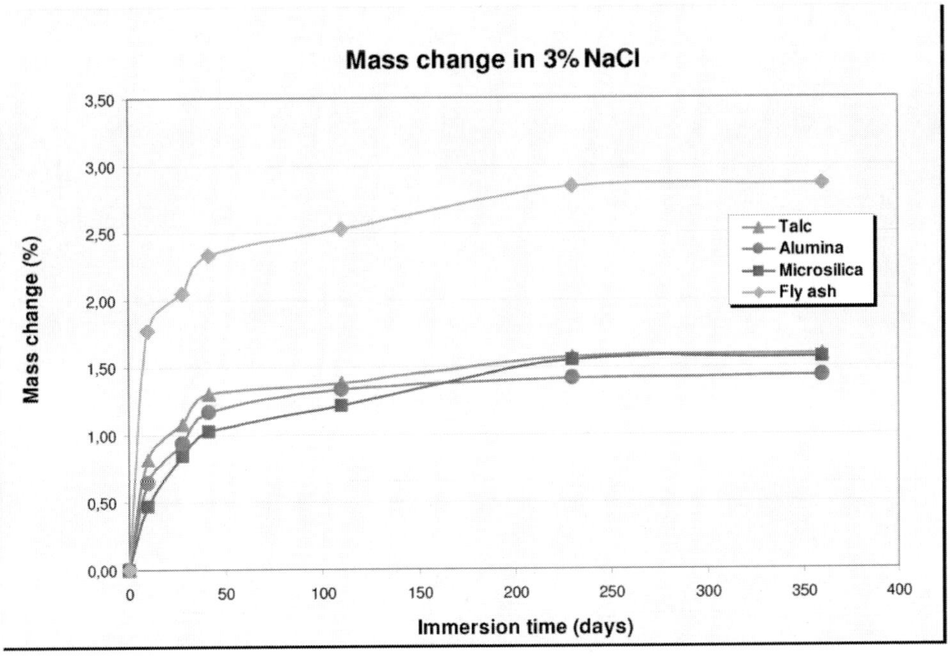

Figure 9. Mass change of sulfur- polymer matrix composite samples with different fillers in 3 % NaCl.

The increase in mass is timed (probably by liquid phase diffusion through open pores) which indicates a conclusion that there is no corrosion for those two samples or it is negligible and limited to the free surfaces of aggregate and filler that are not covered by sulfur. Different corrosion resistance in the solution of sulfuric acid is obviously a consequence of the applied filler and its influence on the structure of sulfur- polymer matrix composite.

Figure 9 presents behavior of sulfur- polymer matrix composite with various fillers in medium concentrated solution of sodium chloride during the period of 360 days.

All sulfur- polymer matrix composite samples exhibited mass increase in sodium chloride solution- the highest was for the sample with fly ash (~ 2.9 %) and quite uniform for the others, whereas the lowest for the sample with alumina (~ 1.4 %). Mass increase is probably a consequence of the liquid phase absorption owing to open surface porosity. The fact that no further mass changes after a period of 200 days for all investigated samples are occurring leads to such a conclusion. It can be concluded that, no matter how small differences in corrosion resistance of sulfur- polymer matrix composite samples in sodium chloride solution, they are the result of applied fillers. Similar behavior of sulfur- polymer matrix composite affected by other corrosive agents whose aqueous solutions are neutral (all inorganic salts and organic substances soluble in water) can be expected.

b) Compressive Strength Change of Sulfur- Polymer Matrix Composite under Acid and Salt Influence

The obtained results, given in the following diagrams in Figures 10-12, represent durability of sulfur- polymer matrix composite samples depending on the type of filler and aggressive agent expressed through compressive strength change as a function of time.

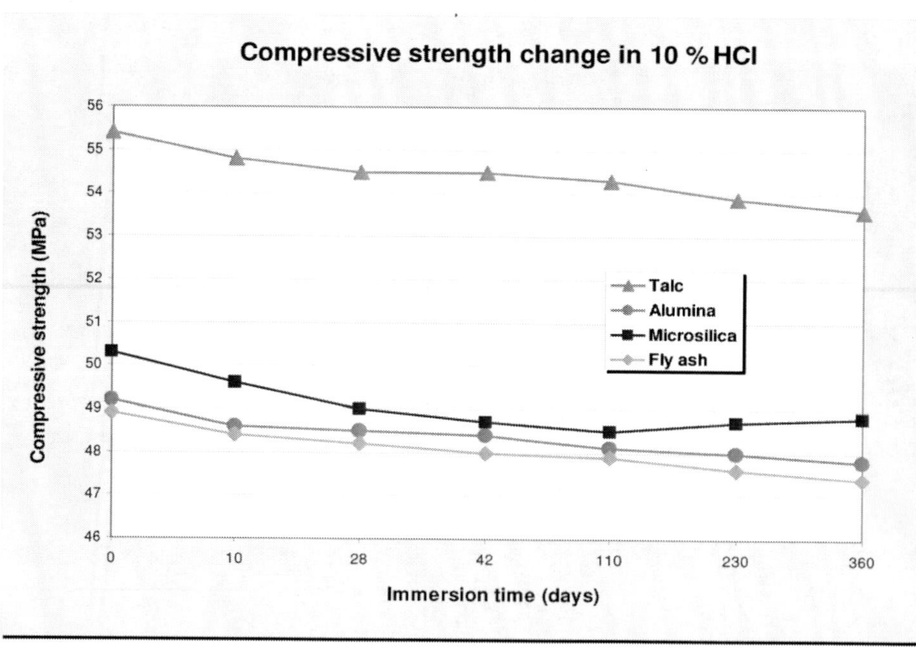

Figure 10. Compressive strength change of sulfur- polymer matrix composite samples with different fillers in 10 % HCl.

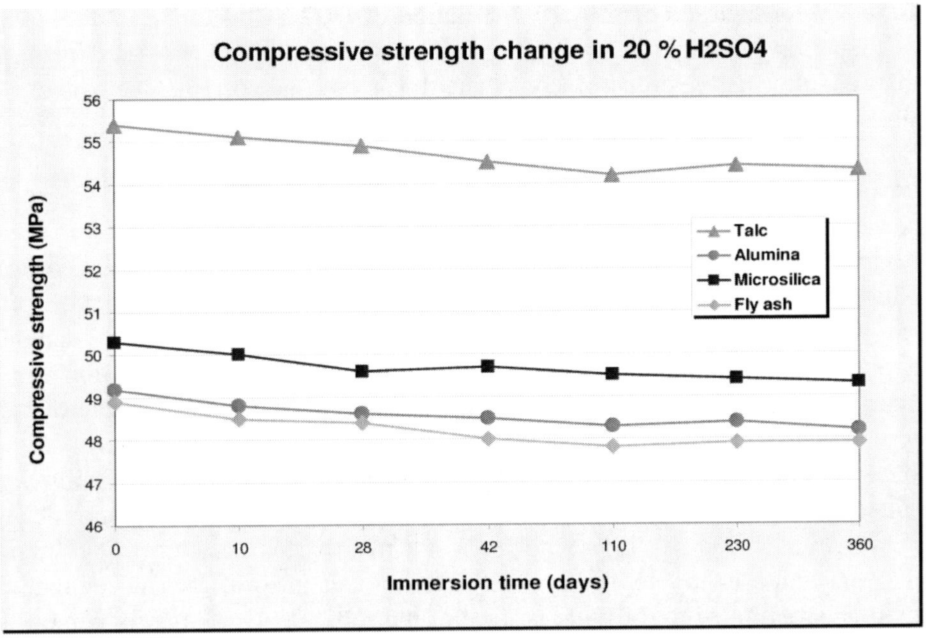

Figure 11. Compressive strength change of sulfur- polymer matrix composite samples with different fillers in 20 % H_2SO_4.

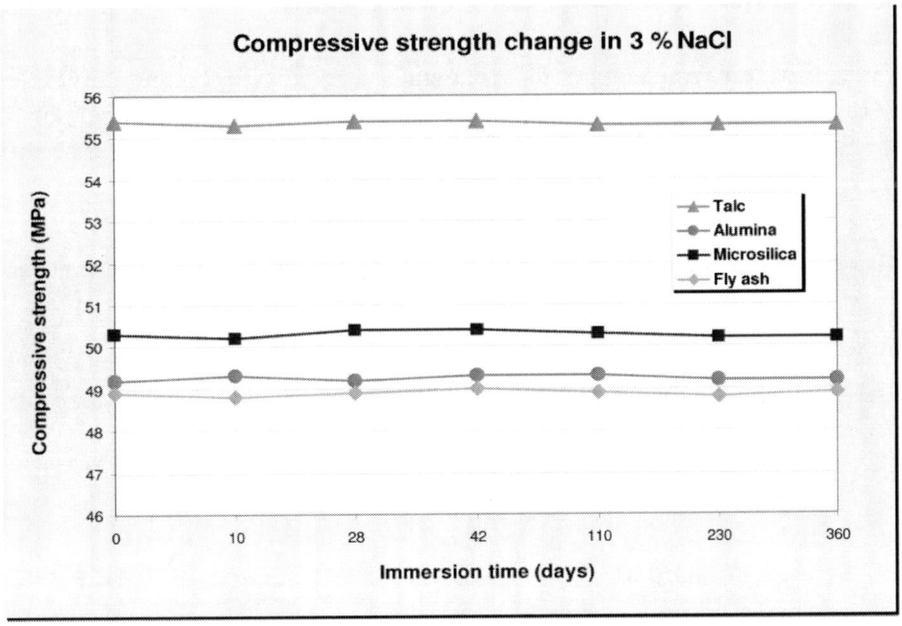

Figure 12. Compressive strength change of sulfur- polymer matrix composite samples with different fillers in 3 % NaCl.

The results showed that behavior of the sulfur- polymer matrix composite samples with different fillers regarding compressive strength changes was quite uniform in the investigated aggressive environments. Generally, all sulfur- polymer matrix composite samples after 360 days lost ~ 3 % of their strength in 10 % solution of HCl, and ~ 2 % in 20 % solution of H_2SO_4. Compressive strength loss after 360 days in 3 % solution of NaCl was negligible for all samples of sulfur- polymer matrix composite. It can be concluded that the type of filler did not exhibit significant influence on strength degradation [25].

In all three cases of chemical attack, found compressive strength loss of the treated sulfur- polymer matrix composite samples can be explained by increased porosity because mechanical strength is dependent on the defects in composite microstructure. Porosity is related to the movement of chemical substances into and out of the material and consequently affects durability of composite, as porosity is connected to many deterioration processes driven by the transport properties of composite. In sulfur- polymer matrix composite, the majority of the matrix is composed of sulfur coated materials (aggregate and filler) and sulfur accumulated in the voids between particles. During the acid attack, dissolving of basic and amphoteric oxides [Eq. (2)-(5)] caused increase in porosity. Acid penetration and therefore corrosion was limited to the surface and open pores that were not coated by sulfur, which in turn resulted in slight porosity increase and slight compressive strength reduction. During the salt attack, increased porosity is probably a result of a partial detachment between sulfur, aggregate and filler owing to sodium chloride crystals growth. Since sodium chloride penetration is only limited to the outer surface of the sulfur- polymer matrix composite [18], porosity increase was very slight and hence compressive strength loss was practically insignificant.

c) Apparent Porosity Change of Sulfur- Polymer Matrix Composite under Acid and Salt Influence

The obtained results, given in the following diagrams in Figures 13-15, represent durability of sulfur- polymer matrix composite samples depending on the type of filler and aggressive agent expressed through apparent porosity change as a function of time.

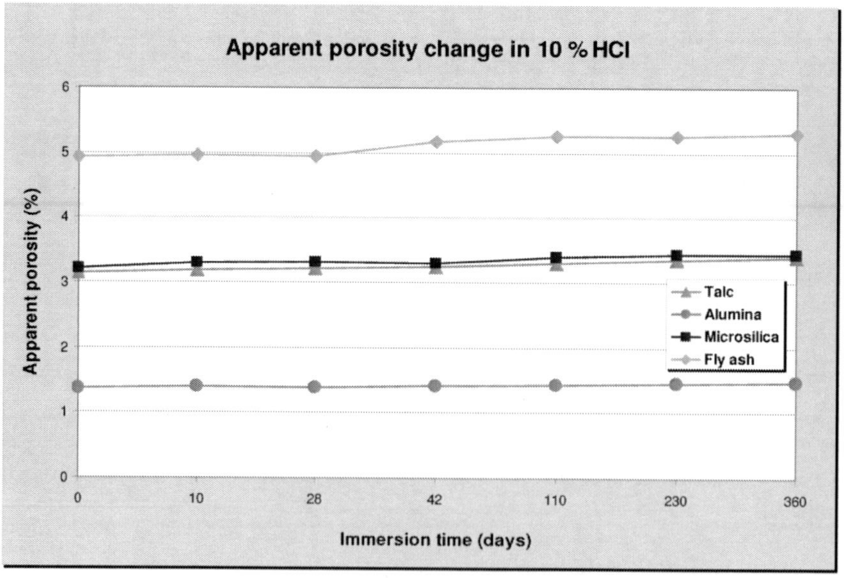

Figure 13. Apparent porosity change of sulfur- polymer matrix composite samples with different fillers in 10 % HCl.

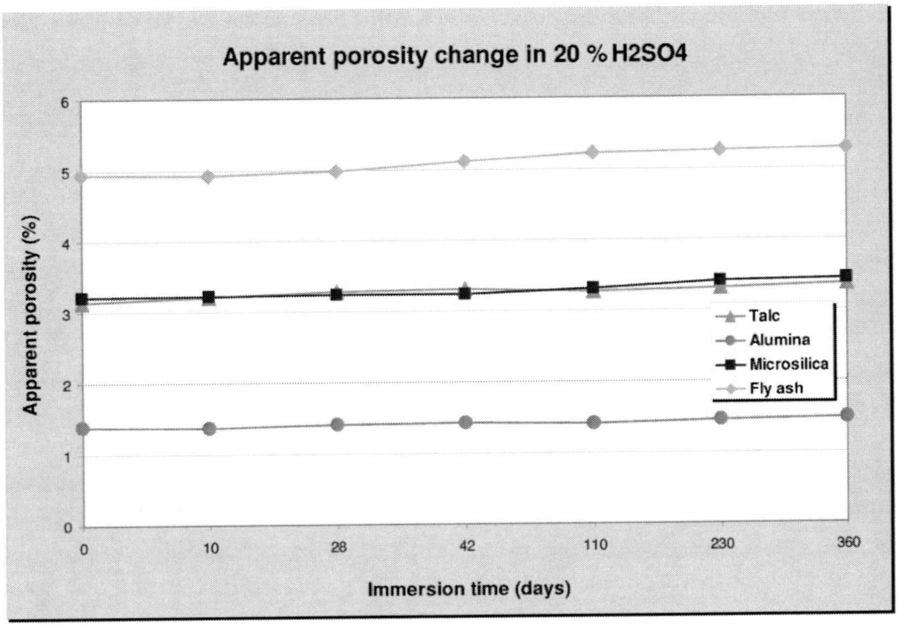

Figure 14. Apparent porosity change of sulfur- polymer matrix composite samples with different fillers in 20 % H_2SO_4.

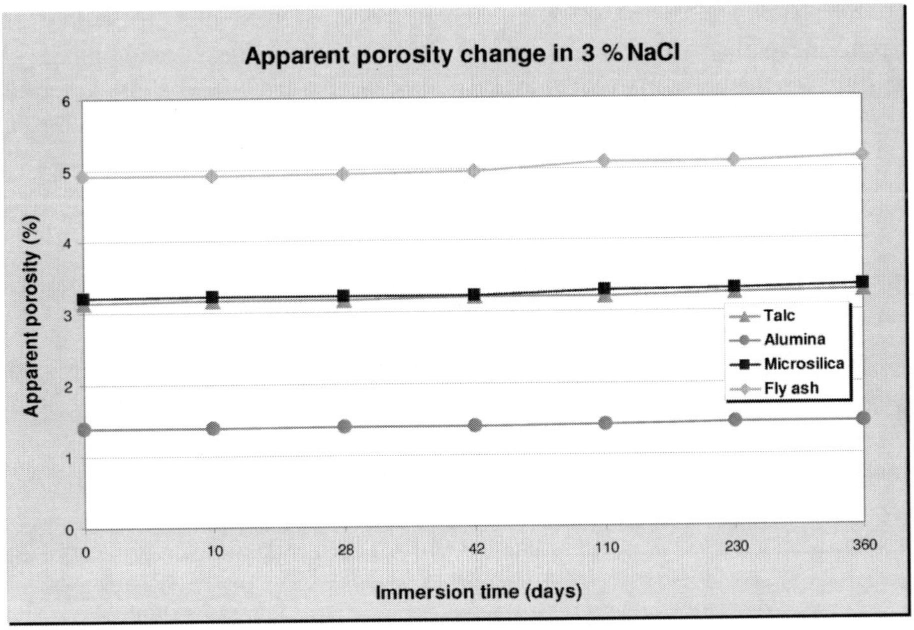

Figure 15. Apparent porosity change of sulfur- polymer matrix composite samples with different fillers in 3 % NaCl.

As it was expected, all samples of sulfur- polymer matrix composite exhibited increase of apparent porosity after a year of treatment in the aggressive environments.

It is obvious that the apparent porosity values were the highest for all samples after a year of immersion in HCl, a bit lower for the samples treated in H_2SO_4, and significantly lower for the samples from NaCl.

The samples treated in HCl underwent the greatest apparent porosity increase of 7.3- 8.3 %, depending on filler. Apparent porosity increase of 6.5- 7.2 % was for the samples with different fillers after treatment in H_2SO_4. In both acids, the samples with talc exhibited the highest apparent porosity changes, unlike the samples with alumina whose changes were the lowest. The difference between those ending values was ~1 % in HCl, and ~ 0.7 % in H_2SO_4. Apparent porosity increase of only ~ 4.5 % happened for the samples treated in NaCl. Since mutual differences of these changes among the samples were negligible, only 0.3 % between the highest and the lowest values, it can be considered that they were not caused by the type of filler.

Based on the obtained results, it can be concluded that apparent porosity increment tendency of sulfur- polymer matrix composite samples with different fillers is in accordance with compressive strength loss due to contact with the aggressive agents.

Image analysis was applied to the sulfur- polymer matrix composite samples in order to monitor the level of destruction during durability testing. Macro image analysis was realized with taking into account the total surface of the samples in order to monitor damage distribution at the surface. For the materials used in engineering, it is very important to analyze the whole sample surface, not only its part, as it is difficult to determine which part is representative for predicting the sample behavior. Similar approach was applied in image analysis of the samples subjected to thermal shock or cavitation erosion [26- 28].

Diagrams presenting non-damaged surface area (P_{nd}) and surface destruction level (P/ P_0) of sulfur- polymer matrix composite samples with alumina in 10 % HCl are given in Figure 16.

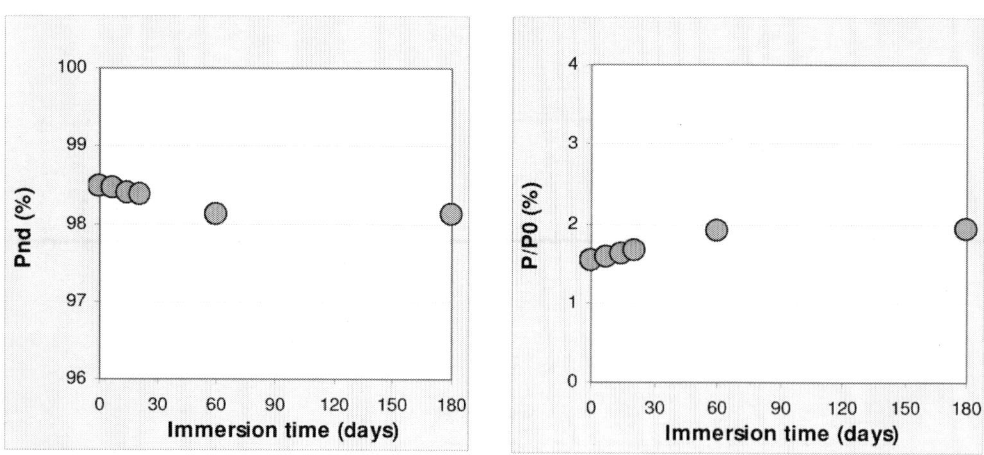

a) b)

Figure 16. a) Non-damaged surface area of sulfur- polymer matrix composite samples with alumina in 10 % HCl versus immersion time; b) Surface destruction level of sulfur- polymer matrix composite samples with alumina in 10 % HCl versus immersion time.

The level of surface destruction is defined as a ratio of damaged surface area (P) to original surface area before the acid resistance testing (P_0) [29-32].

Certain damage was present on sulfur- polymer matrix composite samples with alumina before the testing in 10 % HCl. Surface destruction level before the testing was 1.5 %. The level of surface destruction increased with the time spent in the acid environment. This change was very slight; the value of 1.9 % was reached after 180 days.

Detected surface changes of sulfur- polymer matrix composite samples can be connected with apparent porosity increase results previously reported. Acid penetration is limited to the surface and open pores that were not coated by sulfur, which in turn resulted in slight porosity increase. For the period of 180 days, apparent porosity enhanced from 1.38% to 1.46% (Figure 13). This is in accordance with slight surface destruction increase.

At this point, the analysis took a way to make decision on choosing the engineering method in analysis of the material and its properties for the real application. As it was pointed out in the previous text, the porosity was chosen as the material representative property in this case. The idea was to analyze which of the components (ingredients) had influence on porosity during chemical stress duration. The multicriteria decision methods are designed to analyze data within a multicriteria table including:

- a number of actions,
- several criteria.

In this case, the actions were the percentage of the ingredients used for sulfur- polymer matrix composite production and the criteria were the minimum porosity, maximum duration of chemical influence and the maximum percentage of used chemical agents. In mathematical terms, the problem is the following:

$$\max \{ f_1(a), f_2(a), \ldots, f_j(a), \ldots, f_k(a) \mid a \in A \}$$

where A is a finite set of n actions and f_1 to f_k are k criteria, $f_j(a)$ is the evaluation of action a on criterion f_j. There is no objection to consider some criteria to be maximized and others to be minimized but for the sake of simplicity, we will suppose here that all criteria have to be maximized.

The evaluations of the actions on the criteria form a two- way multicriteria table (Table 3):

The expectation of the decision- maker is to identify an action that is optimal for all the criteria at the same time. That is usually impossible, as the criteria are more or less conflicting one with another. The objective of MCDA is thus to identify the best compromise decisions.

In order to achieve this objective, it is essential to have some information about the preferences and the priorities of the decision- maker. This information is not contained in the multicriteria table. Moreover, different decision- makers will have different preferences and priorities. Gathering information about the decision- maker preferences and priorities can be done in many different ways. In the further text, two common ways will be compared and their qualities and limits will be emphasized. First analysis was the ranking of the actions (ingredients) used for the material production, which is presented in Table 4.

Table 3. Multicriteria table

	f_1	f_2	...	f_f	...	f_k
a_1	$f_1(a_1)$	$f_2(a_1)$	...	$f_f(a_1)$	...	$f_k(a_1)$
a_2	$f_1(a_2)$	$f_2(a_2)$	...	$f_f(a_2)$	...	$f_k(a_2)$
⋮	⋮			⋮		⋮
a_i	$f_1(a_i)$	$f_2(a_i)$	...	$f_f(a_i)$	...	$f_k(a_i)$
⋮	⋮	⋮		⋮		⋮
a_n	$f_1(a_n)$	$f_2(a_n)$	...	$f_f(a_n)$	...	$f_k(a_n)$

Table 4. Ranking actions

Rank	Action	Phi	Phi+	Phi-
1	Alumina	0.1667	0.1667	0.0000
2	Talc	0.0833	0.0833	0.0000
3	Fly ash	0.0000	0.0833	0.0833
4	Microsilica	-0.2500	0.0000	0.2500

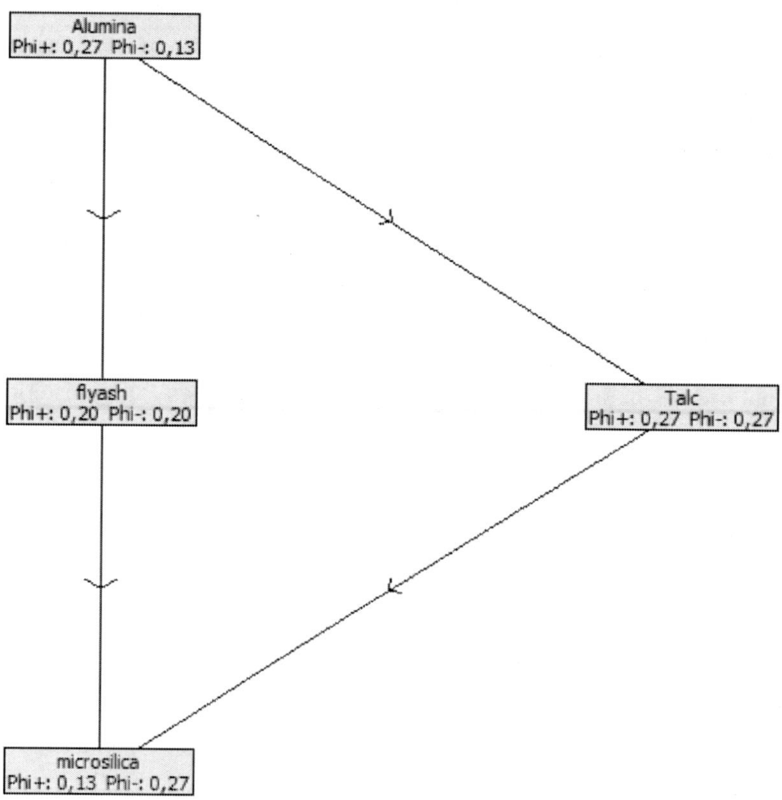

Figure 17. Porosity ranking web.

As it can be seen from the table, the ingredient which had the biggest influence on the porosity variations due to chemical influence was alumina and microslilica was almost inert.

In order to visualize Table 4, the multicriteria web is presented in Figure 17.

The multicriteria analysis in the so called influence space, where the results present the same results as previous with the respect to hydrochloric and sulfuric acids used in the tests are presented in Figure 18. The results present that microsilica is aproximately inert to the chemical stress and has no influence to the variations of porosity during application of the generated sulfur- polymer matrix composite.

This analysis confirms the idea to use the porosity as the representative factor and what is more important to use the apparent porosity for the analysis of behavior of the sulfur-polymer matrix composite under chemical influence.

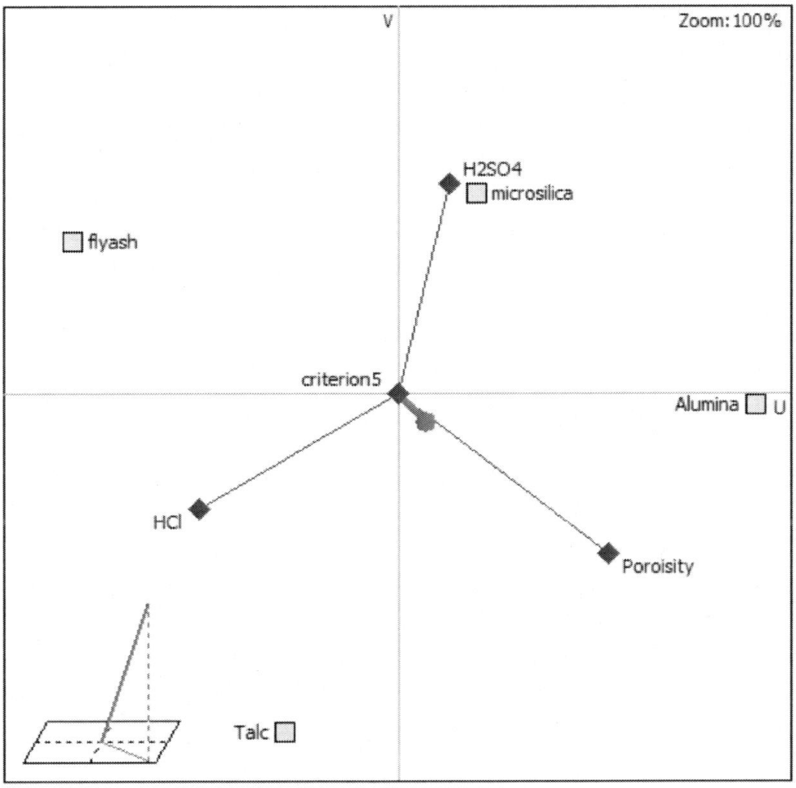

Figure 18. Chemical actions analysis.

CONCLUSION

In this chapter, sulfur- polymer matrix composite samples were produced of the secondary sulfur from the oil refining process, sand as an aggregate, and various fillers: talc, alumina, microsilica and fly ash. Modification of sulfur was performed by cyclic hydrocarbon, dicyclopentadiene.

The behavior of sulfur- polymer matrix composite with different fillers in three aggressive environments (10 % solution of HCl, 20 % solution of H_2SO_4 and 3 % solution of NaCl) as a function of time was the subject of this investigation.

By adequate technological procedure and selection of the initial components, with accent on proper grain size distribution, sulfur- polymer matrix composite with satisfying mechanical strength and good resistance to aggressive solutions was obtained.

Based on the results of analyzing mass and mechanical strength changes of sulfur-polymer matrix composite samples as a function of time spent in various corrosive environments, it was concluded that:

1. In acidic solutions sulfur- polymer matrix composite samples with fly ash and especially with talc showed higher resistance to corrosion regarding mass changes while in saline solution samples with alumina and microsilica were better.
2. Compressive strength loss was higher for sulfur- polymer matrix composite samples after being subjected to corrosive effects of the solution of HCl compared to H_2SO_4 and negligible after treatment in the solution of NaCl but it was completely independent on the type of filler in all aggressive environments.
3. Apparent porosity increase was the highest for the sulfur- polymer matrix composite samples treated in HCl, a bit lower after treatment in H_2SO_4 and the lowest after treatment in NaCl which is connected with the mentioned compressive strength loss.

The level of surface destruction during the testing was monitored using Image analysis in order to obtain information about damage distribution at the surface. Detected surface changes of sulfur- polymer matrix composite were in accordance with the apparent porosity results.

Based on the multicriteria analysis, it was concluded that the ingredient with the biggest influence on the porosity variations of sulfur- polymer matrix composite due to chemical influence was alumina and microslilica was almost inert. The obtained results can be used for finding new possibilities of sulfur- polymer matrix composite applying when high resistance to corrosive effect of the environment is required.

ACKNOWLEDGMENTS

The described research was carried out as a part of the project funded by the Ministry of Education, Science and Technological Development, of the Republic of Serbia under the project number TR 33007. The authors avail themselves of this opportunity to express their gratitude for this support.

REFERENCES

[1] Vlahović, M. & Jovanić, P. (2012). In Microscopy Book Series- Number 5: *Current microscopy contributions to advances in science and technology; Editor,* Méndez-Vilas, A.; *Formatex Research Center, Spain, 2, 1171.*

[2] Vlahović, M., Jovanić, P., Martinović, S., Boljanac, T. & Volkov-Husović, T. (2013). In New Developments in Polymer Composites Research; Editors, Laske, S; Witschnigg, A.; Nova Science Publishers, Inc., USA, 2013. https://www.novapublishers.com/catalog/product_info.php?products_id=46392&osCsid=9e69ae7f0bc1c787bad30b52ee7c0c8d.

[3] Gracia, V., Vàzquez, E. & Cramona, S. (2004). Utilization of by-produced sulfur for the manufacture of unmodified sulfur concrete, International RILEM Conference on the Use of Recycled Materials in Buildings and Structures, Barcelona, Spain, 1054.

[4] Vlahovic, M., Martinovic, S., Boljanac, T., Jovanic, P. & Volkov Husovic, T. (2011). Durability of sulfur concrete in various aggressive environments, *Constr. Build. Mater.*, *25*, 3926.

[5] http://www.shell.com

[6] http://www.sulphurinstitute.org/pub/a0120f9d-b7f3-92f6-a60f-37552901cb41

[7] Abdel-Mohsen, M. O. & El Gamal, M. (2012). Sulfur based hazardous waste solidification, *Environ. Geol.*, *53*, 159.

[8] Abdel-Mohsen, M. O. & El Gamal, M. (2011). Solidification of cement kiln dust using sulfur binder, *J. Hazard. Mater.*, *192*, 576.

[9] Lópeza, F. A., Gázquez, M., Alguacil, F. A., Juan Bolívar, J. P., García-Díaz, I. & López-Coto, I. (2011). Microencapsulation of phosphogypsum into a sulfur polymer matrix: Physico-chemical and radiological characterization, *J. Hazard. Mater.*, *192*, 234.

[10] Kalb, P. D., Adams, J. W. & Milian, L. W. (2001). Sulfur Polymer Stabilization/Solidification (SPSS) Treatment of Mixed-Waste Mercury Recovered from Environmental Restoration Activities at BNL, Brookhaven Science Associates, Upton, Long Island, New York 11973, Formal Report.

[11] Ksiazek, M. (2011). The experimental and innovative research on usability of sulphur polymer composite for corrosion protection of reinforcing steel and concrete, *Composites Part B*, *42*, 1084.

[12] Drop forming method and sulphur polymer concrete process, (2001). United States Patent 6228905.

[13] McBee, W. C., Sullivan, T. A. & Jong, B. W. (1984). Modified Sulfur Concrete Technology. *Proceedings SUFUR-81 International Conference on Sulfur Calgary*, *367*

[14] Blight, L., Currell, B. R., Nash, B.J., Scott, R. A. M. & Stillo, C. (1997). Preparation and Properties of Modified Sulfur Systems.

[15] Đorđević, N., Boljanac, T., Vidojković, V., Vlahović, M., Branković, A. & Martinović, S. (2010). Modification of secondary sulfur from oil refining process by dicyclopentadiene and application of obtained modified sulfur, Patent P-2009/0436.

[16] Mohamed, A. M. O. & El Gamal, M. M. (2009). Hydro-mechanical behavior of a newly developed sulfur polymer concrete, *Cem. Concr. Compos.*, *31*, 186.

[17] Soderberg, A. F. (1983). A New Construction Material. Sudicrete, SUDIC, Calgary.

[18] Mohamed, A. M. O. & El Gamal, M. M. (2010). Sulfur concrete for the construction industry- A sustainable development approach. J. Ross Publishing: U.S.

[19] Banerjee, S. (1998). Monolithic refractories. Westerville: The American Ceramic Society, Singapore: World Scientific Publishing Co. Pte. Ltd.

[20] ACI Committee 548, Guide for Mixing and Placing Sulfur Concrete in Construction, Report 548.2R-93, (American Concrete Institute), Farmington Hills, Mi (1993).

[21] Manufacture of sulfur concrete, (1977). *United States Patent 4025352.*

[22] Vidojković, V., Đorđević, N., Branković, A., Martinović, S., Vlahović, M. & Boljanac, T. (2010). The processing procedure of obtaining sulfur concrete using elemental sulfur produced in the oil refining process and its application, *Patent-2009/0433.*

[23] Methods of testing cement- Part 1: Determination of strength (2008). Serbian standard SRPS EN 196-1, 2008.

[24] Standard Test Method for Density, Absorption, and Voids in Hardened Concrete. (2006). *ASTM C*, 642-06.

[25] Vlahović, M., Boljanac, T., Branković, A., Vidojković, V., Martinović, S. & Đorđević, N. (2010). The influence of filler type on the corrosion stability of the sulfur concrete, *Hemijska industrija., 64*, 129.

[26] Martinović, S., Vlahović, M., Dojčinović, M., Volkov-Husović, T. & Majstorović, J. (2011). Thermomechanical properties and cavitation resistance of a high-alumina low cement castable, *Int. J. Appl. Ceram. Technol., 8*, 1115.

[27] Martinović, S., Vlahović, M., Boljanac, T., Dojčinović, M., Volkov-Husović, T. (2013). Cavitation resistance of refractory concrete: Influence of sintering temperature, *J. Eur. Ceram.* Soc. 33, 7.

[28] Martinović, S., Vlahović, M., Boljanac, T., Majstorović, J. & Volkov-Husović, T. (2014). Influence of sintering temperature on thermal shock behavior of low cement high alumina refractory concrete, *Composites Part B*, *60*, 400.

[29] Vlahović, M., Savić, M., Martinović, S., Boljanac, T. & Volkov-Husović, T. (2012). Use of image analysis for durability testing of sulfur concrete and Portland cement concrete, *Mater. Des. 34*, 346.

[30] Vlahović, M., Jovanić, P., Martinović, S., Boljanac, T. &Volkov-Husović, T. (2013). Quantitative evaluation of sulfur-polymer matrix composite quality, *Composites Part B*, *44*, 458.

[31] Vlahović, M., Martinović, S., Jovanić, P., Boljanac, T. & Volkov-Husović, T. (2014). In Advances in Image Analysis Research; Editor, Echon, R.; Nova Science Publishers, Inc., USA, 147- 170.

[32] Martinovic, S., Vlahovic, M., Boljanac, T., Matovic, B. & Volkov-Husovic, T. (2014). In Advances in Image Analysis Research; Editor, Echon, R.; Nova Science Publishers, Inc., USA, 171- 186.

In: Polymer-Matrix Composites
Editor: Rakesh Kumar

ISBN: 978-1-61122-005-6

Chapter 10

LONG-TERM DURABILITY OF FIBRE-REINFORCED COMPOSITES FOR AUTOMOTIVE APPLICATIONS

Valencia Jacobs*

CSIR Materials Science and Manufacturing, Polymers & Composites,
Port Elizabeth, South Africa
Department of Chemistry, Faculty of Science,
Nelson Mandela Metropolitan University, Port Elizabeth, South Africa
UNESCO-UNISA AFRICA Chair in Nanosciences/Nanotechnology,
University of South Africa, Muckleneuk Ridge, Pretoria, South Africa
Nanosciences African Network, iThemba LABS-National Research Foundation,
Somerset West, Western Cape Province, South Africa

ABSTRACT

In recent years, fibre-reinforced polymer matrix composites have raised great interest due to ever-increasing demand for newer, stronger, stiffer and yet lighter-weight materials in the field of automotive industry. Thus, much effort has been devoted to research and development relating to the long-standing durability of reinforced composites in various environments where these materials are normally utilised. As a result, fibre-reinforced composites are now being rapidly utilised in industries that traditionally used metals or nanofillers to improve the properties of polymer matrix composite materials. The automotive sector is continually seeking products that are inexpensive and more sustainable due to environmental concerns. Owing to their ability to create a positive environmental impact, natural fibres in composite materials are characterized by their biodegradability and decomposability at the end of their lifecycle. These fibres are usually extracted from plants and result in composites with lower specific weight, non-abrasive to mixing and safer handling conditions than the synthetic fibres. This chapter reviews the recent developments, properties and the effect of various conditions under which fibre-reinforced polymer matrix composites are processed.

* Corresponding Author address:CSIR Materials Science and Manufacturing, Polymers & Composites, P.O. Box 1124, Port Elizabeth 6000, South Africa, Email:vjacobs@csir.co.za.

Various types of polymer matrix composites are discussed and their robustness towards specific environmental conditions is further considered.

Keywords: Natural fibres, composites, properties, durability

1. INTRODUCTION

In the past, research and development has seen glass fibres as the most important reinforcing material for polymer composites. Unlike natural fibres, glass fibres are non-renewable and are difficult to dispose at the end of the material lifetime owing to their limitations when thermally recycled by incineration. They are also abrasive, causing damage to processing equipments and further pose health-related challenges during product processing of fibre reinforced parts [1]. Nowadays, ecological concern and environmental safety have led to renewed interest in the utilization of natural fibres as reinforced composites. The reason being, natural fibre possess a combination of good mechanical properties with renewability and biodegradability character [2]. However, there are also some drawbacks in the utilization of natural fibres as reinforcement in the polymer matrix composites. One restriction to the successful exploitation of natural fibres of durable composite application is their hydrophilic characteristic that lowers their compatibility with hydrophobic polymers during composite development.

All natural fibres, whether wood or non-wood types are cellulosic in nature. Their major constituents are cellulose and lignin. However, the amount of cellulose in lignocellulosic material can vary depending on the species and age of the plant. Cellulose is a hydrophilic glucan polymer consisting of 1, 4-β anhydroglucose units which contains hydroxyl groups. The hydroxyl groups form intermolecular and intramolecular hydrogen bonds with the molecule itself and with other cellulose macromolecules. It is this phenomenon that makes all natural fibres hydrophilic in nature [3]. The other limitations in natural fibre-based composites is their low thermal stability, low microbial resistance, susceptibility to rotting as well as quality inconsistancies during processing [4-7].

Owing to the high moisture absorption and elevated temperature, fibre-reinforced composites suffer a great deal of performance degradation that limits the wider use of these materials. Actually, in wet conditions, mechanical performance of composites decreases and therefore the possibility of utilizing these materials in the outdoor necessitate evaluation of their mechanical properties under weathering conditions. To overcome these challenges, previous research indicates that different treatments such as surface modification and chemical treatments have been utilized to improve properties of composites [8-12]. Treatment of fibers with hydrophobic aliphatic and cyclic structures has been attempted. These structures contain reactive functional groups that are capable of bonding to the reactive groups in the polymer matrix. Thus, modification of natural fibers is endeavored to make the fibers hydrophobic and to improve interfacial adhesion between the fiber and the polymer matrix [13-19]. Chemical treatments such as dewaxing (defatting), delignification, bleaching, acetylation, and chemical grafting are used for modifying the surface properties of the fibers and for enhancing its performance. Alternatively, chemical modification through graft copolymerization provides a potential route for significantly altering their physical and mechanical properties [13].

Generally, properties of the fibers, the aspect ratio of the fibers, and the fiber–matrix interface are the important factors that govern the properties of the composites. The surface adhesion between the fiber and the polymer plays an important role in the transmission of stress from the matrix to the fiber and thus contributes toward the performance of the composite. Both thermosets and thermoplastics are attractive as matrix materials for composites. In thermoset composites, a complex system with a large number of components including base resin, curing agents, catalysts, flowing agents, and hardeners is implemented. These processes involve highly chemical resistant cross-linked, three-dimensional network structures with limitations in fibre loading [13]. Unlike thermosets, thermoplastics offer low processing costs with flexible design and ease of molding complex parts. A number of parameters namely, mechanical properties, shape and size, orientation and distribution of reinforcement in the matrix form part of the thermoplastic system [20]. Simple processing techniques such as extrusion and injection molding are utilised for the preparation of these composites.

In the succeeding sections, natural fibre reinforced composite aspects are discussed in detail. Features such as classification of fibres, properties of natural fibres, development and process techniques, are explored. Mechanisms of interfacial interactions that require surface treatments and modification as well as study of compatibility between fibre and matrix in composites are elaborated. Furthermore, the influence of processability on the properties of the composites as well as potential applications in automotive industry is discussed.

2. Natural Fibres

There is no doubt that natural fibres offer a competitive advantage over synthetic fibres, with a possibility of replacing them for various industrial applications. As a result of growing awareness of a combination of global environmental factors, principles of sustainability, industrial requirements and eco-efficiency are being intergrated into the development of the novel generation of materials, products and processes [3, 21]. Increasing environmental concerns over widespread pollution elucidated by non-degradable materials, particularly long-lasting plastics and climate changes resulting from CO_2 emission, promoted a growing tendency towards substitution of synthetic fibre composites. In the past few decades, several reviews and general articles have been dedicated to natural fibres, also known as lignocellulose fibres and their reinforced polymer composites in replacement of synthetic composites.

2.1. Source and Classification of Natural Fibres

Natural fibres are generally grouped into either plants or animals, depending on the source of origin. Plant fibres are principally composed of cellulose and abundantly utilized in composite applications, whilst animal fibres are composed of proteins [22, 23]. These plants can be harvested from the bast or stem fibres, leaf fibres, seed, fruit, wood, cereal straw, and other grass fibres, which classify them as a distinctly renewable resource Figure 1 [3]. In general, there are several types of cellulosic fibres reinforced plastics including flax, hemp, jute and grass. However, many of them are natural waste products, and hence, have the advantage in that they are renewable resources and have market appeal associated with low cost.

While cellulose is the major component which is responsible for the inherent strength and stability of the natural fibre, the non-crystalline matrix phase of the cell wall comprises various compounds, including hemicellulose, lignin and some pectin, all of which form complicated macromolecular networks. The cellulose microfibrils are held together by the hydrogen bonds between the hemicellulose and cellulose forming tight networks. On the other hand, the hydrophobic lignin network affects the properties of the other network as it acts as a coupling agent, increasing the strength of the cellulose-hemicellulose network [1].

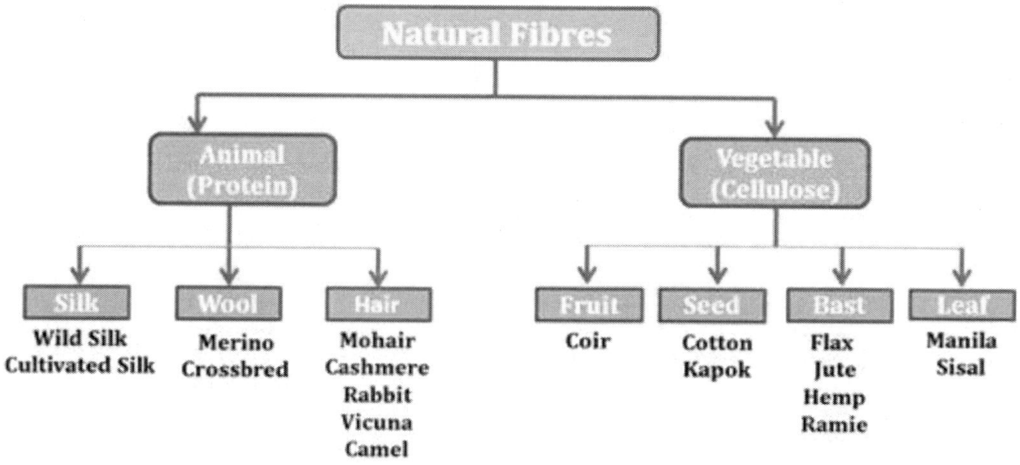

Figure 1. Classification of natural fibres (Reprinted from Ref. [42] Jawaid, M., Khalil. H.P.S.A (2011). Carbohydrate Polymers 86, 1).

Depending on the type of plant, different fibres have different lengths, morphologies and cross-sectional areas that are elucidated by various defects such as microcompressions or voids and cracks.

2.2. Properties of Natural Fibres

The microstructure of natural fibres is quite complex as it comprises hierarchical cell structures. Each elongated cell, with a diameter of approximately 10-20μm is a composite of rigid cellulose microfibrils embedded in a soft lignin and hemicellulose matrix as illustrated in Figure 2. This structure shows the morphology of microfibrils that are helically wound along the hollow fibre axis. In most plant fibres, the cellulose microfibrils are positioned at an angle to the normal fiber axis, namely, microfibrillar angle, thus providing mechanical strength to the fibre cell [1, 3, 24].

Natural fibres are a combination of complex mixtures of organic materials and as a result, a variety of processes leads to differentiated physical, chemical and mechanical properties. In a composite system where natural fibres are utilized as reinforcements, the properties of a composite material are strongly influenced by the properties of its constituents, their distribution and also the quality of interactions among them.

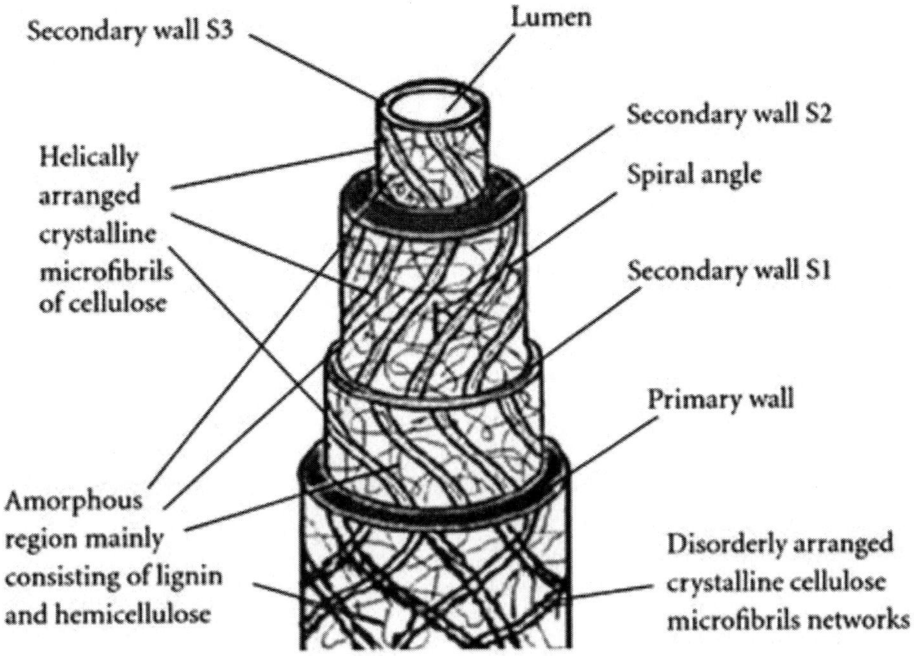

Figure 2. Structural arrangement of plant fibre cell (Reprinted from Ref. [22] Dicker, M.P.M., Duckworth, P.F., Baker, A.B., Francois, G., Hazzard, M.K., Weaver, P.M. (2014). Composites: Part A: Applied Science and Manufacturing 56, 280).

The characteristic values of natural fibres vary from one fibre to another. Some of the characteristic values include cellulose, hemicellulose and lignin content in the fibre. The degree of polymerization of the cellulose and the microfibrils angle fibres as well as thermal resistance of the fibres constitutes the other parts of the fibre. For example, higher degree of polymerization and a lower microfibrillar angle will affect physical properties of the fibre while the variations in the characteristic value exhibit the variations in mechanical properties both along the length of an individual fibre and between fibres. On the other hand, chemical properties of the fibre could be altered by thermal exposure during processing [3].

2.2.1. Physical Properties

In general, natural fibres are characterized by non-uniformity and variation of dimensions, even between individual plants in the same cultivation. Therefore, a number of processing techniques to separate raw materials are required in order to generate fibres suitable for specific end products. Bast or stem fibres, for example, are mainly used in the textile or rope industries because of the length of the fibres. These fibres are normally isolated into fibre bundles, which may contain thousands of single fibres. In contrast, wood is usually separated into single fibres or very small fibre bundles suiting the particular needs of the pulp, paper or board industries. Thus, there are a great number of challenges for selecting fibres in different dimensions and properties [25, 26].

Srinivasa et al. studied the effect of fiber loading on the hardness of composites. The results revealed that the hardness of the composite increased with the increase in the fiber loading. Due to the fact that the hardness is a function of relative fiber volume and modulus, basically, the fibers increase the modulus of composite which in turn increases the hardness of fibre [27]. Previous research indicates examination of physical and mechanical properties on 0, 12, 24, 36 and 48 wt % jute fiber-loaded reinforced epoxy composite using the hand lay-up technique. It was discovered that minimum and maximum void content are in neat epoxy and 12 wt % fiber loading specimens respectively. The fibre loading also had an influence on the hardness, tensile properties and impact strength of the jute-epoxy composites while the flexural strength and inter-laminar shear strength were greatly influenced by the void content of the composites [28]. The fibre surface modifications on high density polyethylene (HDPE) reinforced with continuous henequen fibres was performed using an alkaline treatment and a matrix preimpregnation together with a silane coupling agent. It was revealed that the use of the silane coupling agent improved the degree of fibre matrix adhesion, whereas strength and the stiffness of the composite was increased to approximately 80% [29].

2.2.2. Chemical Properties

The chemical composition of natural fibers varies depending upon the nature of the plant, the source, age of the plant and the extraction method utilized to process the fibres. For example, pineapple leaf fibre is soft and is characterized by high cellulose content while sisal is well known for its exceptional strength. Primarily, fibers contain cellulose, hemicellulose, pectin (Figure 3) and lignin. The properties of each constituent contribute to the overall properties of the fiber. Hemicellulose is responsible for the biodegradation, moisture absorption, and thermal degradation of the fiber as it shows least resistance. Likewise, lignin acts as a cementing material that has a major influence on the structure, properties and morphology of the fibre. Itis also thermally stable and is responsible for the UV degradation. The percentage composition of each of these components varies for different fibers. Generally, the fibers contain 60–80% cellulose, 5–20% lignin, and up to 20% moisture [13]. Other extractives present in plant-sources include pectin and waxes [30-34]. Nevertheless, the chemical composition of some plant fibres are shown in Table 1.

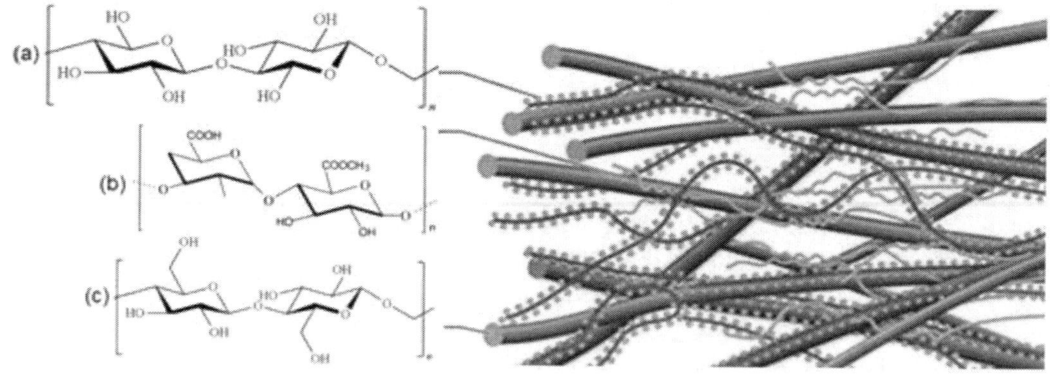

Figure 3. Lignocellulose components a) Hemicellulose; b) Pectin and c) Cellulose.

In the previous research, the effect of chemical treatment on the tensile properties of sisal fibre-reinforced LDPE (low density polyethylene) composites was investigated. Modifications using sodium hydroxide, isocyanate, permanganate and peroxide were carried out to improve the bonding at the fibre-polymer interface. The results showed that the chemical treatments enhanced the tensile properties of the composites with peroxide treatment imparting maximum interfacial interactions [11].

Table 1. Chemical composition of some plant fibres

Fibre	Cellulose (wt %)	Hemicellulose (wt %)	Lignin (wt %)	References
Flax	64.1	16.7	2.0	[33]
Kenaf	36	21	18	[47]
Jute	64.4	12	11.8	[33]
Coir	43	0.3	45	[47]
Banana	60-65	6-19	5-10	[47]
Sisal	65.8	12.0	9.9	[33]
Ramie	68.6	13.1	0.6	[33]
Cotton	82.7-92	2-5.7	0.5-1	[33]

Idicula et al. discovered that there is a significant increase in both thermal conductivity and density values of banana/sisal fibre composites when these fibres were treated with NaOH and Polystyrene maleic anhydride. This shows that the chemical treatment allows a better contact between the components (fibre/matrix) and reduces considerably the thermal contact resistance [35].

2.2.3. Mechanical Properties

The final mechanical behaviour of a composite material depends to a great extent on the adhesion between the reinforcing fibre and the surrounding matrix. The adhesion between two different materials is a function of several factors, among which are the surface roughness and surface polarity [4, 36]. In general, the variability of fibres depends on the environmental conditions as well as the source of fibres. However, the mechanical properties of natural fibres can be affected by various factors that may be introduced during manufacturing, at the service life point and during chemical modification of fibres to improve compatibility in the matrix. As a result, various factors that influence the properties of the fibres include:

- The fiber dispersion in the matrix
- The orientation and geometry of the fibers within the composites
- The fiber volume fraction
- The quality of the interface between the reinforcing fiber and polymeric matrix phase

These parameters are often difficult to control particularly during processing. Previous research indicates that a variety of treatments to enhance the compatibility between fibres and the matrices have been explored to develop composites with better mechanical properties [37]. Earlier research shows that Young's modulus of a flax fibre was estimated by taking into account the composition of the fibre and the evolution of the orientation of the fibrils during a tensile test, using micro-mechanical equations which were compared with experimental results [38].

In another research, the calculation of the elastic properties of cellulose-based natural fibres was performed by means of different mechanical models. The results showed how the modulus in fibre axis decreases with an increasing spiral angle and degree of anisotropy, while shear modulus reached a maximum value at a spiral angle of 45° [39]. Mishra et al. investigated the effect of fibre loading on the development of natural fiber based polymer composites consisting of bidirectional jute fiber mat as reinforcement and epoxy resin as matrix material. The composites were fabricated using hand lay-up technique and were characterized with respect to their physical and mechanical properties. Results showed the significant effect of fiber loading on the mechanical properties of the composites. Also, the formation of voids in the composites was found to be an influencing factor on the mechanical properties of the materials [28].

Researchers investigated biodegradable foams made from potato starch and how natural fibres obtained by extrusion influenced their properties. The effects of varying origins of these fibres on foam properties were studied, as well the relationships between their properties and the foam microstructure. The results revealed that the addition of fibres increased the expansion index while the foam properties were improved. It was also discovered that, as the relative humidity increased, the foam strength decreased. Optimum mechanical properties containing 10 wt% hemp fibre and had a maximal resistance of 4.14 MPa and a modulus of 228 MPa, corresponded to a more compact and dense microstructure [40]. The influence of different process times and temperature on the mechanical properties of unidirectional and multidirectional flax/polypropylene treated with maleic anhydride acid was examined. It was found that the treatment improved the mechanical properties of the composites [41].

3. Reinforcement of Composites

The interest in natural fiber-reinforced polymer composite materials is rapidly growing both in terms of their industrial applications and fundamental research. Generally, the continuous reinforcing fibers of advanced composites mainly constitute to their high strength and stiffness. In the process, interfaces play a critical role in determining the physical and mechanical properties of the reinforced composites. However, reinforcement of fibres is normally given surface treatments to improve their compatibility with the matrix. More often, the most important fibers in current use are glass, graphite, and aramid. Other organic fibers, such as oriented polyethylene, are also becoming important. Polymer matrix composites contain about 60% reinforcing fiber by volume. For instance, unidirectional, high strength graphite/epoxy has over three times the specific strength and stiffness of common metal alloy. Taking into cognisance, specific properties such as modulus of lignocellulosic, the lignocellulosic fibres show values that are comparable to or even better than glass fibres. Lignocellulosic fibres exhibit significantly better elongation at break which will translate in better composite damage tolerance. Low cost and better durability make cellulosic fibre attractive for automotive applications with light weight requirements. However, for better strength and stiffness cellulose fibres like hemp, flax, jute, kenaf and sisal are becoming increasingly prominent in composites production [42].

3.1. The Development of Fibre-Reinforced Composites

Advances in the development of more advanced materials and manufacturing techniques in composites has seen increase to a recognized and well-respected engineering discipline, providing solutions to conventional and challenging applications. Nowadays, fibre-reinforced composites (FRCs) are amongst the most common types of composites utilized in various industrial applications. The current technologies allow designs in various forms with different structural properties, specifiers and end-users that give wide scope of choice on these materials, thus, providing design flexibility as well as low manufacturing and maintenance cost.

Extensive research has been done in the development of the fibres as a function of raw material optimisation, growth-maturity-retting-decortication processes and the use of advanced methods for the preparation and processing of fibres. The purpose of composite materials development is, in fact, to obtain a product able to combine the positive properties of each constituent component. In this sense, lignocellulosic reinforcements have been of growing interest in the last few years due to their environmentally friendly aspect [43]. In fact, environment legislation as well as consumer demands are all increasing the manufacturing requirements of these materials and final products to consider the virtual impact of the products at all stages of their life cycle. The latter aspect validates the renewed interest that natural reinforcements have gained, especially as a glass fibre substitute in the automotive industries [32, 37, 44]. Furthermore, the development of specific manufacturing technologies has to be addressed in order to obtain reliable products based on these materials.

The main advantages of using natural reinforcements in composite materials can be attributed to the fact that they can be easily eliminated from the environment after the life cycle and their lower weight contributes to their strength and stiffness as opposed to glass. Generally, industrial production of these composites requires minimum energy while the usage of CO_2 and emission of O_2 offer environmental benefits. Consequently, this low cost production with user-friendly and less hazard processes makes natural fibre reinforcement composite materials more attractive to developing countries. These materials, however, do have some drawbacks that prevent their optimal usage at industrial level. In some other cases, the unpredictable environmental conditions have a negative influence on properties of the composites. For example, the swelling of fibres that is elucidated by the materials high absorption tendency, lower durability and impact strength pose challenges in fabrication of composites. Some of the advantages and disadvantages of natural reinforcement composites are tabulated below in Table 2 [37, 45].

Table 2. Advantages and disadvantages natural fibre reinforcements in composite materials [32]

Advantages	Disadvantages
Biodegradable	Variable quality
Renewable	Hygroscopic
Low cost	Poor strength
Hazard-free	Flammable
User-friendly	Expensive
Thermal resistant	Lack of durability
Light specific weight	Limited thermal exposure

There are, however, some benefits relating to properties of fibre reinforced composites, for instance, materials derived from agriculture are emerging as promising substitutes for conventional plastics. Thermoplastic starch has a low initial cost (US$ 0.25–0.6/kg) and wide availability; thus, it is the focus of a great interest in the production of disposable products [40]. The processing of natural fiber thermoplastic composites involves extrusion of the ingredients at melt temperatures followed by shaping operations such as injection molding and thermoforming. Fiber–fiber interactions as well as fiber–matrix interactions play a crucial role in determining the properties of such composites. Mostly, challenges relating to properties of fibre reinforced composites are elucidated from a lack of proper and effective reinforcement system resulting from poor adhesion at the fiber matrix interface. Another major challenge is the agglomeration of cellulose fibers as a result of poor dispersion in a hydrophobic polymer matrix and thus poses difficulties in achieving a uniform distribution of fiber in the matrix. Previous research indicates numerous alternative surface treatment methods to improve compatibility of fibre-matrix systems [13, 40].

Alternatively, thermosets are processed by simple sequential processing techniques such as hand layup and spraying, compression, transfer, resin transfer, injection, compression injection, and pressure bag molding operations. These materials are impregnated with the thermosetting resins and then exposed to high temperature for curing to take place [13].

3.1.1. Fibre Dispersion in the Matrix

Traditionally, the fundamental requirements for obtaining an optimal composite with superior properties are to attain a good fibre matrix system. Prominently, the parameters that normally influence the adhesion of fibres in the matrix are fiber-fibre interaction and fibre length. To overcome such challenges, many researchers have conducted investigations relating to pre-treatments of fibres before processing and several methods have been reported to improve fibre dispersion and interfacial interaction between fibre and matrix [46-48]. It is well known that lignocellulosic fibres are hydrophilic in nature and incompatible to hydrophobic polymeric matrices resulting in poor adhesion between fibre and matrices [49].

Researchers studied composites of cellulosic fibres with polypropylene, polyethylene, and polystyrene and found that the bad dispersion of fibres results in their agglomeration into knotty masses, leading to composites with poor final properties [50].

Earlier studies indicate that composites containing 30 wt% of chemical pulps, thermomechanical pulp (TMP) and wood flour was produced by compounding and injection moulding. Fibre morphologies, dispersion of the fibres and mechanical performance of the composites were also investigated. The results indicated that fibre length was reduced during melt processing steps, reduction being higher with longer fibres. Comparing the wood fibre types, TMP fibres provided the highest improvement in mechanical properties in polylactic acid composites with uniform fibre dispersion [51]. However, other parameters that can improve fibre adhesion include power input, processing time and shortened fibre length. In the previous research, composite samples made out of polylactic acid (PLA) matrix reinforced with olive pit powders, were manufactured and mechanically characterized. The outcomes indicated that with filler loading, an increase in the tensile modulus but a decrease of the flexural strength may be due to the poor interfacial bonding between olive pit powder and PLA. Other alternatives to overcome fibre adhesion challenges and significantly improved mechanical properties of the composites, is the pretreatment of the fiber with suitable additives prior to processing [52]. Also, the processing parameters in the case of thermoplastic composites hamper the properties of the composites.

Report on the effect of processing parameters such as mixing time, rpm, and temperature in compounding polypropylene/wood composites indicated that tensile strength varied nonlinearly with rpm, mixing temperature and time, leading to improved properties of the composites [13, 53].

3.1.2. Fibre Matrix Adhesion

In composite technology, natural fibres have been favourably utilized as reinforcements in the fabrication of materials with optimum strength and superior qualities. To obtain such qualities, it is fundamental to improve the compatibility between the matrix and the hydrophilic cellulose-based natural fibres [11, 54]. In most cases, the incorporation of hydrophilic natural fibers in polymers leads to heterogeneous systems with inferior properties that are elucidated by the lack of adhesion between the fibers and the matrix. Hence, the treatment of fibers for improved adhesion is a critical step in the development of such composites. Generally, moisture absorption in natural fibres is high due to the presence of hydroxyl and other polar groups on the surface and throughout fibre. This phenomenon leads to poor wettability of the polymers and weak interfacial bonding between fibres and hydrophobic polymers [26]. To encourage proper adhesion of the matrix onto the fibres, various treatments of fibers may be explored such as bleaching, grafting of monomers, acetylation and others. Alternatively, use of a compatibilizer or a coupling agent for effective stress transfer across the interface can also be utilized. The compatibilizer can be polymers with functional groups grafted onto the chain of the polymer. On the other hand, the coupling agents are tetrafunctional organometallic compounds based on silicon, titanium, and zirconium and are commonly known as silane, zirconate, or titanate coupling agents [13].

Virtually, the properties of the polymer matrix composite depend on the matrix, the reinforcement, and the interphase. Consequently, there are many variables to consider when designing a polymer matrix composite. These include not only the types of matrix and reinforcement but also their relative proportions, the geometry of the reinforcement, and the nature of the interphase. During processing, a strong interracial bond makes polymer matrix composite more rigid, but brittle, while a weak bond decreases stiffness, but enhances toughness. If the interracial bond is not at least as strong as the matrix, debonding can occur at the interphase under the influence of certain loading conditions. To maximize the properties of the polymer matrix composite, the suitable coupling agent is often intermediate between the strong and weak limits. The character of the interracial bond is also critical to the long-term durability of the polymer matrix composite, playing a key role in fatigue properties, environmental behavior, and resistance to hot or wet conditions.

In an advanced composite system, toughness is a complicated function of the matrix, fiber, and interphase, as well as the geometry of reinforcement. Moreover, mechanical properties are highly symbiotic. For example, cracking associated with shear stresses may result in a loss of stiffness. Also, impact damage can seriously reduce the compressive strength of polymer matrix composites. Furthermore, compressive and shear properties can be seen to relate strongly to the toughness of the matrix, and to the strength of the interfacial bond between matrix and fiber. Earlier research indicate studies on surface modification of cellulosic fibres in depth and concluded that most promising approach of chemical modification seemed to be the one that gave rise to continuous covalent bonds between fibre surface and matrix [55]. Recently, Wilhelmy plate method was utilised to measure the perimeter of regenerated cellulose as a function to calculate Interfacial Failure Shear Stress (IFSS) [56].

3.1.3. Processability

Given the variety of fibers and different matrices from which polymer matrix composites can be fabricated, the subject of polymer matrix composite manufacturing is an extremely broad one. However, more than any other single area, low cost manufacturing technologies are the basis for products that can be processed as advanced composites. Fundamentally, polymer matrix composites are divided into two categories, namely, reinforced palstics and so-called advanced composites. The difference between the two is based on the level of mechanical properties, particularly, strength and stiffness. Reinforced plastics are relatively inexpensive and consist of polyester resins reinforced with glass fibres. On the other hand, advanced polymers constitute a combination of fibre and matrix that yield superior strength and stiffness.

Some of the basic steps of fibre processing are shown in Figure 4. Depending on the process, these steps may occur distinctly or simultaneously. For instance, the starting material for many polymer matrix composites is a prepreg; i.e., a fiber tape or cloth that has been preimpregnated with resin and partially cured. Unlike the previous technique, pultrusion, impregnation, forming, and curing are done in one continuous process. Fundamentally, in the aerospace sector, advanced composite structures are commonly fabricated by the slow and labor-intensive process of hand lay-up of prepreg tapes. Hand lay-up involves placement of sheets or tapes of prepreg on a tool (the contoured surface that defines the shape of the finished part).

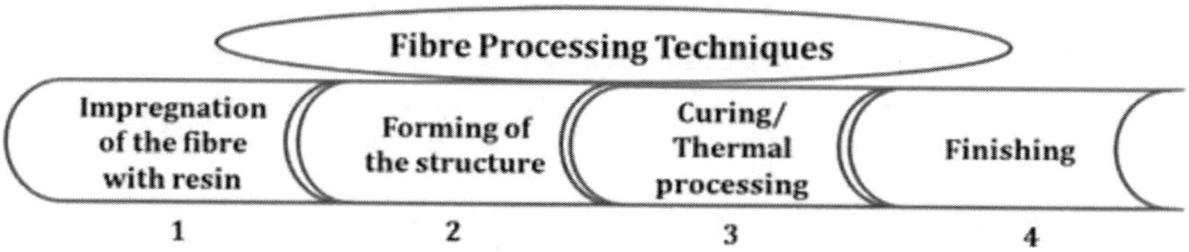

Figure 4. Processing techniques of natural fibres.

Earlier research indicates that pressing on laboratory scale was done with the use of a heated press at varied temperatures and in a mold. After 12 min pre-heating, the samples were pressed at ±220°C (205°C for maleic acid anhydride modified polypropylene) during 2.5 min and then cooled for 6 min. On pilot scale, tests with three different temperatures (190, 200 and 210°C) in combination with two pressing times (3.5 and 5 min) were carried out. Thereafter the composites were cold pressed during 5 min. The optimum conditions were discovered as 200°C– 3.5 min combination, and were utilised for further production of panels [41]. In another study, the low density polyethylene-sisal composites were prepared by a solution mixing technique followed by extrusion as reported by Joseph et al. [57].

The developmentof coir composite and their mechanical properties were evaluated. A comparison between qualitative evaluation of the interfacial properties of coir /epoxy and glass fibers was conducted from fracture surfaces using scanning electron micrographs [58]. Further research on the characteristics of the coir stack was investigated. It was discovered that the length of the fibers was in the range between 8 and 337 mm. The fibers with the length range of 15~145 mm was 81.95% of all measured fibers while the weight of fibers with the length range of 35~225 mm accounted for 88.34% of all measurements. The average fineness of the coir fibers was 27.94 tex [59].

The design and manufacture of composite material was examined using three Jamaican natural cellulosic fibers. The study revealed that bagasse from sugar cane, banana trunk from banana plant and coconut coir from the coconut husk were subjected to standardized tests. These methods comprises ash and carbon content, water absorption, moisture content, tensile strength, elemental analysis and chemical analysis [60].

Insulating material for construction were prepared from four fibers derived from banana-trees (leaf, trunk) and coconut-tree (husk, fabric) before their incorporation in cementitious matrices. Thermal degradation as a function of temperature of pyrolysis was studied between 200 and 700 °C under nitrogen gas flow. The solid residues obtained were analyzed by classical elemental analysis; Fourier Transform Infra Red (FTIR) spectroscopy and the morphology of the materials were observed by Scanning Electron Microscopy (SEM). The study revealed that there is a relation between botanical, chemical composition with both localization of fiber in the tree and type of tree. Also, the decomposition of banana fibers with increasing temperature of pyrolysis and the rough, hollow morphology of the fibres was reported [61].

3.1.4. Moisture Content of Fibres

It is well known that natural fibres are hydrophilic in nature and this is elucidated by the presence of cellulose bearing three alcohol groups per glucan repeating units that eventually imparts water absorption characteristic of the fibre reinforced composite material. Subsequently, water absorption poses negative effect on the strength of the composites as it causes fibre swelling which leads to delamination and suface roughness of the end product. The influence of moist environmental conditions can also affect the durability of the composite by facilitating the growth of fungus and bacteria [22, 62, 63]. It was discovered that for 40 wt% jute/epoxy composites, water absorption stabilized at 17.5 wt% of total specimen [64]. The water absorption of epoxy composites reinforced with carbon nanofillers was investigated. The study revealed that the water uptake induces the plasticization of the resin and improvement of the postcuring reaction. The mechanical properties of the composites were also affected by the absorption of both type I and II bound water molecules. As a result, the mechanical strength dropped due to plasticization whereas after saturation, the mechanical strength was regained owing to the influence of secondary crosslinked network [65]. Flax/polypropylene composites treated with maleic anhydride-polypropylene copolymer (MAPP), as compatibilizer, reduced equilibrium water uptake of the composites. The decrease was attributed to the improved interfacial adhesion that prevents interference of water molecules into the modified composites [66-68]. Others relate this behavior to some fibre hydrophilic –OH groups reacting with maleic anhydride to form ester links that contributes to lower water uptake [69]. Some reports state that water molecules could affect the composite in one of the following ways [70]:

i. Diffuse to matrix and fibre causing degradation of both constituents
ii. Penetrate to the interphase between matrix and fibre leading to de-bonding
iii. Transport through deformations such as pores resulting in accumulation of stress

Researchers studied the cellulosic fibre based hybrid composites and discovered that hybrid arrangement tends to inhibit the absorption of moisture into the composite and pack arrangement of fibre tend to fill up the voids which form during the formation of the composite. This is elucidated by the hydrophilic nature of cellulosic materials, enabling the composites to take up a high amount of moisture from the surrounding environment. During manufacturing of thermoset based composites, the mechanical properties of the composites are decreased due to voids that are affecting fibre–matrix bonding [42]. Further studies were done on the influence of water absorption on the tensile strength of kenaf, jute and abaca. It was found that tensile strength of flax and coir decreased significantly while hemp was notably degraded. On the other hand, abaca had the highest absorption moisture capacity (164 wt%) which escalated the swelling ability of the fibre up to 245.85% [22, 71].

3.1.5. Thermal Stability

Basically, manufacturing processes regarding cellulosic fibres can be limited due to low thermal stability, and thus restricts the use of the composites to low temperature applications. In general, low thermal stability increases the possibility of cellulosic degradation and the likelywood of emissions of volatile materials that could adversely affect the composite properties [22]. Relatively low degradation temperatures of cellulosic reinforcements, in comparison with the process temperatures of polymeric matrices, represents the most important limit for their use in plastics with the conventional processing equipment. In spite of other advantages, the thermal properties of cellulosic fibres pose a drawback when compared to synthetic fibres. However, the main influential properties of cellulosic fibres for composite materials are thermal expansion and thermal degradation. Regarding thermal expansion only few data are available in the scientific literature with values of the corresponding coefficient ranging in the order of 3105, while glass fibres are in the order of 5 106. On the other hand, thermogravimetric and differential scanning calorimetric analysis demonstrate that noticeable mass loss in natural fibres start at about 200-220°C while; irreversible degradation of the fibres occurs above this temperature [72]. To overcome such challenges, the thermal stability of the cellulose-based fibres can be improved with several physical and chemical treatments.

Numerous reports on research activity regarding the thermal degradation of cellulosic fibres for composites have been reported in the scientific literature. For instance, the thermal behaviour of untreated and acetylated jute fibreswas determined. Also, decomposition temperatures of the hemicellulose and α-cellulose showing the missing of the first peak and the increase of the second peak temperature for treated fibres, was proven to be caused by higher thermal stability [73]. The effect of acetylated and non-acetylated sisal fibres on the thermal degradation of blends of polypropylene and polyolefins was examined. The studies revealed that the starting temperature of the decomposition of the fibre, whether treated or not, was maintained almost at the same level whereas acetylated fibre stability was found to increase with a higher activation energy value. A change in the degradation mechanisms with two stages degradation for the treated fibres was also observed [74].

Thermogravimetric analysis, TGA, of untreated and different surface modified jute fibres was also performed with the realisation of composites based on a Biopol matrix. It was found that all surface modified fabrics show the same or better thermal stability compared to untreated fabrics [75]. In another interesting study, the introduction of TGA of *Hildegardia populifolia* fibre into poly(propylene carbonate) matrix led to a slightly improved thermo-oxidative stability of the composite [76].

The longitudinal Young's modulus of the fibre was determined to exceed the transverse fibre modulus by as much as an order of magnitude in certain temperature regimes. The fibres exhibited negative thermal expansion coefficients along their lengths; however, large positive thermal expansion coefficients, similar in magnitude to many polymeric materials, have been estimated in the transverse directions [32]. The thermophysical properties of the banana/sisal fibre reinforced composite induced a decrease on the thermal conductivity of the composite. The study indicated that thermal conductivity of polymeric matrix seemed to be more important than the banana/sisal fibre [35].

In the research for the thermal behavior of oil palm fiber reinforced phenol-formaldehyde resin, it was discovered that the thermal stability of the composites increased after chemical treatment. Consequently, maximum thermal stability was obtained for resin treated with peroxide phenol-formaldehyde composites in comparison to fiber treated phenol-formaldehyde composites [77].

3.2. Surface Modification of Fibres

A natural fiber reinforced composite with an outstanding combination of properties is not impossible nowadays. Use of proper processing techniques, physical treatments and chemical modification as well as use of coupling agents can lead to composites with superior properties for set applications. These modifying strategies have been generally reviewed recently [78, 79]. For instance, treatment of natural fibers is beneficial in order to improve the water resistance of fibers, enhance the wettability of the natural fiber surface and promote interfacial adhesion.

Physical treatments such as electronic discharge in the different media relating to plasma and corona technologies may create a hydrophilic or hydrophobic fiber surface by changing the surface energy to consequently increase the compatibility of the treated fiber with the polymer matrices. These surface treatments only modify a very shallow surface of cell walls and do not penetrate the fibre itself and thus the hygroscopic characteristics of fibers is not affected. Therefore, use of chemical modification is explored to provide the means of permanently altering the nature of fiber cell walls. These processes include either grafting polymers onto the fibers or crosslinking of the fiber cell walls or by using coupling agents [13, 80].

Extensively coupling agents have been used for natural fibre reinforced composites containing maleic anhydride such as maleated polypropylene (MAPP) or maleated polyethylene (MAPE) [29-31]. The anhydride groups of the copolymers may react with the surface hydroxyl groups of natural fibers forming ester bonds whilst the other end of copolymer entangles with the polymer matrix due to their similar polarities. Various methods of modifying the surface of the cellulosic fibers have been explored in an effort to improve the interaction that occurs at the interface between cellulosic fibres and the matrix [81,82]. It is apparent that the stronger the molecular interactions that occur at the interface between cellulose fibers and polymer matrix, the stronger the resulting interfacial adhesion and the optimal stress transfer efficiency. However, the strongest adhesion can be achieved when covalent bonds are formed at the interface between cellulose fibers and coupling agent as well as molecular entanglement between coupling agent and the polymer itself [83]. Owing to their ability to induce better flow of the molten polymer during processing, coupling agents have the ability to enhance fibre dispersion, while melt elasticity and melt strength can improve in the resulting polymer composites [84].

The study on the effects of different treatments on the fibre–matrix compatibility in terms of surface energy and mechanical properties of composites has been reported. Thecomposites were prepared using two kinds of flax fibres (natural flax and flax pulp) and polypropylene. The applied treatments were maleic anhydride (MA), maleic anhydride-polypropylene copolymer (MAPP) and vinyl trimethoxy silane (VTMO). The results indicated that the three treatments reduced the polar component of the surface energy of the fibre. The highest mechanical properties were obtained for composites containing MAPP, whilst there were no significant values obtained for MA and VTMO-treated fibre that gave similar values to that for the untreated ones [4].

The influence of different process times and temperatures on the mechanical properties of Unidirectional (UD) and Multidirectional (MD) flax/polypropylene was investigated. The unidirectional composites of boiled flax combined with MAA-PP show the best mechanical properties, whereas, MD composites showed no significant improvements on properties of the composites [41]. Sisal-low density polyethylene composites were subjected to different fibre treatments such as alkali, isocyanate, permanganate and peroxide. Alkali treated fibre composites showed better tensile properties than untreated composites due to their rough surface topography and increased aspect ratio [11, 31]. Extensively, treatments of fibre for surface modification may include bleaching process that utilise oxidizing agents such hydrogen peroxide. Other processes comprise grafting of monomer by initiation, propagation, termination and dimerization stages as shown below [13, 18-19]:

1. Bleaching Process

$H_2O_2 \rightarrow HOO^- + H^+$ ------ (i)

2. Grafting of Monomer on Cellulose Molecule

- Formation of radical molecule

$Cell—H + R^\bullet \rightarrow Cell^\bullet + R—H$ ------ (ii)
$(R^\bullet = OH^\bullet \text{ or } SO^\bullet_4)$

- Grafting the monomer

$Cell^\bullet + M \rightarrow Cell—M^\bullet$ ------ (iii)
$Cell—M^\bullet + M \rightarrow Cell—M^\bullet_2$ ------ (iv)
|
$Cell—M^\bullet_{n-1} + M \rightarrow Cell—M^\bullet_n$ ------ (v)

$Cell—M^\bullet_n + Cell—M^\bullet_n \rightarrow$ graft copolymer ------ (vi)

$2Cell^\bullet \rightarrow$ dimerized polymer ------ (vii)

Generally, the most common process of bleaching the cellulose-based fibres uses hydrogen peroxide, sodium hypochlorite or calcium hypochlorite as reducing agents.

Interestingly, the modification of the fibre surface with a multifunctional monomer can be exploited in secondary reactions (i.e., radical grafting) with the polymer matrix and thus providing stable bonding network between the fibres and the matrix. Numerous studies on the chemical treatment or surface modification of fibres to improve adhesion between fibre and matrix have been explored and reported [85-88].

4. Properties of Fibre-Reinforced Composites

It is well known that the matrix properties in the composite structure determine the resistance of the polymer matrix composite to most of the degradative processes that eventually cause failure of the structure. Basically, these processes include impact damage, delamination, water absorption, chemical attack, and high-temperature creep. Thus, the matrix is typically the weak link in the polymer matrix composite structure. In most cases the matrix phase of commercial polymer matrix composites is classified as either thermoset or thermoplastic. Owing to their escalated demand that led to rapidly heightened cost, thermoplastic materials are being increasingly used for various applications [89]. This situation made it necessary to use low cost fillers as a means of reducing the cost of the end product. However, the widely used inorganic fillers, such as glass fibre and mica are very expensive compared to natural fibres [90]. On the other hand, processes utilised to produce natural fibre-reinforced thermoplastic composites are cost effective when compared to the original thermoplastics which makes it possible to meet any future shortage of thermoplastic composites. To their added advantage, the use of natural fibre in thermoplastic composites is highly beneficial, because the strength and stiffness of the plastics can be altered for better performance. Besides, lack of good interfacial adhesion and hydrophillicity made the use of natural fibre reinforced composites less attractive [91]. This challenge can be overcome by subjecting these fibres to physical and chemical treatments [11].

4.1. Thermoset Polymers in Composites

Thermosetting refers to a natural fibre reinforcement impregnated by cross-linking resins to produce semi-permanent mouldings, which is an irreversible process. The resins used may be saturated polyester, poly urathaepoxy, poly vinyl ester or phenolic resin. In a thermosetting plastic, the polymer is made of atomic chains which are characterized by many bonds between different chains. A linear wire like chain is the normal structure for all polymers; bonds between chains are called cross-links and they stabilize the plastic shape (Figure 5) [92]. However, in natural fibre reinforced composites, the most commonly used polymers are polypropylene, polyamide and polyester.

The matrix phase plays a crucial role in the performance of polymer composites. Both thermosets and thermoplastics are attractive as matrix materials for composites. In thermoset composites, formulation is complex because a large number of components are involved such as base resin, curing agents, catalysts, flowing agents, and hardeners. These composite materials are chemically cured to a highly cross-linked, three-dimensional network structure which is supported by a high content of fibre which is subsequently aligned for better enhancement. As a result, their three-dimensional crosslinked structure is characterized by dimensional stability, high-temperature resistance, and good resistance to solvents [13].

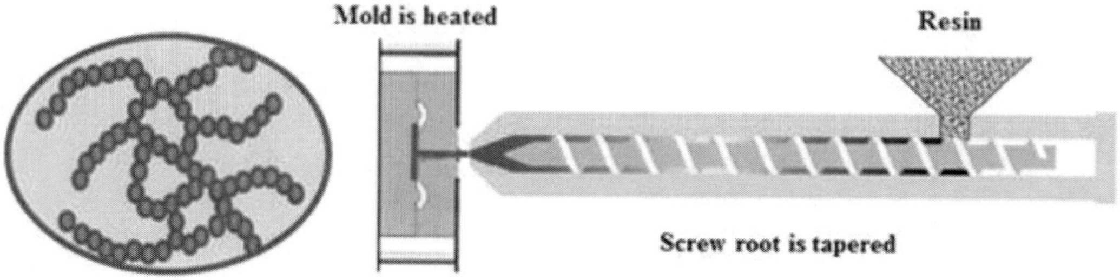

Figure 5. Structural arrangement of plant fibre cell.

Previous research indicates that improved strength retention properties of sisal/polyester composites both in humid and dry conditions were ascribed to treatments with silane, titanate, and zirconate coupling agents and N-substituted methacrylamide. As a result of these effects, the mechanical properties of the composites were high [94]. The development and mechanical behaviour of cementious fibre and powder was determined. The effect of 0% to 1% silane treatment on the bond characteristics of the fibre–matrix of ultra-high-performance concrete (UHPC) was examined. The results revealed that the incorporation of fibres and powders by sand substitution increased the bending and tensile strength by 32%, while the substitution of silica fume by the powder led to an increase of 37% in flexural and tensile strengths. Also, enhanced fibre–matrix interfacial properties were obtained due to bond characteristics of fibres in the UHPC matrix whereas the fibre pullouts were improved by the treatment using vinyl trimethoxy silane (VTMO) [95].

Earlier reseach shows flexural strength (97 MPa) and elastic modulus (6.5 GPa) were obtained when 30% fiber content was utilized in banana/ polyester composites [96]. Similarly, the curing studies on oil/palm fiber reinforced rubber composites were conducted. The results indicated that the presence of bonding agents prolonged the curing time whereas the mechanical properties were better due to improved interfacial bonding as evidenced by SEM studies [97]. Recently, the nanotubes were dispersed in dimethyl formamide, ethanol and acetone by sonication before treatment with epoxy. The resulting composite was cured at 50°C for 10min and then for 24hrs at room temperature. It was discovered that while the dispersion properties were improved, unprocessed traces of solvent had negative impact on the mechanical properties of the composites [98].

4.2. Thermoplastic Polymers in Composites

Thermoplastics deals with a moulding incorporating synthetic polymer fibres or resins as binders/matrices, which can be modified by the application of heat, as a consequence, offer many advantages over thermoset polymers. One of the advantages of thermoplastic matrix composites is their low processing costs. Another benefit is their design flexibility and ease of molding complex parts. Simple methods such as extrusion and injection molding are used for processing of these composites. In thermoplastics most of the work reported so far uses polymers such as polyethylene, polyvinyl chloride, polypropylene, polystyrene, and poly(vinyl chloride) [13, 99-101]. This is mainly because the processing temperature is restricted to low temperatures to avoid thermal degradation of the natural fibers. They consist of long, discrete molecules that melt to a viscous liquid at the processing temperature, typically 260 to 3710 °C, and, after forming, are cooled to an amorphous, semicrystalline, or crystalline solid. The degree of crystallinity has a strong effect on the final matrix properties.

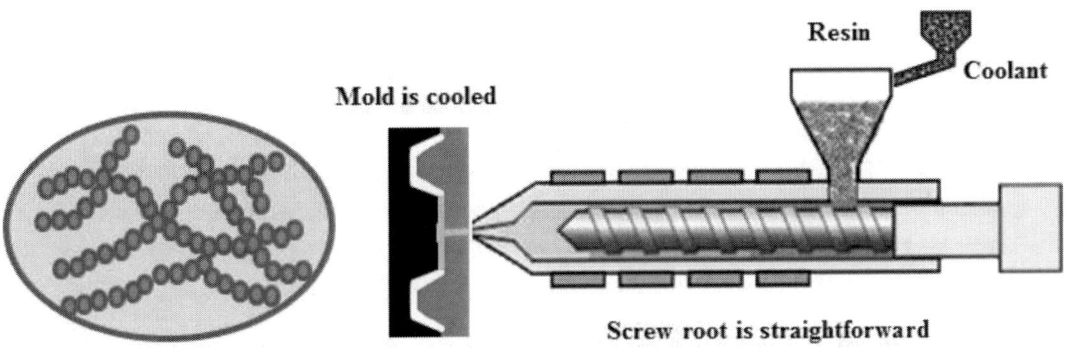

Figure 6. Structural arrangement of plant fibre cell.

Unlike the curing process of thermosetting resins, the processing of thermoplastics is reversible, and, by simply reheating to the process temperature, the resin can be formed into another shape if desired. Although thermoplastics are generally inferior to thermosets in high-temperature strength and chemical stability, they are more resistant to cracking and impact damage. Thus, making thermoplastic matrices attractive to high-volume industries such as the automobile sector. Currently, thermoplastics are used primarily with discontinuous fiber reinforcements such as chopped glass or carbon/ graphite. However, there is great potential for high-performance thermoplastics reinforced with continuous fibers.

In an interesting study, the mechanism of in-situ mechanical reinforcement of blends of two thermotropic liquid crystalline polymers (TLCPs), with brittle and ductile matrix materials was studied using injection molded technique. More-over, wide angle X-ray diffraction measurements, as well as determination of the fiber-moduli, revealed that during spinning not only a macroscopic elongation of the fibrils was achieved, but also a considerable molecular orientation within the TLCP domains [102]. Sodium hydroxide, isocyanate, and peroxide treatments were utilized to determine their influence on the properties of sisal/polyethylene composites. It was discovered that the enhancement in the properties was due to the bonding between the fibre and the polyethylene matrix [11]. Studies indicate that electron beam processed polypropylene/wood fibre composites were significantly enhanced than those of traditionally manufactured composites. This was attributed to creation of active sites both in the polymer and the fibres and thus interfacial interactions were encouraged for improved adhesion [103].

The mechanical properties of natural fibres (sisal, kenaf, hemp, jute and coir) reinforced polypropylene composites were studied using compression moulding via a film stacking method. Kenaf, hemp and sisal composites showed comparable tensile strength and modulus results but in impact properties hemp appears to out-perform kenaf. The tensile modulus, impact strength and the ultimate tensile stress of kenaf reinforced polypropylene composites were found to increase with increasing fibre weight fraction [104].

5. Applications in Automotive Industry

Increasing competition and innovation in automobile sector is enforcing advancement of the existing products and replacement of old products by new and cutting-edge materials for various components. These efforts are undertaken to ensure the comfort of the user while utilization of the safer and lighter materials are borne into these products. Being one of the user-friendly materials, natural-fibre-reinforced composites are established products in today's automotive industry with all major original equipment manufacturer's (OEM) having models with interior or even exterior trim parts made from flax, hemp, sisal or wood fibres formed with a thermoset or thermoplastic resin system. Nowadays, current thermoset resin systems feature epoxy, polyurethane, acrylic and phenolic binders, which are all derived from fossil resources. The automotive industry is widely viewed as being the industry in which the greatest volume of advanced composite materials could be used in the future [105-108]. Because the industry is mature and highly competitive, the principal motivation for introducing polymer matrix composites has economic benefits. Nevertheless, the automotive industry continues unabated to be interested in saving weight as it pursues the conflicting goals of larger automobiles and higher fuel efficiency.

In 1980, the first interior door panel using natural fibres, was manufactured by the Institute of Structural Mechanics in Germany. Subsequently in 2000, Audi came out with the A2 midrange vehicle which was the first mass produced car with an all aluminum body. Taking advantage of the properties of natural fibres such as lightweight and dimensional stability, a hybrid of flax/sisal reinforced polyurathanes was utilized to make door trim panels [109]. The same year saw Daimler Chrysler spending $1.5 billion in environmentally friendly production processes and products for enhanced materials. Succeeding years have seen significant growth in the overall usage of natural fibres, particularly, flax, kenaf. sisal and hemp due to their economic and functional properties [110-113].

Fundamentally, the properties such as high stiffness with light weigth per unit area makes natural fibres key materials for the automotive applications. Other attractive characteristics of cellulosic fibres are their ease of cycling, 30-40% lighter than glass fibre, good thermal and acoustic insulating properties. In the early years, the most predominantly users of components reinforced with natural fibres were Germany, with BMW, Audi and Daimler Chrysler as leading examples. Nowadays, basically all manufactures incorporate natural fibre composites in their vehicles [75].

Virtually, natural fibre reinforced composites used in automobile applications continues to gain momentum with typical applications ranging from door panels, rear-window shelves and car roofs, to sparewheel covers and noise absorber panels. Multidirectional flax/polypropylene composites were made on laboratory scale and on pilot scale in order to develop a car rear-window shelf that could compete with the conventional used ones. Different kinds of flax/polypropylene non-wovens were used as starting material and pressed under varying process parameters. The effects on the mechanical and physical properties of fibre reinforced composites were studied and their impact on the automobile sector is discussed below [3, 41, 42].

Strength	***Weaknesses***
▪ *Reduction cost, natural fibre is 25-50% cheaper than glass fibre* ▪ *Reduction in weight, 9kg of synthetic based panel performance is equivalent to 5kg natural based panel* ▪ *Safer crash behaviour in accident tests* ▪ *Good thermal insulation and acoustic properties*	▪ *Natural fibres are hydrophilic in nature* ▪ *Lack of proper adhesion between the fibre and the matrix* ▪ *Invariable properties due to source and processing techniques* ▪ *Low impact strength due fibre defects*
Threats	***Opportunities***
▪ *Inferior to glass due to lower inherent strengths and stiffness* ▪ *Lack of interfacial interactions due to hygroscopic character* ▪ *Variable quality*	▪ *Drivers of "green" composites due to safer and user-friendly characteristics* ▪ *Potential to produce bio-friendly, fuel efficient and low polluting vehicles.* ▪ *Cost and weight reduction for automotive industry*

In the earlier research, applications of natural fibre reinforced composites have been explored in the automotive sector [75, 108, 109]. For instance, Johnson Controls, Inc is using natural fibre reinforced polypropylene to manufacture dashboards and door panels for Daimler Chrysler [114]. In their manufacturing processes, Ford utilizes between 5 to13kg of natural fibres per vehicle whereas BMW processes about 8kg to over 13kg, taking into cognizance that the amounts include wool and cotton [115]. Reports indicate that researchers successfully developed a car prototype with all body panels made from jute fibre reinforced composites [116]. A hybrid of flax/sisal reinforced epoxy composite was used to design door panels of E-Class Mercedes-Benz and the significant weight reduction of 20% was obtained [117].

CONCLUSION

Increasing concerns and awareness of recent environment conservation regulations and the impact of human activity on the environment is currently the driving force for developing materials that are sustainable, more ecologically sound, cost effective and are purely produced from renewable resources. The possibility to use natural fibres as reinforcements for composite materials offers an adequate reply to these considerations. The use of natural fibres present a useful technique for developing natural fibre based composites, which are lightweight, biodegradable and can have properties that can be tailored for their specific application.The potential use of natural fibres as reinforcements is greatly hampered because of their hydrophilic nature. Even though these composites have some disadvantages, advantages such as combination of the useful properties of the two different materials, cheaper manufacturing cost, versatility, make them useful in various fields of engineering and technology. Moreover, their specific mechanical properties appear to be competitive when compared to the other conventional fibres. Using different mechanisms of processing, surface modifications and chemical treatments can improve the properties of composites for better performance. In the automotive sector, natural fibre reinforced composites have offered solutions to pressures for production of fuel efficient, low polluting and lighter vehicles.

References

[1] Stamboulis, A., Baillie, C.A., (2000). Environment durability of flax fibres and their composites based on polypropylene matrix. *Appl. Compos. Mater.* 7, 273.

[2] Kalia, S., Dufrense, A., Cherian, B.M., Kaith, B.S., Avérous, L., Njuguna, J., Nassiopoulos, E., (2011). Cellulose-based bio- and nanocomposites: A review. *Int. J. Polym. Sci.* 35. doi:10.1155/2011/837875.

[3] Mohanty, A.K., Misra, M., Drzal, L.T., Selke, S.E., Harte, B.R., Hinrichsen, G., (2005). Natural fibres, biopolymers, and biocomposites: An introduction. Natural fibres, biopolymers, and biocomposites. Mohanty, A.K., Misra, M., Drzal, L.T., eds., *CRC Press*, Boca Raton.

[4] Cantero, G., Arbelaiz, A., Llano-Ponte, R., Mondragon, I., (2003). Effects of fibre treatment on wettability and mechanical behavior of flax/polypropylene composites. *Compos. Sci. Technol.* 63, 1247.

[5] Kokta, B.V., Raj, R.G., Daneault C., (1989). Use of wood flour as filler in polypropylene: studies on mechanical properties. *Polymer-Plastics Technology and Engineering* 28, 247.doi: 10.1080/03602558908048598.

[6] Joseph, P.V., Joseph, K., Thomas, S., (1999). Effect of processing variables on the mechanical properties of sisal fibre reinforced polypropylene composites. *Compos. Sci. Technol.* 59, 1625.

[7] Felix, J.M., Gatenholm, P., (1991). The nature of adhesion in composites of modified cellulose fibers and polypropylene. *J. Appl. Polym. Sci.* 42, 609.

[8] Marcovich, N.E., Reboredo, M.M., Aranguren, M.I., (1998). Dependence of the mechanical properties of woodflour–polymer composites on the moisture content. *J. Appl. Polym. Sci.* 68, 2069.

[9] Mwaikambo, L.Y., Martuscelli, E., Avella, M., (2000). Kapok/cotton fabric–polypropylene composites. *Polym. Test.* 19, 905.

[10] Hornsby, P.R., Hinrichsen, E., Tarverdi, K., (1997). Preparation and properties of polypropylene composites reinforced with wheat and flax straw fibres. *J. Mater. Sci.* 32, 1009.

[11] Joseph, K., Thomas, S., Pavithran,C., (1996). Effect of chemical treatment on the tensile properties of short fibre-reinforced polyethylene composites. *Polymer* 37, 5139.

[12] Sydenstricker, T.H.D., Mochnaz, S., Amico, S.C., (2003). Pull-out and other evaluations in sisal-reinforced polyester biocomposites. *Polym. Test.* 22, 375. doi:10.1016/S0142-9418(02)00116-2.

[13] Saheb, D.N., Jog, J.P., (1999). Natural Fiber Polymer Composites: A Review. *Advances in Polym. Technol.* 18, 351.

[14] Rowell, R.M., Clemson, C.M., (1992). Proceedings of the 26th International Particleboard/Composites Symposium, WA 251.

[15] Bledzki, A.K., Reihmane, S., Gassan, J., (1996). Properties and modification methods for vegetable fibers for natural fiber composites. *J. Appl. Polym. Sci.* 59, 1329.

[16] Mohanty, A.K., Singh, B.C. (1987). Redox-initiated graft copolymerization onto modified jute fibers. *J. Appl. Polym. Sci.* 34, 1325.

[17] Muzumdar, P., Sanyal, S., Dasgupta, B., Shaw, S.C., Guha, R., (1994). *Indian J. Fibre Techn.* 19, 286.

[18] Samal, R.K., Sahoo, P.K., Samantaray, H.S., (1986). *Journal of Macromolecular Science-Reviews in Macromolecular Chemistry and Physics* C26, 81.

[19] Mani, P., Satyanarayan., K.G., (1990). Effects of the surface treatments of lignocellulosic fibers on their debonding stress. *J. Adhes. Sci. Technol.* 4, 17.

[20] Arbelaiz, A., Fernández, B., Ramos, J.A., Mondragon, I., (2006). Thermal and crystallization studies of short flax fibre reinforced polypropylene matrix composites: Effect of treatments. *Thermochimica Acta* 440, 111.

[21] Monteiro, S.N., Calado, V., Rodriguez, R.J.S., Margem, F. M., (2012). Thermogravimetric stability of polymer composites reinforced with less common lignocellulosic fibers – an overview. *J. Mater. Res. Technol.* 1, 117.

[22] Dicker, M.P.M., Duckworth, P.F., Baker, A.B., Francois, G., Hazzard, M.K., Weaver, P.M. (2014). Green composites: A review of material attributes and complementary applications. *Composites: Part A : Applied Science and Manufacturing* 56, 280.

[23] Zini, E., Scandola, M., (2011). Green composites: an overview. *Polym. Compos.* 32, 1905.

[24] Bos, H.L., Mussig, J., van den Oever, M.J.A., (2006). Mechanical properties of short-flax-fibre reinforced compounds. *Composites: Part A* 37, 1591.

[25] Wirawan, R., Zainudin, E.S., Sapuan, S.M., (2009). Mechanical properties of natural fibre reinforced PVC composites: A Review. *Sains Malaysiana* 38, 531.

[26] Olesen, P., Plackett, D., (1999). Perspective on the performance of natural plant fibres. Natural Fibres Performance Forum, May 27-28. Copenhagen.

[27] Srinivasa C.V., Bharath K.N., (2011). Impact and hardness properties of areca fibre-epoxy reinforced composites. *J. Mater. Environ. Sci.* 2, 351.

[28] Mishra, V., Biswas, S., (2013). Physical and mechanical properties of bi-directional jute fiber epoxy composites. *Procedia Engineering* 51, 561.

[29] Herrera-Franco P.J., Valadez-González, A., (2004). Mechanical properties of continuous natural fibre-reinforced polymer composites. *Composites Part A*: *Applied Science and Manufacturing* 35, 339.

[30] Hon, D.N.S. (1991). Wood and Cellulosic Chemistry. Marcel Dekker, Inc. 665.
[31] Bledski, A.K., Gassan, J., (1999). Composites reinforced with cellulose based fibres. *Prog. Polym. Sci.* 24:221.
[32] Biagiotti, J., Puglia, D., Kenny, J.M., (2004). A review on natural fibre-based composites-Part I. *J. Natural Fibers* 1, 37.
[33] Gassan, J., Bledzki, A.K., (1996). Modification methods on nature fibers and their influence on the properties of the composites. *J. Eng. Appl. Sci.* 2, 2552.
[34] http://csls-text.c.u-tokyo.ac.jp/large_fig/c_fig11_2.html.
[35] Idicula, M., Boudenne, A., Umadevi, L., Ibos, L., Candau, Y., Thomas, S., (2006). Thermophysical properties of natural fibre reinforced polyester composites. *Comp. Sci. Technol.* 66, 2719.
[36] Woodhams R.T., Thomas G., Rodgers D.K., (1984). Wood fibers reinforcing fillers for polyolefins. *Polym. Eng. Sci.* 24, 1166.
[37] Brouwer, W.D., (2000). Natural fibre composites: Where can flax compete with glass? *Sampe Journal* 36, 18.
[38] Baley, C., (2002). Analysis of the flax fibres tensile behaviour and analysis of the tensile stiffness increase. *Composites Part A: Applied Science and Manufacturing (Incorporating Composites and Composites Manufacturing)* 33, 939.
[39] Gassan, J., Chate, A., Bledzki, A.K., (2001). Calculation of elastic properties of natural fibers. *J. Mater. Sci.* 36, 3715.
[40] Bénézet, J., Stanojlovic-Davidovic, A., Bergeret, A., Ferry, L., Crespy, A., (2012). Mechanical and physical properties of expanded starch, reinforced by natural fibres. *Ind. Crops Prod.* 37, 435.
[41] Van de Velde, K., Kiekens P., (2003). Effect of material and process parameters on the mechanical properties of unidirectional and multidirectional flax/polypropylene composites. *Composite Structures* 62, 443.
[42] Jawaid, M., Khalil. H.P.S.A (2011). Cellulosic/synthetic fibre reinforced polymer hybrid composites: A review. *Carbohydr. Polym.* 86, 1.
[43] Rowell, R.M., Sanadi, A.R., Caulfiel, D.F., Jacobson. R.E., (1997). Utilization of natural fibers in plastic composites: Problems and opportunities. In: Lignocellulosic-Plastic Composites. Ed. by A.L. Leao, F.X. Carvalho and E. Frollini. Sao Paulo, Brazil: Universidade de Sao Paulo Press: 23-52.
[44] Kenny, J.M., (2001). Natural fibre composites in the European automotive industry. Proceedings of the Sixth International Conference on Woodfibre-Plastic Composites, May 15-16, Madison, Wisconsin, 9-12
[45] Peijs, T., Jacobs, M.J.N., Lemstra, P.J., (2000). High performance polyethylene fibres in Chou, T.W., Kelly, A., Zweben, C., (eds) Comprehensive Composites, vol.1 Oxford: Elsevier Science Publishers Ltd, 263-303.
[46] Dubois, P., Alexandre, M., Hindryckx, F., Jerome, R., (1998). Polyolefin based composites by polymerization-filling technique. *Polymer Reviews* 38, 511.
[47] Jones, R. M. (1975). Mechanics of composite materials. New York: McGraw-Hill.
[48] Zhang, Y., Rodrigue, D., Aiȉ^t-Kadi, A. (2004). Polyethylene–kevlar composite foams II: Mechanical properties. *Cellular Polymers* 23, 61.
[49] Rials, T. G., Wolcott, M. P., Nassar, J. M., (2001). Interfacial contributions in lignocellulosic fiber-reinforced polyurethane composites. *J. Appl. Polym. Sci.* 80, 546.

[50] Gatenholm, P., Bertilsson, H., Mathiasson, A., (1993). Effect of chemical composition of interphase on dispersion of cellulose fibers in polymers. I. PVC-coated cellulose in polystyrene. *J. Appl. Polym. Sci.* 49, 197.

[51] Peltola, H., Pääkkönen, E., Jetsu, P., Heinemann. S., (2014). Wood based PLA and PP composites: Effect of fibre type and matrix polymer on fibre morphology, dispersion and composite properties. *Composites: Part A* 61, 13.

[52] Koutsomitopoulou, A.F., Bénézet, J.C., Bergeret, A., Papanicolaou G.C., (2014). Preparation and characterization of olive pit powder as a filler to PLA-matrix bio-composites. *Powder Technology* 255, 10.

[53] Takase, S., Shiraishi, N. J., (1989). Studies on composites from wood and propylene. II. *J. Appl. Polym. Sci.* 37, 645.

[54] George, J., Sreekala, M.S., Thomas, S., (2001). A review on interface modification and characterization of natural fiber reinforced plastic composites. *Polym. Eng. Sci.* 41, 1471.

[55] Belgacem, M.N., Gandini, A., (2005). The surface modification of cellulose fibres for use as reinforcing elements in composite materials. *Composites Interfaces* 12, 41.

[56] Karlsson, J.O., Gatenholm, P., Blachot, J.F., Peguy, A., (1996). Improvement of adhesion between polyethylene and regenerated cellulose fibers by surface fibrillation. *Polymer Composites* 7, 300.

[57] Joseph, K., Thomas, S., Pavithran, C. and Brahmakumar, M., (1993). Tensile properties of short sisal fiber-reinforced polyethylene composites. *J. Appl. Polym. Sci.* 47, 1731.

[58] Harish, S., Michael, P.S., Bensely, A., Lal, D.M, Rajadurai, A., (2008). Mechanical property evaluation of natural fiber coir composite. *Materials Characterization* 60, 44.

[59] Wang, W., Huang, G., (2009). Characterization and utilization of natural coconut fibers composites. *Materials & Design* 30, 2741.

[60] Jústiz-Smith, N.G., Virgo, G.J., Buchanan, V.E., (2008). Potential of Jamaican banana, coconut coir and bagasse fibers as composite materials. *Material Characterization* 59, 1273.

[61] Bilba, K., Marie-Ange Arsene, M.-A., Ouensanga, A., (2007). Study of banana and coconut fiber botanical composition, thermal degradation and textural observations. *Bioresour. Technol.* 98, 58.

[62] Dittenber, D.B., GangaRao, H.V. (2012). Critical review of recent publications on use of natural composites in infrastructure. *Composites Part A: Applied Science and Manufacturing* 43, 1419.

[63] Mohanty, A.K., Misra, M., Drzal, L.T., (2001). Surface modifications of natural fibers and performance of the resulting biocomposites: an overview. *Composite Interfaces* 8, 313.

[64] Masoodi, R., Pillai, K.M., (2012). A study on moisture absorption and swelling in bio-based jute-epoxy composites. *J. Reinf. Plastics Compos.* 31, 285.

[65] Prolongo, S.G., Gude, M.R., Urena, A., (2012). Water uptake of epoxy composites reinforced with carbon nanofillers. *Composites Part A: Applied Science and Manufacturing* 43, 2169.

[66] Avella, M., Bozzi, C., dell'Erba, R., Focher, B., Marzetti, A., Martuscelli, E., (1995). Steam-exploded wheat straw fibers as reinforcing material for polypropylene-based composites. Characterization and properties. *Die Angewandte Makromolekulare Chemie* 233, 149.

[67] Thwe, M.M., Liao, K., (2003). Durability of bamboo-glass fiber reinforced polymer matrix hybrid composites. *Comp. Sci. Technol.* 63, 375.

[68] George, J., Bhagawan, S.S., Thomas, S., (1998). Effects of environment on the properties of low-density polyethylene composites reinforced with pineapple-leaf fibre. *Comp. Sci. Technol.* 58, 1471.

[69] Rana, A.K., Mandal, A., Mitra, B.C., Jacobson, R., Rowell, R., Baneerjee, A.N., (1998). Short jute fiber-reinforced polypropylene composites: effect of compatibilizer. *J. Appl. Polym. Sci.* 69, 329.

[70] Zhuang, R.-C.; Liu, J.-W.; Plonka, R.; Huang, Y.-X., Mäder, E., (2011). $NaBF_4$ as a hygrothermal durability enhancer for glass fibre reinforced polypropylene composites. *Comp. Sci. Technol.* 71, 1461.

[71] Symington, M.C., Banks, W.M., West, O.D., (2009). Tensile testing of cellulose based natural fibers for structural composite applications. *J. Compos. Mater.* 43, 1083.

[72] Wielage, B., Lampke, Th., Marx, G., Nestler, K., Starke, D., (1999). Thermogravimetric and differential scanning calorimetric analysis of natural fires and polypropylene. *Thermochimica Acta* 337, 169.

[73] Rana, A.K., Basak, R.K., Mitra, B.C., Lawther, M., Banerjee, A.N., (1998). Studies of acetylation of jute using simplified procedure and its characterization. *J. Appl. Polym. Sci.* 64, 1517.

[74] Albano, C., González, J., Ichazo, M., Kaiser, D., (1999). Thermal stability of blends of polyolefins and sisal fibre. *Polym. Degrad. Stab.* 66, 179.

[75] Mohanty, A.K., Khan, M.A., Sahoo, S., Hinrichsen, G., (2000). Effect of chemical modification on the performance of biodegradable jute yarn-Biopol® composites. *J. Mater. Sci.* 35, 2589.

[76] Li, X.H., Meng, Y.Z., Wang. S.J., Rajulu. A.V., Tjong, S.C., (2004). Completely biodegradable composites of poly(propylene carbonate) and short, lignocellulose fiber *Hildegardia populifolia*. *Journal of Polymer Science Part B: Polymer Physics* 42, 666.

[77] Agarwal, R., Saxena, N.S., Sharma, K.B., Thomas, S., Sreekala, M.S., (2000). Effect of different treatments on the thermal behavior of reinforced phenol-formaldehyde polymer composites. *J. Appl. Polym. Sci.* 78, 603.

[78] Netravali, A.N., (2007). Fiber/Resin Interface Modification in "Green" Composites. In: Bhatta-charyya SFaD, Editor. Handbook of Engineering Biopolymers Homopolymers, Blends and Composites. Munich: Hanser, 847-68.

[79] Lu, J.Z., Negulescu, I.I., Wu, Q., (2005). Maleated wood-fibre/high-density-polyethylene compo- sites: coupling mechanisms and interfacial characterization. *Composite Interfaces* 12, 125.

[80] Pracella, M., Chionna, D., Anguillesi, I., Kulinski, Z., Piorkowska, E., (2006). Functionalization, compatibilization and properties of polypropylene composites with hemp fibres. *Comp. Sci. Technol.* 66, 2218.

[81] De Bruijn J.C.M., (2000). Natural fibre mat thermoplastic products from a processor's point of view. *Applied Composite Materials* 7, 415.

[82] Karmaker, A.C., Youngquist, J.A., (1996). Injection molding of polypropylene reinforced with short jute fibers. *J. Appl. Polym. Sci.* 62, 1147.

[83] Huda, M.S., Drzal, L.T., Misra, M., Mohanty, A.K., (2006). Wood-fibre reinforced poly(lactic acid) composites: evaluation of the physicomechanical and morphological properties. *J. Appl. Polym. Sci.* 102, 4856.

[84] Xie, Y., Hill C.A.S., Xiao, Z., Militz, H., Mai, C., (2010). Silane coupling agents used for natural fiber/polymer composites: a review. *Composites Part A: Applied Science and Manufacturing* 41, 806.

[85] Cech, V., Prikryl, R., Balkova, R., Vanek, J., Grycova, A., (2003). The influence of surface modifications of glass on glass fiber/polyester interphase properties. *J. Adhes. Sci. Technol.* 17, 1299.

[86] Li, R., Ye, L., Mai, Y.-W., (1997). Application of plasma technologies in fibre-reinforced polymer composites: a review of recent developments. *Composites Part A: Applied Science and Manufacturing* 28, 73.

[87] Kalia, S., Kaith, B.S., Kaur, I., (2009). Pretreatments of natural fibers and their application as reinforcing material in polymer composites-a review. *Polym. Eng. Sci.* 49, 1253.

[88] Xie, Y., Hill, C.A.S., Xiao, Z., Militz, H., Mai, C., (2010). Silane coupling agents used for natural fiber/polymer composites: a review. *Composites Part A: Applied Science and Manufacturing* 41, 806.

[89] Folkes, M.J., Short Fibre Reinforced Thermoplastics, John Wiley & Sons, New York, 1982, 5148.

[90] Lightsey, G.L., (1983). Organic Fillers for Thermoplastics. *Polym. Sci. Technol.* 17, Plenum Press, New York.

[91] Bisanda, E.T.N., Ansell, M.P., (1991). The effect of silane treatment on the mechanical and physical properties of sisal-epoxy composites. *Comp. Sci. Technol.* 41, 165.

[92] http://www.whatischemistry.unina.it/en/thermosetting.html

[93] http://www.daltonelectric.com/News/TGID-080511.htm

[94] Singh, B., Gupta, M., Verma, A., (1996). Influence of fiber surface treatment on the properties of sisal-polyester composites. *Polym. Compos.* 17, 910.

[95] Sebaibi, N., Benzerzour, M., Abriak, N. E., Binetruy, C., (2012). Mechanical properties of concrete-reinforced fibres and powders with crushed thermoset composites: The influence of fibre/matrix interaction. *Construction and Building Materials* 29, 332.

[96] Zhu, W.H., Tobias, B.C., Coutts, R.S.P., (1995). Banana fibre strands reinforced polyester composites. *J. Mater. Sci. Lett.* 14, 508.

[97] Ismail, H.R., Rosnah, N. (1997). Curing characteristics and mechanical properties of short oil palm fibre reinforced rubber composites. *Polymer* 38, 4059.

[98] Lau, K-t., Lu, M., Chun-ki, L., Cheung H-y., Sheng, F-L., Li, H-L., (2005). Thermal and mechanical properties of single-walled carbon nanotube bundle-reinforced epoxy nanocomposites: The role of solvent for nanotube dispersion. *Comp. Sci. Nanotechnol.* 65, 719.

[99] Beshay, A.D., Kokta, B.V., Maldas, D., Daneault, C., (1985). Use of wood fibers in thermoplastic composites: ii. polyethylene. *Polym. Compos.* 6, 261.

[100] Maiti, S.N., Singh, K., (1986). Influence of wood flour on the mechanical properties of Polyethylene. *J. Appl. Polym. Sci.* 32, 4285.

[101] Raj, R. G., Kokta, B. V., Maldas, D., Daneault, C. (1989). Use of Wood Fibers in Thermoplastics: VII. The Effect of Coupling Agents in Polyethylene Wood Fiber Composites. *J. Appl. Polym. Sci.* 37, 1089.

[102] Crevecoeur, G., Groeninckx, G., (1990). Morphology and mechanical properties of thermoplastic composites containing a thermotropic liquid crystalline polymer. *Polym. Eng. Sci.* 30, 532.

[103] Czvikovszky, T., (1994). Electron-beam processing of wood fibre reinforced polypropylene. *Periodica Polytechnica, Mechanical Engineering.& thinsp* 38, 209.
[104] Wambua, P., Ivens, J., Verpoest, I., (2003). Natural fibres: can they replace glass in fibre reinforced plastics? *Comp. Sci. Technol.* 63, 1259.
[105] Larbig, H., Scherzer, H., Dahlke, B., Poltrock, R., (1998). Natural-fiber reinforced foams based on renewable resources for automotive interior applications. *Journal of Cellular Plastics* 34, 361.
[106] Marsh, G., (2003). Next step for automotive materials. *Materials Today* 6, 36.
[107] Holbery, J., Houston, D., (2006). Fiber-reinforced polymer composites in automotive applications. *Journal of the Minerals, Metals and Materials Society* 58, 80.
[108] Zah, R., Hischier, R., Leão, A.L., Brown, I., (2007). Curaua fibers in automobile industry – A sustainability assessment. *Journal of Cleaner Production* 15, 1032.
[109] Suddell, B.C., Evans, W.J., (2005). Chapter 7 Natural Fiber Composites in Automative Applications, In: Natural Fibers, Biopolymers and Biocomposites, Mohanty, (Ed.), Taylor &Francis Group, Florida, pp. 237–297, ISBN 084931741X.
[110] Bledzki, A.K., Mamun, A.A., Faruk, O., (2007). Abaca fibre reinforced polypropylene composites and comparison with jute and flax fibre polypropylene composites. *eXPRESS Polymer Letters* 1, 755.
[111] Puglia, D., Biagiotti, J., Kenny J.M., (2004). A review on natural fiber based composites Part II: Application of natural reinforcements in composite materials. *Journal of Natural Fibers* 1, 23.
[112] Kumar, R., Obrai, S., Sharma, A., (2011). Chemical modifications of natural fiber for composite material. *Pelagia Research Library Der Chemica Sinica*, 2, 219.
[113] Jacob, M., Thomas, S., Varughese, K.T., (2004). Mechanical properties of sisal/oil palm hybrid fibres reinforced natural rubber composites. *Composites Science and Technology* 64, 955.
[114] 'Green' door trim panels use polypropylene and natural fibres, 2000. *Plastic Technology* 46, 27.
[115] Taylor, A., Case study on fibres in composite materials, e.g. hemp in automotive applications, 4th meeting of government-Industry forum on Nonfood uses of Crops, GIFNFC 4/4 Fibres in Composite Material. Jan. 22 2002, DTI Conference Centre, London.
[116] Al-Qureshi, H.A., (2001). Automobile leaf springs from composite materials. *Journal of Materials Processing Technology* 118, 58.
[117] Schuh, K.L., Cunningham, D.J., (2004). Rhizome and the mind: describing the metaphor. *Semiotica* 149, 325.

In: Polymer-Matrix Composites
Editor: Rakesh Kumar
ISBN: 978-1-61122-005-6

Chapter 11

REVIEW ON EROSIVE WEAR PERFORMANCE OF NATURAL FIBER REINFORCED THERMOSET POLYMER-MATRIX COMPOSITES

Sandeep Kumar*
Centre for Polymer Science and Engineering,
Indian Institute of Technology Delhi, Hauz Khas,
New Delhi, India

ABSTRACT

Recently, natural fiber reinforced polymer-matrix composites have been fabricated by manual techniques (open mould / hand-lay-up) for the evaluation of erosive wear performance as a function of various operating and material variables such as impingement angle, impact velocity, stand-off distance, erodent size, erodent flow rate, reinforcement content as well as their mechanical and thermal response. Various types of natural fibers such as bagasse, bamboo, chicken feather, coir, jute, lantana–camera etc. have been used for the reinforcement purpose in polymer-matrix composites to improve their erosive wear properties. It is observed that polymer composites reinforced with bamboo fibers and jute fibers show maximum erosion resistance as compared to other reported natural fibers reinforced composites. The design of experiments approach based on Taguchi method have been generally used to obtain an optimal set of operating and material variables to minimize the erosive wear rate of composites. The morphological characterization of the eroded surface by using scanning electron microscopy (SEM) has been accepted to understand the possible mechanistic attributes responsible for erosive wear in air medium (solid particle impingement). Further, it can be deciphered that the combinations of natural and synthetic (traditional) fiber as reinforcement phase in polymer-matrix composites improve resistance to erosive wear against natural fiber reinforced composites.

Keywords: Erosive wear, Composites, Natural fiber

---

* Corresponding Author address: Cenre for Polymer Science and Engineering, IIT Delhi, New Delhi-16, India. Email: *sanju_chemiit@yahoo.co.in*

1. Introduction

In the recent years much attention and curiosity has been brought into the development of recyclable, environmentally sustainable and bio-compatible polymer composite materials. An additional pressure is being imposed continuously on the manufacturers of polymeric materials and their end/finish products in several countries due to environmental legislation as well as consumer demands. Materials manufacturers have to consider the environmental impact of their end-use products at the entire stages of their life cycle, including recycling and ultimate disposal. The extensive and significant interests in the field of development of recyclable and biodegradable or partially biodegradable polymeric composite materials have been recently developed. Therefore, research or investigation related to polymer composites containing natural fibers has now attracted much attention. The virgin polymers based on petroleum products and three dimensional cross-linked thermoset polymers disclose a harmful environmental potential due to poor biodegradation. Therefore, alternative materials that are compatible with the environment are required for structural, packaging and tribological applications instead of virgin or cross-linked thermoset polymers. The alternative materials can be prepared by incorporation of natural fiber in polymer matrix without compromising the physical, mechanical and thermal performance. Also natural fibers absorb CO_2 during their growth and they help to balance carbon-content in the environment. The additional advantage of natural fiber is that their supply can be unlimited and they are much easier to handle as compared to traditional glass fiber and carbon fibre.

The variation in mechanical properties of natural fibres depends on their growth conditions and plants or animals part from which fiber is extracted. It is difficult to predict the mechanical properties of natural fibre reinforced composites due to variety of the extraction process, chemical composition and shape of fibre, fibre strength and ability to adhere with polymer matrix.

"Wear" is related to mechanical, chemical and thermal interactions between two dynamically interacting mating solid surfaces. That may lead to progressive loss, damage, deformation, alteration in shape, size and geometry of the surface and material as a whole. Wear may be broadly referred as a system response. Among the various form of wear, viz, abrasive, adhesive, fatigue, fretting, reciprocating, corrosion and erosion, erosive wear process poses specific challenges with respect to applications, materials and operational environments [1-2]. Erosion wear can be defined as a loss of materials due to repeated impact of erodent with the eroding materials surface in gas (solid particle erosion) or liquid (slurry erosion) medium. Erosion wear is a widely encountered mechanism in industry perspectives, e.g. the erosive wear in piping and pumping equipment associated with the movement of slurries, helicopter rotor blades, high speed vehicles, wind shields operating in desert environments, radomes and surfing boats, desert structure and housing. Typically, the performance requirements of composites in harsh/extreme and dusty environments are very complex. In this regard, solid particle erosion resistant polymeric composites used in several industrial and marine applications may be of high relevance. Solid particle erosion manifests itself in many facets such as thinning of components, surface degradation, surface roughening, macroscopic scooping appearance, which may eventually lead to reduction in functional life of structures. Hence, solid particle erosion has been considered as a serious problem being responsible for many failures in engineering applications. The influence of various experimental variables such as impact velocity, impingement angle (α), erodent type, erodent shape, erodent hardness and erodent size as well as target material properties such as strength, crystallinity, cross link density, ductility, reinforcement content and arrangement on solid particle erosion of polymers and their composites is comprehensively reviewed [3-5]. The rate of erosive wear is dependent upon a multitude of factors; broadly more than twenty. The characteristics of the erodent such as their hardness, shape, impingement angle, and impact velocity are the primary factors along with the properties of the surface being eroded. In fiber-reinforced polymer composites, the major factors that influence the erosion behaviour are fiber or filler content [6-9], fiber type [10-13], fiber orientation [10, 14-19] and fabrication technique of composites [11, 16, 20-24] as well as extent of interfacial interaction between fiber and matrix [10] and intrinsic materials properties [20, 25-26].

It is well known that the mechanical and thermal properties of natural fiber reinforced polymer composites are poor as compared to traditional fiber reinforced polymer composites. Thus, mechanical and thermal properties of natural fiber reinforced polymer composites are not the part of discussion here. On the other hand, the operational factors such as impingement angle [3-26], impact velocity [6-14], test temperature [14, 24], mass flow rate [15, 21, 27-28], erodent type and erodent size [29-31], erodent density [32] and stand-off distance [33] control the erosion rate of composites. It is also found that the tensile strength has no influence on the erosion resistance of materials [3] but it is required for structural applications. The ductile mode of erosion failure is characterized by maximum erosion rate at low impingement angle, typically $15° < \alpha < 30°$. On the other hand, if maximum erosion rate occurs at normal impact ($\alpha = 90°$), the erosion mode behavior of material is brittle. Fiber reinforced polymer composites generally exhibit semi-ductile mode of erosion behaviour which is characterized by maximum erosion rate at an intermediate impingement angles, typically $45° < \alpha < 60°$ [6-19]. However, the above classification is not absolute as the erosion behavior of materials strongly depends upon the experimental conditions and composition of target materials.

The present review focuses on the progress of natural fiber in the development of polymer composites to maximize erosive wear resistance and an overview on the problem faced by solid particle erosion of polymer matrix composites. We also discuss in this chapter how erosion wear is a multi-parametric problem.

2. Types of Commonly Natural Fibers and Thermoset Resins

Natural fibers are much better than traditional fibers (such as glass fiber and carbon fiber) because of their low cost, combustibility, low density, low coefficient of thermal expansion (CTE), high specific strength, renewability, non-abrasivity, non-toxicity, defrostability, abundance, and ultimately biodegradability. Natural fiber with biodegradation property can be categorized on the basis of their original source as:

- Fibers obtained from plant/vegetable (e.g., sisal, jute, bagasse, coir dust etc.)
- Fibers derived from animal species (e.g., sheep wool, chicken feather etc.)
- Fibers from bird / aqueous species (e.g., fish (Lobea Rohita) flakes)

The composition of natural fiber includes cellulose, hemicelluloses, lignin, pentasans, waxes and ashes etc. [34]. The mechanical properties in the natural fibers are mainly provided by the cellulose content [34]. The varieties of natural fibers are available and a continuous discovery of new natural fiber is still proceeding in this field. Natural fibers can be obtained from leaf or stem and provide different mechanical and physical properties. Table 1 shows the mechanical properties of commonly used natural fibers for composite fabrication. These fibers have poor strength compared to conventional glass fiber (tensile strength~3400MPa) whereas poor elastic modulus compared to conventional carbon fiber (elastic modulus~240GPa).

Table 1. Properties of commonly used natural fibers for composites fabrication

Natural fiber	Density (gm/cm^3)	Tensile Strength (MPa)	Elastic Modulus (GPa)	Elongation (%)	Ref.
Bamboo	0.6-1.16	140-610	35.9-46	1.3	[34-36]
Sugarcane	1.25	170-290	15-19	-	[34, 37]
Kenaf	1.32-1.45	260-930	53	-	[34, 38-39]
Banana	1.35	54-754	7-20	1.3-3	[34-35, 40]
Palm oil	1.45	200-250	3.2	25	[34, 41]
Cotton	1.5-1.6	400	5.512.6	7.0-8.0	[42]
Jute	1.3	393-773	26.5	1.5-1.8	[42-43]
Flax	1.5	500-1500	27.6	2.7-3.2	[42-43]
Hemp	1.47	550-900	70	2-40	[42, 36]
Ramie	N/A	400–938	61.4–128	3.6-3.8	[42]
Sisal	1.5	511–635	9.4–22	2.0-7	[42-44]
Coir	1.2	175-593	4.0–6.0	30	[42, 45]
Softwood kraft pulp	1.5	1000	40	4.4	[42]
Pineapple	2.5	170	72	2.4	[43, 46]
Date palm frond (DPF)		4.28-135.04	0.23-1.15	0.69-3.39	[41]
Agave Americana	1.20	5.69	-	7.07	[47]

Thermoset resins are usually viscous liquids or low melting point solids in their initial form and are used to produce finished goods by curing reactions, which takes place by the use of a catalyst/accelerator/cross-linking agent and heat/pressure/vacuum or a combination of these. Once cured, this cross-linked structure cannot be reshaped after applying heat and pressure. The primary functions of the cured resin matrix in the thermoset polymer composites are given as:

1. To transfer stress between the reinforcing fibers.
2. To hold the fibers together and give them shape as per design consideration.
3. To protect the fibers from external mechanical and environmental damage.

The most common thermosetting resins used in the composites manufacturing industry are unsaturated polyesters, epoxies, vinyl esters and phenolics and their characteristics are presented in the Table 2 [48-50].

Table 2. Characteristics of commonly thermoset resin used for composite fabrication [45-47]

Resin	Characteristics
Polyester	Low cost
	Extreme processing versatility
	Long history of performance
	Emit odor
Vinyl ester	Similar to polyester with better handling & faster cure
	Excellent mechanical (Tensile Strength & Stiffness) & fatigue properties
	Excellent water, corrosion and chemical resistance
Epoxy	Excellent mechanical properties
	Good fatigue resistance
	Low shrinkage
	Good heat, UV and corrosive liquid resistance
	Emit little odor
Phenolics	Excellent fire retardancy with low smoke & toxicity emissions
	High Strength at high temperatures
	Excellent thermal insulation & sound damping properties
	High creep & temperature resistance
Polyurethane	Superior toughness & abrasion resistance
	Good impact resistance
	Rapid curing
	Good surface quality

On the basis of fiber and resin characteristics or properties, one can use the combinations of these two types of raw materials to produce composite materials for required specific applications. A range of processing techniques for thermoset resins is available to achieve the desired finished product. Further, these resins can be modified by means of formulations or chemically tailored to provide process compatibility with desired properties. The properties of typical thermoset resin used for fabrication of natural fiber reinforced composites are given in Table 3.

Table 3. Properties of typical thermoset polymers resin (Reprinted from Ref. [42] Ku, H., Wang, H., Pattarachaiyakoop, N., Trada, M., (2011). Composite Part B 42, 856)

Property	Polyester resin	Vinyl ester resin	Epoxy
Density (g/cm^3)	1.2–1.5	1.2–1.4	1.1–1.4
Elastic modulus (GPa)	2–4.5	3.1–3.8	3–6
Tensile strength (MPa)	40–90	69–83	35–100
Elongation (%)	2	4–7	1–6
Cure shrinkage (%)	4–8	N/A	1–2
Water absorption (24 h@20°C)	0.1–0.3	0.1	0.1–0.4

There are various problems in using natural fibers as a reinforcement phase in polymer matrix composites. Some of them are listed below:

- Poor dimensional stability due to water absorption
- Poor fire resistance
- Enormous variability
- Non-uniformity of diameter
- Lack of fibre-matrix adhesion
- Variety of treatment of improving interfacial strength
- Low modulus of elasticity
- Poor strength
- Difficulty of getting single filament

These disadvantages of natural fiber generate variations in the results of erosive wear performance of natural fiber reinforced polymer matrix composites. It is expected that the erosion rate would be faster if aging of these composites is fast with the external environment.

3. Review on Erosive Wear Performance of Natural Fiber Reinforced Composites

With regard to the tribological applications, natural fiber reinforced thermoset polymer matrix composites encounter different types of wears in realistic situations. The erosion wear process is a combinations of abrasive (sliding), fatigue, adhesive and corrosion wear.

Mostly abrasive wear studies have been done by using pin-on-disc wear tester. The epoxy based composites reinforced with baggase fiber (obtained from the waste sugarcane) showed better abrasion resistance than that of lantana-camara fiber (obtained from flowering shrubs) reinforced epoxy composites [51-52]. The polyester matrix based hybrid composites reinforced with combination of pine-bark plant fiber and cement-kiln-dust showed improvement in the sliding wear behavior against the composites reinforced with only pine-bark plant fibers [53]. Borassus fruit fiber reinforced epoxy composites with fiber length of 5mm revealed good abrasion resistance [54]. The natural fiber reinforced polymer composites, which show good resistant to abrasive, adhesive, fatigue and corrosion wear, can also be expected to give good results for erosion wear.

Though a huge amount of research work is reported on the erosion wear studies of polymer matrix composites containing traditional fiber such as glass fiber and carbon fiber, however, there are few reports available on solid particle erosion wear studies of composites containing natural fiber. The dimensionless erosion rate is the ratio of the mass loss of the eroded materials to the mass of erodent causing erosion of material. The overview of erosive wear performance of natural fiber reinforced composites at room temperature is presented in Table 4. It is known that impingement angle, impact velocity, erodent size, erodent feed rate, fiber content, type of polymer matrix, fiber/matrix interface are the major parameters for evaluation of erosive wear performance.

The impingement angle (α) is defined as the angle between the eroded surface and the trajectory of the erodent immediately before impact. If maximum erosion rate is obtained at low impingement angle ($<30^o$), it is known as ductile mode of erosion wear. If maximum erosion wear occur at normal impact (i.e., impingement angle of 90°), then it is brittle mode of erosion wear.

Table 4. Synopsis of the erosive wear performance of natural fiber reinforced composites at room temperature (RT) with silica sand as an erodent

Matrix	Fiber (Type, content)	Erodent size, shape & feed rate	Angle (°)	Velocity (m/s)	Peak erosion Angle (°)	Key issues addressed	Ref.
Epoxy	Roving bidirectional bamboo mat (10, 20, 30, 40 wt.-%) (Dia=2.5mm)	(125-300μm), irregular square pyramidal, (4.7 g/min)	15, 30, 60, 90	30, 52, 88	75 (relatively brittle failure)	Effect of impingement angle and erodent velocity	[55]
Epoxy	Bagasse fiber (10, 15, 20 wt.-%), L*=10mm, Dia=1mm)	(150-250μm), irregular shape, (0.572 g/min)	30, 45, 60, 90	48, 70, 82, 109	90 (brittle failure)	Effect of fiber loading, impingement angle, impact velocity	[56]
Epoxy	Short feather fiber (20-wt-%) L*=10-15mm,	(300μm, 500 μm, 800μm), spherical, N/A	30, 45, 60, 75, 90	N/A	90 (brittle failure)	Effect of impingement angle on erosion rate	[57]
Polyvinyl alcohol (PVA)	Date palm leaf fiber, (10, 20, 28, 30 40 wt.-%) L*=15mm, W*=1mm	200±50μm, Angular, (1.467g/min)	15, 30, 45, 60, 90	48, 70, 82, 109	45 (semi-ductile)	Effect of fiber content, impingement angle, impact velocity on erosion rate, Taguchi analysis	[62]
Epoxy	Jute mat (62 wt.-%)	200±50μm, Irregular, (4 g/min)	30, 45, 60, 90	48, 70, 82	45 (semi-ductile)	Effect of impingement angle on erosion rate at various velocity	[58]
Epoxy	Short Lantana-Camara, (10, 20, 30, 40 wt.-%)	200±50μm, Angular, (1.467g/min)	30, 45, 60, 90	48, 70, 82, 109	45 (semi-ductile)	Effect of impingement angle and fiber content on erosion rate at various velocity	[59]
Epoxy	Coir-dust fiber (20, 40, 60 wt.-%)	(200μm, 300 μm, 400μm, 500μm, 600 μm), N/A, (0.650 g/min)	30, 45, 60, 75, 90	34, 48, 60, 78, 92	90 (brittle failure)	Effect of impingement angle, content, impact velocity on erosion rate	[60]
Epoxy	Cross-plied jute fiber (20, 30, 40 wt.-%)	200 μm, Cubical, (10g/min)	15, 30, 45, 60, 75, 90	32, 44, 58	60 (semi-ductile)	Effect of impingement angle, fiber content on erosion rate, Taguchi analysis	[66]
Epoxy	Short fish (Lobea Rohita) flakes (5, 10, 15 wt-%) (L*=6-8mm, W*=1mm)	(300μm, 500 μm, 800 μm), spherical, N/A	15, 30, 45, 60, 75, 90	32, 44, 58	90 (brittle failure)	Effect of impingement angle, on erosion rate, Taguchi & ANN model for erosion rate	[76]

*where L and W represent the length and width of fiber.

It was found that with the addition of bamboo fiber to the epoxy matrix the maximum erosion rate shift from an impingement angle of 60° to 75° which indicate the brittle mode of erosion wear compared to epoxy matrix alone (Figure 1). If velocity of erodent is very low (< critical velocity) then the stresses genreated at target surface during collision of erodent to the target material surface are insufficient for plastic deformation to occur and wear proceeds by surface fatigue wear mechanisms. If velocity of erodent is higher than critical velocity, the eroded composite material deform plastically on impact of erodent on target material surface.

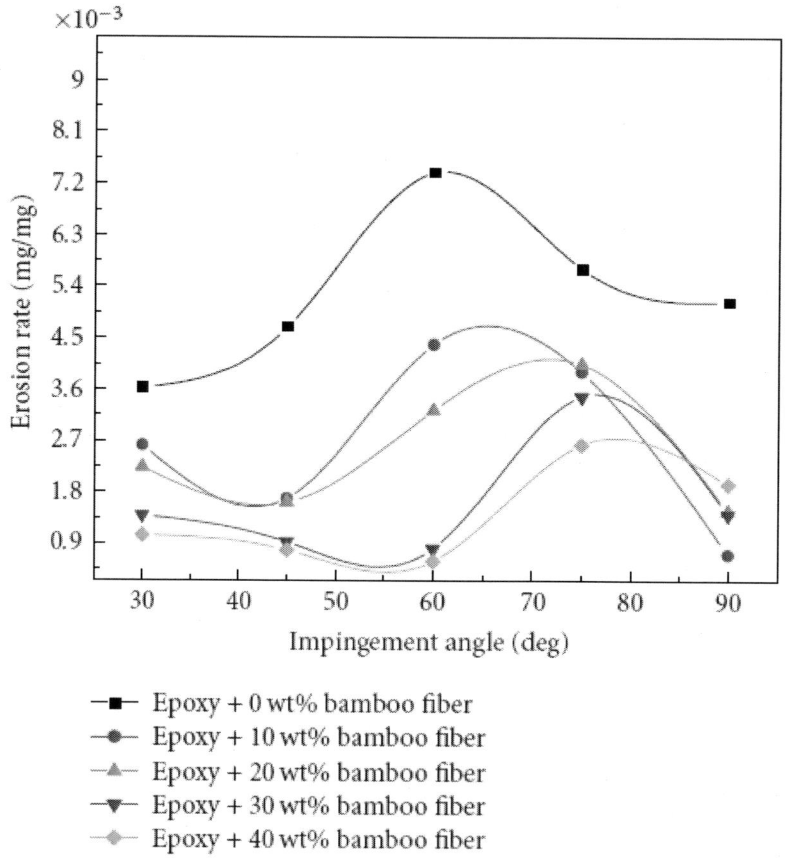

Figure 1. Effect of impingement angle on the erosion wear rate of the composites (Reprinted from Ref. [55] Gupta, A., Kumar, A., Patnaik, A., Biswas, S., (2011) International Journal of Polymer Science, Article ID 592906).

There was no significant change in the erosion rate of bamboo fiber reinforced epoxy composites in the velocity range of 35m/s to 45m/s and it increased considerably after applying velocity of 45 m/s [55]. The erosion wear behavior with erodent size revealed maximum erosion rate at erodent size of 250μm. The dimensionless erosion rate (g/g) has been found in the range of ~0.8×10^{-3} to 8×10^{-3} for these composites [55].

The bagasse fiber reinforced epoxy composites have been investigated for erosion wear behavior (or pattern) as a function of impingement angle at various impact velocities of 48m/s, 70m/s, 82m/s, 109m/s at constant stand-off distance of 10mm and erosion time of 10 min and erodent size of 150-250μm. The erosion rate increased with increase of impingement angle upto normal impact in all experiments indicating brittle mode of erosion wear. The magnitude of erosion rate also increased with increasing impact velocity in these composites. The erosion rate (g/g) was found in the range of 0.1×10^{-3} to 10×10^{-3}.

Figure 2 shows the pulverization of fiber as a dominant erosion mechanism in bagasse fiber reinforced composites [56]. The short poultry feather fiber reinforced epoxy composites also revealed maximum erosion rate at normal impingement angle indicating again brittle mode of erosion behavior [57]. The maximum erosion rate for jute fabric reinforced epoxy composite was found at 45° impingement angle indicating semi-ductile mode of erosion behavior ($45<\alpha<60$). The erosion rate (g/g) was observed in the range of 1.5×10^{-4} to 10×10^{-4} in these composites [58]. The lantana-camara fiber reinforced composites also showed maximum erosion rate at 45° impingement angle for all compositions and velocities indicating again semi-ductile behavior of materials [59].

In bio-waste reinforced composites, it was also found that the erosion rate decreased with increasing the coir dust concentration [60]. The erosive wear mechanism is dominated by the fiber content which is soft in nature compared to the hard polymer matrix material. The erosion rate of coir dust fiber reinforced polymer composites increased with increasing the impingement angle and attained a peak value at an impingement angle of 90°, indicating brittle type of erosion failure. The magnitude of erosion rate of these composite increased with increasing impact velocity. This may be due to increase in the tangential component of the impact force. The improved erosion resistance was also obtained for short fish flake reinforced epoxy composites compared to neat epoxy resin matrix. The variation of erosion rate with impingement angle at constant erodent size of 800μm, impact velocity of 32m/s, stand-off distance of 200mm showed continuous increment and attained maximum value at normal impact (i.e., 90° impingement angle) indicating brittle mode of erosion failure [61]. The impingement angle versus erosion rate at constant parameters (impact velocity=32 m/s, stand-off distance=100 mm, and erodent size=200 mm) showed peak erosion rate at 60° impingement angle. The date palm leaf reinforced polyvinylalcohol composites showed peak erosion rate at 45° impingement angle indicating ductile mode of erosion failure [62].

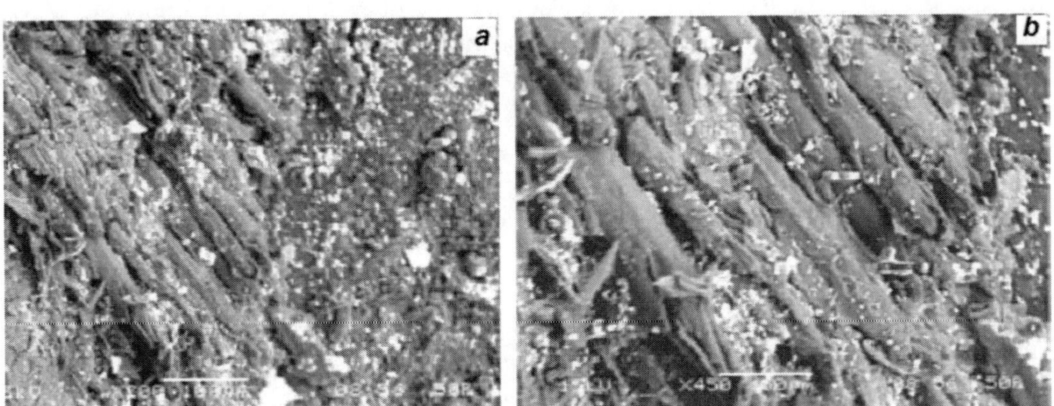

Figure 2. SEM micrographs (a) eroded at 90° (b) magnified view of surface eroded at 90° (Reprinted from Ref. [56] Mishra, P., Acharya S. K., (2010) International Journal of Physical Sciences, 5, 109).

4. Erosive Wear Behavior of Hybrid Composites Containing Natural Fiber

The content of copper slag was varied from 5 wt.-% to 15wt.-% in epoxy/short bamboo fiber (45 wt.-%) based hybrid composites for the study of erosive wear performance. The variation of impingement angle with erosion rate showed that the peak erosion rate was more intensely shifted towards 60° impingement angle with the increase in the concentration of copper slag. The results indicate semi-ductile erosion mode of failure in epoxy/short bamboo fiber/copper slag based hybrid composites. The composites without copper slag, i.e. short bamboo fiber reinforced epoxy composites, showed peak erosion rate at 45° impingement angle. The erosion rate increased (~8-12%) with increase of erodent temperature (from 35°C to 140°C). The epoxy polymer matrix based hybrid composite containing 15 wt.-% of copper slag and 45 wt.-% of short bamboo fiber showed maximum erosion resistance [63]. The bamboo fiber (BF) mat reinforced epoxy hybrid composites with red mud (filler) in the range of 10wt.-% to 20wt.-% showed peak erosion rate at 60° impingement angle. The erosion rate increased ~8-10 % with increase in erodent temperature from 30°C to 60°C in these hybrid composites. [64].

Figure 3 shows the eroded surface morphology of bi-directional bamboo-fiber reinforced composites filled with red mud [64]. The brittle mode of erosion failure mechanisms of these composites mainly includes the fiber breakage, protruding of fiber and fiber-matrix debonding etc. The four ply stacking sequence of woven jute fiber and glass fiber (jute-jute-glass-jute, jute-jute-glass-glass, jute-glass-jute-glass, jute-glass-glass-jute, glass-jute-jute-glass) laminate composites are also used to study erosive wear behavior. The peak erosion rate at 45° impingement angle in these composites indicated semi-ductile mode of erosion failure. The stacking sequence of fibers such as jute-glass-glass-jute in epoxy based hybrid composite offered maximum erosion resistance at all velocities [58].

The peak erosion rate of short bamboo fiber (BF)/E-glass fiber (1:1 ratio) reinforced epoxy based hybrid composites was found at 60° impingement angle. The boost up in impact velocity showed increase in erosion rate and 15 wt.-% of fiber content in epoxy matrix showed maximum erosion resistance [65]. The 20 wt.-% of SiC filler in jute fiber (40 wt.-%) reinforced epoxy based hybrid composites also confirmed improvement in erosive wear resistance and the peak erosion rate was shifted from an impingement angle of 60° (for unfilled composites) to 75° (for SiC filled hybrid composites) [66]. The alkali treated bamboo fiber and cement-by-pass-dust (CBPD) reinforced epoxy based hybrid composites revealed marginal changes in the erosion rate with impact velocity as compared to untreated bamboo fiber and CBPD reinforced epoxy based hybrid composites. The peak erosion rate was shifted from an impingement angle of 45° (untreated composites) to 60° (treated composites) in these hybrid composites [67]. The peak erosion rate for epoxy based hybrid composites reinforced with rice husk and E-glass fiber is observed at 60° impingement angle and maximum resistance to erosive wear is observed for 15 wt.-% of rice husk filled hybrid composites [68].

The interesting feature is that the composites containing only natural fiber acquire maximum erosion rate at an impingement angles of 45°, 60°, 75° and 90° whereas hybrid composites containing combination of natural fiber and traditional fiber typically show maximum erosion rate at an impingement angles of 45° and 60°. If maximum erosion rate in any composites takes place at lower impingement angle, the abrasive wear is the prominent mechanism in material removal process. On the other side, the kinetic energy of erodent is more prominent to material removal process if maximum erosion rate is observed at normal impact (i.e. impingement angle of 90°). The fatigue wear also takes part in erosion (material removal) process when erodent size increases. In this case, the erosion rate increased with increase in the erodent size if other operating conditions are constant. From erosion wear results, it was observed that the abrasion resistance of composites reinforced with bagasse fiber, fish fiber, chicken feather fiber and coir dust are much better than that of other natural fiber reinforced composites.

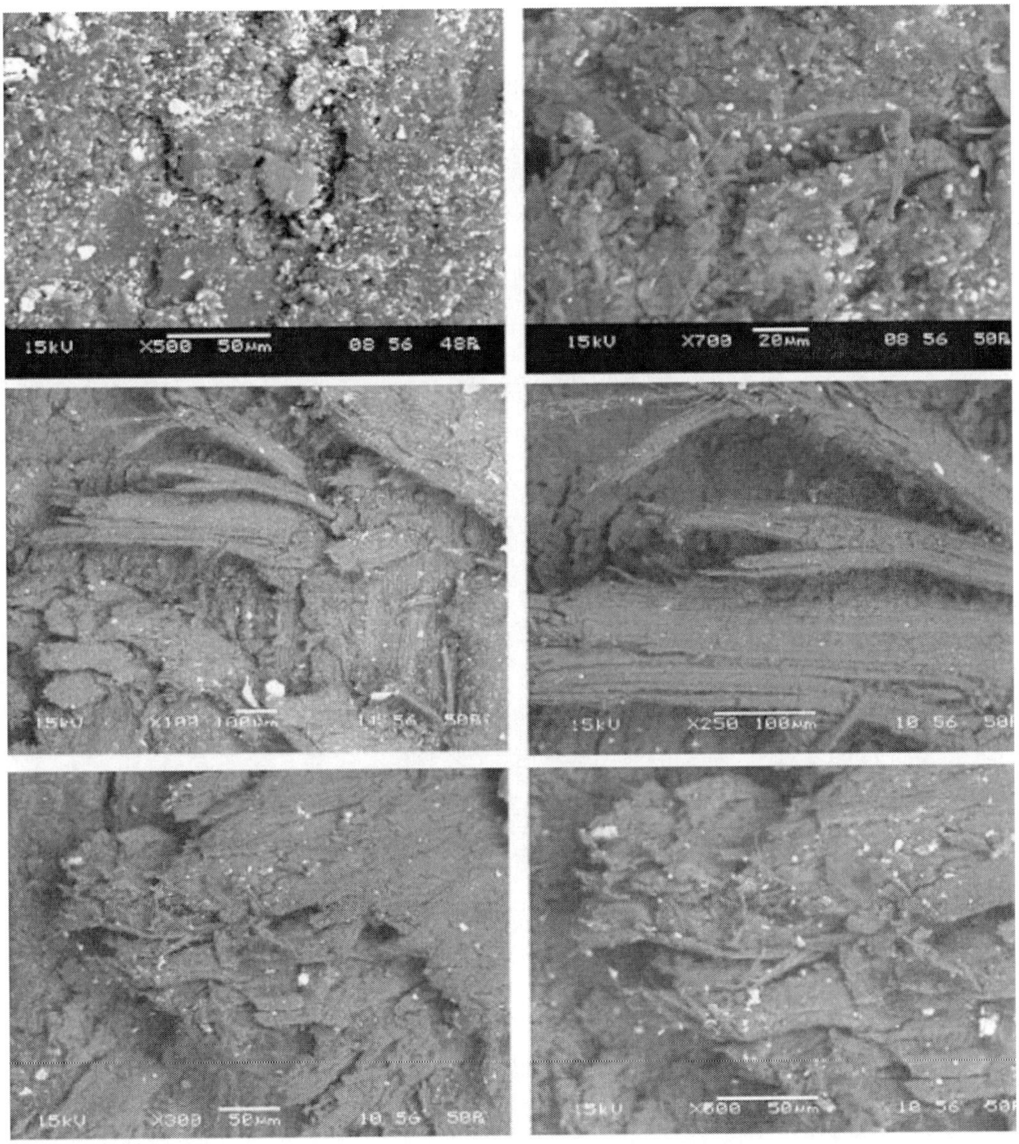

Figure 3. SEM micrographs of the eroded bamboo–epoxy composites filled with red mud (Reprinted from Ref. [64] Biswas, S., Satapathy A., (2010) Materials and Design, 31, 1752).

Table 5. Synopsis of the erosive wear performance of hybrid composites containing natural fiber as one component at room temperature (RT) with silica sand as an erodent

Materials	Erodent size, shape & feed rate	Angle (°)	Velocity (m/s)	Peak erosion Angle (°)	Key issues addressed	Ref
Epoxy+short BF (45 wt.-%) +copper slag (5, 10, 15 wt.-%)	N/A	45, 60, 75, 90	35, 45, 55, 65	60	Effect of copper slag and erodent temperature on erosion rate, Taguchi analysis	[63]
Epoxy+ jute-Jute-glass-Jute, Jute-Jute-glass-glass, Jute-glass-Jute-glass, Jute-glass-glass-jute, Glass-jute-jute-glass	200±50μm, Irregular, (4 g/min)	30, 45, 60, 90	48, 70, 82	45	Effect of impingement angle on erosion rate at various velocity	[58]
Epoxy+BF mat (*Dia: 1.5 mm) +red mud (10-20 wt.-%)	Square Pyramidal, (300μm, 450 μm, 600μm), (10g/min)	15, 30, 45, 60, 75, 90	43, 54, 65	60	Effect of impingement angle, red mud content and erodent temperature on erosion rate, Taguchi analysis	[64]
Epoxy+ short BF/E-GF (1:1 ratio) (15, 30, 45 wt.-%)	Pyramidal, 125μm, N/A	45, 60, 75, 90	35, 45, 55, 65	60	Effect of impingement angle, fiber content and impact velocity on erosion rate, Taguchi analysis	[65]
Epoxy+40 wt.-% jute fiber mat + SiC (10, 20 wt.-%)	200μm, Cubical, (10 g/min)	15, 30, 45, 60, 75, 90	32, 44, 58	60	Effect of impingement angle, SiC content on erosion rate, Taguchi analysis	[66]
Epoxy+bi-directional bamboo fiber mat (40 wt.-%)+Cement by-pass dust (CBPD) (10, 20 wt.-%)	(125μm, 250μm, 350μm), Pyramidal, N/A	30, 45, 60, 75, 90	45, 55, 65	45 (untreated), 60 (treated)	Effect of impingement angle, CBPD content, impact velocity, alkali treatment of bamboo fiber on erosion rate, Taguchi analysis	[67]
Epoxy+E-glass fiber mat (40 wt.-%)+Rice husk (5, 10, 15 wt.-%)	(40μm, 60μm, 80μm, 100μm),N/A, (10g/min)	15, 30, 45, 60, 75, 90	30, 40, 50, 60	60	Effect of filler and impingement angle on erosion rate, Taguchi analysis	[68]

*where Dia represents the diameter of fiber.

5. Design of Experiments for Erosive Wear Performance

The erosive wear is a complex multi-parametric problem and it is well presented in Figure 4. The conventional approach such as One Variable at A Time (OVAT) is not sufficient to evaluate the erosion performance of materials. Therefore, the design of experiments approach based on Taguchi method is generally used to characterize erosive wear performance of composites on a platform over a wide spectrum of variables to identify the set of operating variables or combination of variables to which a composite is most resistant within a tolerance limit of 95%. This method is used to minimize the experimental time, cost and to obtain optimal parameters setting for getting low erosion rate of material. This approach has been successfully applied by many researchers to solve the multi-parametric erosive wear related problems [61-68].

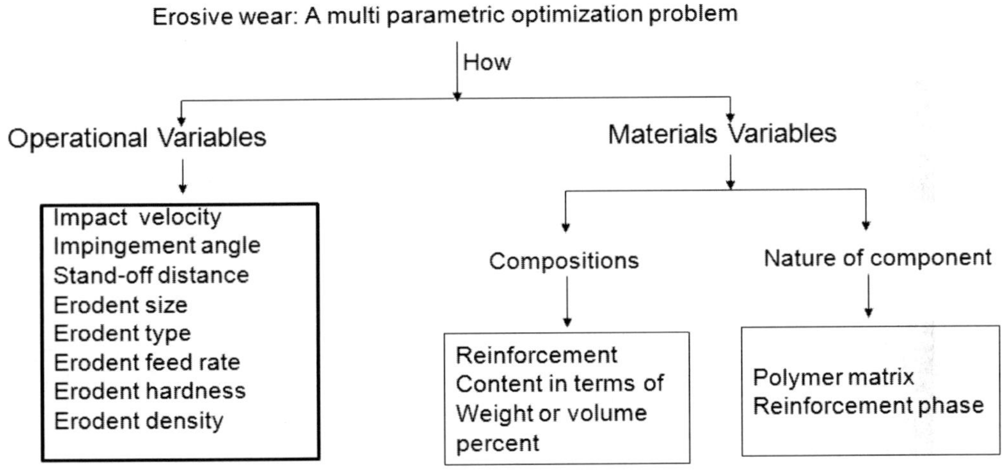

Figure 4. Multi-paramteric problem for evaluation of erosion performance of composites.

To select an appropriate orthogonal array for design of experiments, the total degrees of freedom need to be computed [69]. The degrees of freedom are defined as the number of comparisons between process parameters that need to be made to determine which level is better and specifically how much better it is [70-72]. In all cases, degrees of freedom can be computed by parameters levels. Once the required degrees of freedom for all variables (parameters) are known, the next step is to select an appropriate orthogonal array to fit the specific task. For example, five parameters at four levels have 15 degree of freedom and thus L_{16} orthogonal array are paramount fit for experimental run. Furthermore, the erosion wear parameters-levels are assigned to columns and each row represents the experimental conditions. The experimental runs must be conducted in such a way that the levels of control factors (parameters) should be occured an equal number of times in an orthogonal array. Moreover, the results of experimental runs are reported in the completed design layout. Analysis of the influence of parameters on erosion wear rate and design of orthogonal array are performed by using MINITAB software [73-75]. The design of experiments approach based on Taguchi method has been used to study the erosion characteristics of dried flakes of fish scales reinforced epoxy composites [76]. The overview of input control parameters, assigned orthogonal design and optimal setting for minimization of wear rate for various composite systems are presented in Table 6. The optimal settings of parameters-levels have a great significance in the industry where these materials can be used for transportation of sea water, coal and other erosive particles.

Table 6. Overview of parameters, orthogonal design and optimal control setting

Materials	Para-meters	Levels	Orthogonal Design	Optimal settings	Ref.
Epoxy+ Short feather fiber	4	3	L_9	Velocity of impact=32m/s Impingement angle=30° Erodent size=300μm Stand-off distance=160mm	[54]
Epoxy+ Short fish (Lobea Rohita) flakes	5	3	L_{27}	Impact velocity=58m/s, Fiber content =5wt.-%, Impact angle=30°, Stand-off-distance=120mm, erodent size=300μm	[58]
Epoxy+SBF (45 wt.-%) +Copper slag filler	5	4	L_{16}	Impact velocity=35m/s, Filler content = 15wt.-%, Impact angle=90°, Stand-off-distance=85mm, erodent temperature= 140°C	[60]
Epoxy+BF mat (dia:1.5 mm)+red mud filler	6	3	L_{27}	Velocity of impact=43m/s Filler loading =10 wt.-% Erodent temperature=40°C Impingement angle=90° Stand-off-distance=75mm Erodent size=450μm	[61]
Epoxy+ short BF/E-GF (1:1 ratio)	5	4	L_{16}	Impact velocity=35m/s, Fiber content =30wt.-%, Impact angle=90°, Stand-off-distance=55mm, erodent temperature=105°C	[62]
Epoxy+40 wt.-% jute fiber mat+ SiC filler	4	3	L_9	Velocity of impact=32m/s Impingement angle=30° Erodent size=300μm Filler loading =20 wt.-%	[63]
Epoxy+bi-directional bamboo fiber mat (40 wt.-%)+Cement by-pass dust (CBPD) filler	5	3	L_{27}	Impact velocity=45m/s, Filler content(wt.-%)=20 (untreated), 10 (treated) Impact angle=60°, Stand-off-distance=65mm, erodent size(μm)= 350 (untreated), 250 (treated)	[64]
Epoxy+E-glass fiber mat (40 wt.-%)+Rice husk filler	4	4	L_{16}	Velocity of impact=50m/s Impingement angle=45° Erodent size=40μm Filler loading =15 wt.-%	[65]

6. Mechanism of Erosive Wear

Scanning electron microscopy (SEM) is generally used to find out the wear mechanism of composite materials. The erosion mechanisms relevant to the material removal process with respect to the variety of polymer resins and nature of natural fibers have been found different. After experimental run, the SEM photomicrographs of wear composite sample represent the mechanisms of material removal process. The details of wear mechanism as reported in several literatures are presented in Table 7 for various composite systems. These data help in easy understanding of erosion wear process in natural fiber reinforced polymer matrix composites.

Table 7. Erosion wear mechanisms based on natural fiber reinforced polymer composites

Material	Erosion Failure	Ref.
Epoxy/bamboo fiber	Microploughing, transverse shearing, stripping, fibrillation of fiber, fiber-matrix debonding, fiber fracture & protrusion	[55]
Epoxy/baggase fiber	Intensive debonding, breakage of fiber, fiber microcutting leads to damage and pulverization	[56]
PVA/DPL	Microploughing, microcutting, plastic deformation, microcracking, matrix chipped-off	[62]
Epoxy/lantana-camara	Microcutting, microploughing, fiber breaking and bending, intensive debonding.	[59]
Epoxy/coir dust	Increament in the pulverization process with increasing the impingement angle, debonding and fiber breakage, microcracking, microcutting, pulverization	[60]

CONCLUSION

Currently, more research work related to erosive wear is needed for natural fiber reinforced polymer composites because of lack of systematic studies and availability of less discussion on the effect of void content, geometry and shape on the erosion rate. During the last few years, research has been conducted to utilize the natural fiber for development of erosive wear resistant composites. The bamboo fiber and jute fiber reinforced composites present the maximum erosion resistance as the higher erodent feed rate and impact velocity is used in erosion experiments. The polymer matrix composites based on fibers obtained from animal and aqueous species such as fish show brittle mode of erosion failure whereas polymer composites based on fibers obtained from plants show brittle to semi-ductile mode of erosion failure. Taguchi method is used to find out the optimal parameters-levels setting for minimization of erosion rate of polymer matrix composites.

REFERENCES

[1] Stachowiak, G.W., (2005). *Wear-materials mechanisms and practice*. John Wiley & Sons.

[2] Arai, K., Tsuda, K., Hojo, H., (1987). Erosion damage of cylindrical plastic specimen by slurry flow. *J. Mater. Sci. Soc. Jpn.* 24, 90.

[3] Patnaik, A., Satapathy, A., Chand, N., Barkoula, N.M., Biswas, S., (2010). Solid particle erosion wear characteristics of fiber and particulate filled polymer composites: A review. *Wear* 268, 249.

[4] Barkoula, N.M., Karger-Kocsis, J., (2002). Review-Processes and influencing parameters of the solid particle erosion of polymers and their composites. *J. Mater. Sci.* 37, 3807.

[5] Harsha, A.P., Tewari, U.S., Venkatraman, B., (2003). Solid particle erosion behaviour of various polyaryletherketone composites. *Wear* 254, 693.

[6] Suresh, A., Harsha, A.P., Ghosh, M.K., (2009). Solid particle erosion studies on polyphenylene sulfide composites and prediction on erosion data using artificial neural networks. *Wear* 266, 184.

[7] Zhou, R., Lu, D.H., Jiang, Y.H., Li, Q.N., (2005). Mechanical properties and erosion wear resistance of polyurethane matrix composites. *Wear* 259, 676.

[8] Kulkarni, S.M., Kishore, K., (2001). Influence of matrix modification on the solid particle erosion of glass/epoxy composites. *Polym. Polym. Compos.* 9, 25.

[9] Miyazaki, N., Takeda, T., (1993). Solid particle erosion of fiber reinforced plastics. *J. Compos. Mater.* 27, 21.

[10] Miyazaki, N., Hamao, T., (1996). Effect of interfacial strength on erosion behavior of FRPs. *J. Compos. Mater.* 30, 35.

[11] Qian, D., Bao, L., Takatera, M., Kemmochi, K., Yamanaka, A., (2010). Fiber reinforced polymer composites materials with high specific strength and excellent solid particle erosion resistance. *Wear* 268, 637.

[12] Suresh, A., Harsha, A.P., Ghosh, M.K., (2009). Solid particle erosion of unidirectional fibre reinforced thermoplastic composites. *Wear* 267, 1516.

[13] Harsha, A.P., Jha, S.K., (2008). Erosive wear studies of epoxy-based composites at normal incidence. *Wear* 265, 1129.

[14] Drensky, G., Hamed, A., Tabakoff, W., Abot, J., (2011). Experimental investigation of polymer matrix reinforced composite erosion characteristics. *Wear* 270, 146.

[15] Tewari, U.S., Harsha, A.P., Häger, A.M., Friedrich, K., (2002). Solid particle erosion of unidirectional carbon fibre reinforced polyetheretherketone composites. *Wear* 252, 992.

[16] Suresh, A., Harsha, A.P., Ghosh, M.K., (2008). Solid particle erosion behavior of high performance thermoplastic polymers. *J. Mater. Sci.* 43, 1757.

[17] Kim, A., Kim, I., (2009). Solid particle erosion of CFRP composite with different laminate orientations. *Wear* 267, 1922.

[18] Tewari, U.S., Harsha, A.P., Häger, A.M., Friedrich, K., (2003). Solid particle erosion of carbon fibre-and glass fibre-epoxy composites. *Compos. Sci. Technol.* 63, 549.

[19] Kumar, R., Zhang, L., (2009). Aligned ramie fiber reinforced arylated soy protein composites with improved properties. *Compos. Sci. Techn.* 69, 555.

[20] Lins, V.F.C., Branco, J.R.T., Diniz, F.R.C., Brogan, J.C., Berndt, C.C., (2007). Erosion behavior of thermal sprayed, recycled polymer and ethylene-methacrylic acid composite coatings. *Wear* 262, 274.

[21] Zahavi, J., Schmitt, G.F., (1981). Solid particle erosion of polymeric coatings. *Wear* 71, 191.

[22] Roy, M., Vishwanathan, B., Sundararajan, G., (1994). The solid particle erosion of polymer matrix composites. *Wear* 171, 149.

[23] Rajesh, J.J., Bijwe, J., Tewari, U.S., Venkataraman, B., (2001). Erosive wear behavior of various polyamides. *Wear* 249, 702.

[24] Friedrich, K., (1986). Erosive wear of polymer surfaces by steel blasting. *J. Mater. Sci.* 21, 3317.

[25] Veinthal, R., Tarbe, R., Kulu, P., Kaeerdi, H., (2009). Abrasive erosive wear of powder steels and cermets. *Wear* 267, 1838.

[26] Balu, P., Kong, F., Hamid, S., Kovacevic, R., (2013). Finite element modelling of solid particle erosion in AISI 4140 steel and nickel-tungsten carbide composite materials produced by the laser-based powder deposition process. *Tribol. Int.* 62, 18.

[27] Krishnamoorthy, P.R., Seetharamu, S., Sampathkumaran, P., (1993). Influence of the mass flux and impact angle of the abrasive on the erosion resistance of materials used in pulverized fuel bends and other components in thermal power stations. *Wear* 165, 151.

[28] Johansson, S., Ericson, F., Akeschweitz, J., (1987). Solid particle erosion-a statistical method for evaluation of strength properties of semiconducting materials. *Wear* 115, 107.

[29] Chevallier, P., Vannes, A.B., Forner, A., (1995). New parameters in erosion for study of bulk materials and coatings. *Wear* 186-187, 210.

[30] Akbarzadeh, E., Elsaadawy, E., Sherik, A.M., Spelt, J.K., Papini, M., (2012). The solid particle erosion of 12 metals using magnetite erodent. *Wear* 282-283, 40.

[31] Williams, J.H., Lau, E.K., (1974). Solid particle erosion of graphite-epoxy composites. *Wear* 29, 219.

[32] Macchini, R., Bradley, M.S.A., Deng, T., (2013). Influence of particle size, density, particle concentration on bend erosive wear in pneumatic conveyors. *Wear* 303, 21.

[33] Patnaik, A., Tejyan, S., (2012). A Taguchi approach for investigation of solid particle erosion response of needle-punched nonwoven reinforced polymer composites: Part II. *J. Ind. Text.*, doi: 10.1177/1528083712460264.

[34] Rassiah, K., Ahmad, M.M.H.M., (2013). A review on mechanical properties of bamboo fiber reinforced polymer composite. *Australian Journal of Basic and Applied Sciences* 7, 247.

[35] Okubo, K., Fujii, T., Yamamoto, Y., (2004). Development of bamboo-based polymer composites and their mechanical properties. *Composites: part A* 35, 377.

[36] Amada, S., Untao, S., (2001). Fracture properties of bamboo. *Composites: part B* 32, 451.

[37] El-Tayeb, N.S.M., (2008). A study on the potential of sugarcane fibers/polyester composite for tribological applications. *Wear* 265, 223.

[38] Sen, T., Reddy, H.N.J., (2011). Various industrial applications of hemp, kenaf, flax and ramie natural fibers. *Int. Journal of Innovation, Management and Technology* 2, 192.

[39] Holbery, J., Houston, D., (2006). Natural-fiber-reinforced polymer composites in automotive applications, *Journal of the Minerals, Metals and Materials Society (JOM)* 11, 80.

[40] George, J., Sreekala, M.S., Thomas, S., (2001). A review on interface modification and characterization of natural fiber reinforced plastic composites. *Polymer Engineering and Science* 41, 1471.

[41] Tlijani, M., Gouadria, A., Benyounes, R., Durastanti, J.F., Mazioud, A., (2013). Study and optimization of palm wood mechanical properties by alkalization of the natural fiber. *International Journal of Sciences: Basic and Applied Research* 11, 30.

[42] Ku, H., Wang, H., Pattarachaiyakoop, N., Trada, M., (2011). A review on the tensile properties of natural fiber reinforced polymer composites. *Composites: Part B* 42, 856.

[43] Nabi, S.D., Jog, J. P., (1999). Natural fiber polymer composites, a review. *Advances in Polymer Technology* 18, 351.

[44] John, M.J., Francis, B., Varughese, K.T., Thomas, S., (2008). Effect of chemical modification on properties of hybrid fiber biocomposites. *Composites: Part A*39, 352.

[45] Sanjeevamurthy, G.C., Srinivas, G.R., (2012). Sisal/Coconut coir natural fibers – epoxy composites: water absorption and mechanical properties. *International Journal of Engineering and Innovative Technology (IJEIT)* 2, 166.

[46] Joseph, K., Filho, R.D.T., James, B., Thomas, S., de Carvalho, L.H., (1999). A review on sisal fiber reinforced polymer composites. *R. Bras. Eng. Agríc. Ambiental, Campina Grande* 3, 367.

[47] Mylsamy, K., Rajendran, I., (2011). Influence of fibre length on the wear behaviour of chopped agave americana fibre reinforced epoxy composites. *Tribol. Lett.* 44, 75.

[48] http://www.catamaransailingphuket.com/EPOXY-VINYL%20ESTER.pdf.

[49] http://www.azom.com/Details.asp?ArticleID=997.

[50] http://www.mdacomposites.org/mda/psgbridge_CB_Materials2_Resins.html#Polyurethane.

[51] Raghavendra, G., Acharya, S.K., Deo, C.R., Mishra, P., (2012). Fabrication-Modelling and analysis on tribological performance of natural composites using Taguchi Approach. *Procedia Engineering* 38, 2635.

[52] Mishra, P., (2012) Statistical analysis for the abrasive wear behavior of bagasse fiber reinforced polymer composites. *Int. J. of Applied Research in Mechanical Engineering* 2, 2231.

[53] Patnaik, A., Satapathy, A., Dwivedy, M., Biswas, S., (2010). Wear behavior of plant fiber (pine-bark) and cement kiln dust-reinforced polyester composites using taguchi experimental model. *J. Compos. Mater.* 44, 559.

[54] Boopathi, L., Sampath, P.S., Mylsamy, K., (2012). Influence of fiber length in the wear behavior of borassus fruit fiber reinforced epoxy composites. *International Journal of Engineering Science and Technology* 4, 4119.

[55] Gupta, A., Kumar, A., Patnaik, A., Biswas, S., (2011). Effect of different parameters on mechanical and erosion wear behavior of bamboo fiber reinforced epoxy composites. *International Journal of Polymer Science* doi:10.1155/2011/592906.

[56] Mishra, P., Acharya, S.K., (2010). Solid particle erosion of bagasse fiber reinforced epoxy composite. *International Journal of Physical Sciences* 5, 109.

[57] Rao, V.A., Satapathy, A., Mishra, S.C., (2007). Polymer composites reinforced with short fibers obtained from poultry feathers. *Proceedings of International and INCCOM-6 Conference, IIT Kanpur, UP, India.*

[58] Patel, B.C., Acharya, S.K., Mishra, D., (2011). Effect of stacking sequence on the erosive wear behavior of jute and glass fabric reinforced epoxy composite. *International Journal of Engineering, Science and Technology* 3, 213.

[59] Deo, C., Acharya, S.K., (2009). Solid Particle Erosion of Lantana Camara Fiber-Reinforced Polymer Matrix Composite. *Polymer-Plastics Technology and Engineering* 48, 1084.

[60] Aireddy, H., Mishra, S.C., (2011). Tribological behavior and mechanical properties of Bio-waste reinforced polymer matrix composites. *Journal of Metallurgy and Materials Science* 53, 139.

[61] Satapathy, A., Patnaik, A., Pradhan, M.K., (2009). A study on processing, characterization and erosion behavior of fish (Labeo-rohita) scale filled epoxy matrix composites. *Mater. Des.* 30, 2359.

[62] Mohanty, J.R., Das, S.N., Das, H.C., Mahanta T.K., Ghadei, S.B., (2014). Solid particle erosion of date palm leaf fiber reinforced polyvinyl alcohol composites. *Advances in Tribology*, http://dx.doi.org/10.1155/2014/293953.

[63] Biswas, S., (2014). Erosion wear behaviour of copper slag filled short bamboo fiber reinforced epoxy composites. *IACSIT International Journal of Engineering and Technology* 6, 91.

[64] Biswas, S., Satapathy, A., (2010). A comparative study on erosion characteristics of red mud filled bamboo–epoxy and glass–epoxy composites. *Mater. Des.* 31, 1752.

[65] Biswas, S., Xess, P.A., (2012). Erosion wear behaviour of bamboo/glass fiber reinforced epoxy based hybrid composites. *International Journal of Mechanical and Industrial Engineering (IJMIE)* 1, 79.

[66] Jha, A.K., Mantry, S., Satapathy, A., Patnaik, A., (2010). Erosive wear performance analysis of jute_epoxy_SiC hybrid composites. *J. Compos. Mater.* 44, 1623.

[67] Gupta, A., Kumar, A., Patnaik, A., Biswas, S., (2012). Effect of filler content and alkalization on mechanical and erosion wear behavior of cbpd filled bamboo fiber composites. *Journal of Surface Engineered Materials and Advanced Technology* 2, 149.

[68] Rout, A.K., Satapathy, A., (2012). Study on mechanical and tribo-performance of rice-husk filled glass-epoxy hybrid composites. *Mater. Des.* 41, 131.

[69] Taguchi, G., Konishi, S., (1987). Taguchi methods orthogonal arrays and linear graphs: Tools for quality engineering. *Amer. Supplier Inst.* 35.

[70] Ross, P.J., (1988). Taguchi techniques for quality engineering-loss function, orthogonal experiments, parameter and tolerance design. New York: McGraw-Hill. 10-50.

[71] Bendell, A., Disney, J., Pridmore, W.A., (1989). Taguchi methods: applications in world industry. UK: IFS Publications. 20.

[72] Glen, S.P., (1993). Taguchi method: a hands-on approach. Addison-Wesley.

[73] Rout, A., Satapathy, A., Mantry, S., Sahoo, A., Mohanty, T., (2012). Erosion wear performance analysis of polyester-GF-Granite hybrid composites using the Taguchi Method. *Procedia Eng.* 38, 1863.

[74] Patnaik, A., Satapathy, A., (2009). Mahapatra, S.S., Dash, R.R., Tribo-performance of polyester hybrid composites: Damage assessment and parameter optimization using Taguchi design. *Mater. Des.* 30, 57.

[75] Biswas, S., Satapathy, A., (2009). Tribo-performance analysis of red mud filled glass-epoxy composites using Taguchi experimental design. *Mater. Des.* 30, 2841.

[76] Satapathy, A., Patnaik, A., Pradhan, M.K., (2009). A study on processing, characterization and erosion behaviour of fish (Labeo-rohita) scale filled epoxy composites. *Mater. Des.* 30, 2359.

EDITOR CONTACT INFORMATION

Dr. Rakesh Kumar
Assistant Professor
BIT Mesra
Patna Campus, Patna
India
Tel: 9709474061
E-mail: krrakesh72@gmail.com

INDEX

B

C

D

E

H

I

N

O

P

Q

R

S

T

U

V

W

X

Y

Z